SYNTHETIC JETS

SYNTHETIC JETS
Fundamentals and Applications

Edited by
Kamran Mohseni
Rajat Mittal

CRC Press is an imprint of the
Taylor & Francis Group, an **informa** business

CRC Press
Taylor & Francis Group
6000 Broken Sound Parkway NW, Suite 300
Boca Raton, FL 33487-2742

First issued in paperback 2019

ISBN-13: 978-1-4398-6810-2 (hbk)
ISBN-13: 978-0-367-37836-3 (pbk)

Library of Congress Cataloging-in-Publication Data

Synthetic jets : fundamentals and applications / editors, Kamran Mohseni, Rajat Mittal.
pages cm
Includes bibliographical references and index.
ISBN 978-1-4398-6810-2 (hardcover : alk. paper) 1. Axial flow. 2. Jets--Fluid dynamics. I. Mohseni, Kamran. II. Mittal, Rajat.

TA357.5.A95S96 2015
621.6'91--dc23 2014024202

Visit the Taylor & Francis Web site at
http://www.taylorandfrancis.com

and the CRC Press Web site at
http://www.crcpress.com

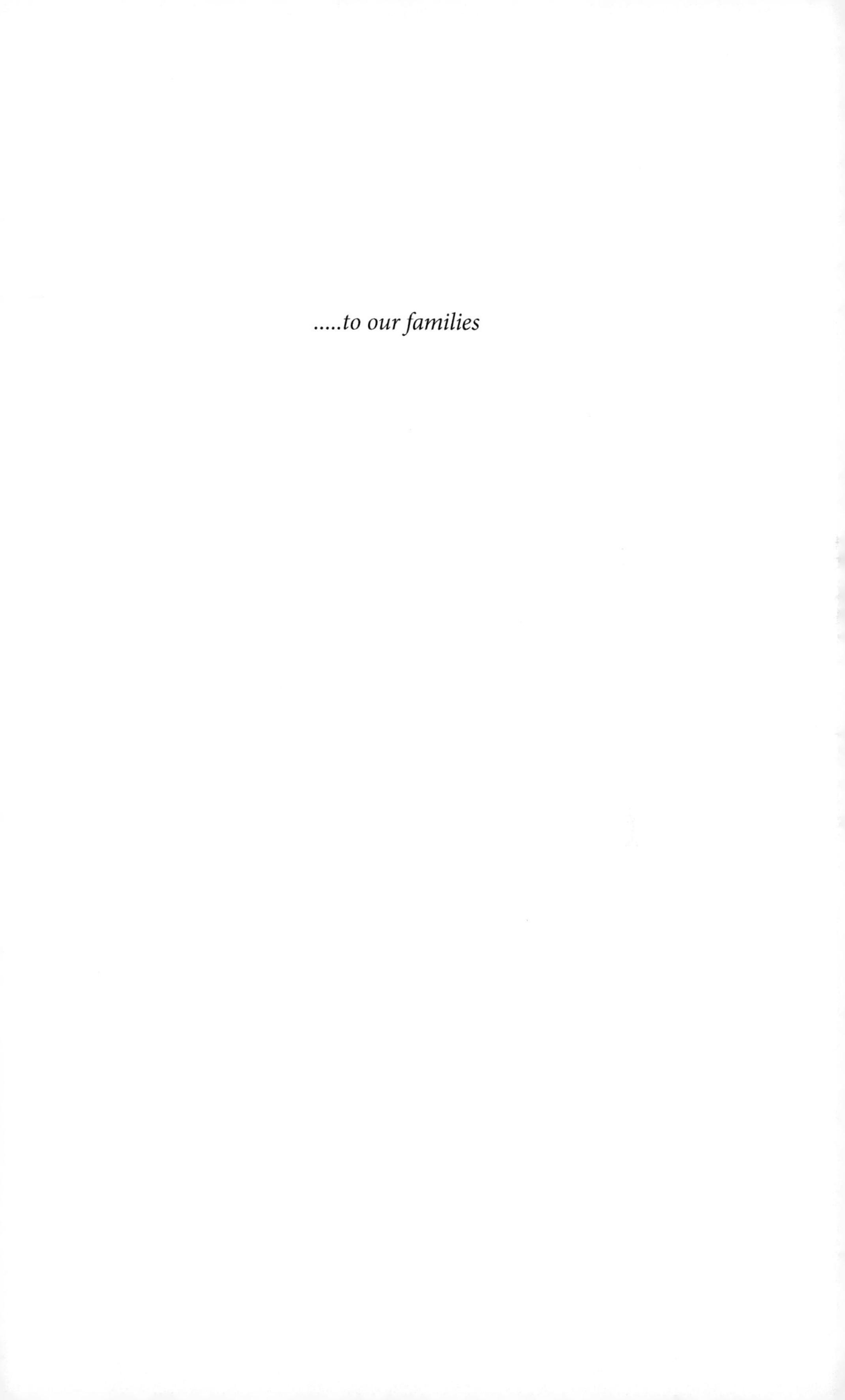

.....to our families

Contents

Preface

Synthetic jets have been the subject of intense research and development for over two decades and have in this time been deployed in a wide variety of applications, ranging from separation and turbulence control to electronic cooling and propulsion. These jets are characterized by the fact that they are entirely synthesized from the surrounding fluid and introduce no net mass into the external flow. This fact eliminates the need for fluidic circuitry and enables self-contained, compact designs that integrate power and actuation and even control electronics. Although synthetic jets do not inject net mass, they do impart momentum, energy, and vorticity, and in doing so, effect local as well as global modifications in the external flow. The ability to prescribe and/or modify the jet frequency and amplitude as well as tailor the geometry of the jet opens up a vast parameter space that can be exploited in a diverse range of applications. This large operational space is, however, associated with a rich cornucopia of fluid dynamic mechanisms, and these mechanisms have to be understood well, in order to make effective use of these actuators in any given application.

Although the topic of synthetic jets may have matured to some degree, the scientific knowledge associated with these actuators remains scattered across hundreds of journal articles and conference papers. The objective of this book is to compile in one place fundamental as well as applied knowledge of these fluidic actuators. By providing a concise survey of the fundamental principles and analysis techniques, and a few selected applications, this book offers a treatment of the subject that should serve as a useful starting point for students, researchers, and technologists interested in this topic.

Kamran Mohseni
University of Florida

Rajat Mittal
The Johns Hopkins University

Editors

Kamran Mohseni, who received his PhD in Mechanical Engineering from the California Institute of Technology in 2000, is professor and W.P. Bushnell endowed chair in the Department of Mechanical and Aerospace Engineering and in the Department of Electrical and Computer Engineering at the University of Florida in Gainesville. He is also the director of the Institute for Networked Autonomous Systems there. He was previously associate professor of Aerospace Engineering Sciences, University of Colorado at Boulder, as well as courtesy professor in the Departments of Electrical Engineering and Mechanical Engineering, and an affiliated faculty of the Department of Applied Mathematics. He was also a founding fellow of the Renewable and Sustainable Energy Institute (RASEI), a joint partnership with the National Renewable Energy Laboratory (NREL) in the Department of Energy. He earned his doctorate from the California Institute of Technology in Mechanical Engineering and his masters from Imperial College London in Aeronautics and Applied Mathematics. His research interests include bioinspired unmanned aerial and underwater vehicles, vehicle system dynamics and control, mobile sensor networking, and fluid dynamics. He is an associate fellow of the American Institute of Aeronautics and Astronautics and a member of other professional societies, including the American Physical Society, American Society of Mechanical Engineers, Society for Industrial and Applied Mathematics, Institute of Electrical and Electronics Engineers, and American Geophysical Union.

Rajat Mittal is professor in the department of mechanical engineering at the Johns Hopkins University in Baltimore, Maryland. He earned his PhD in applied mechanics from the University of Illinois at Urbana–Champaign and his masters in aerospace engineering from the University of Florida in Gainesville, Florida. He was a postdoctoral researcher in the Center for Turbulence Research at Stanford University before joining the University of Florida's department of mechanical engineering as an assistant professor. He was then appointed as faculty in the department of mechanical and aerospace engineering at The George Washington University in Washington, DC. His research focuses on computational fluid dynamics, low Reynolds number aerodynamics, biomedical flows, active flow control, biomimetics and bioinspired engineering, and fluid dynamics of locomotion. He is a fellow of the American Society of Mechanics Engineers as well as the American Physical Society, and an associate fellow of the American Institute of Aeronautics and Astronautics.

Contributors

Michael Amitay
Department of Mechanical, Aerospace and Nuclear Engineering
Rensselaer Polytechnic Institute
Troy, New York

Shawn Aram
Department of Mechanical Engineering
The Johns Hopkins University
Baltimore, Maryland

Mehmet Arik
Department of Mechanical Engineering
Özyeğin University
Istanbul, Turkey

Louis N. Cattafesta
Department of Mechanical Engineering
Florida State University
Tallahassee, Florida

John Farnsworth
Department of Mechanical, Aerospace and Nuclear Engineering
Rensselaer Polytechnic Institute
Troy, New York

Mike Krieg
Department of Mechanical and Aerospace Engineering
University of Florida
Gainesville, Florida

Rajat Mittal
Department of Mechanical Engineering
The Johns Hopkins University
Baltimore, Maryland

Kamran Mohseni
Department of Mechanical and Aerospace Engineering and Department of Electrical and Computer Engineering
University of Florida
Gainesville, Florida

Matias Oyarzun
Department of Mechanical and Aerospace Engineering
University of Florida
Gainesville, Florida

Reni Raju
Department of Mechanical and Aerospace Engineering
The George Washington University
Washington, DC

Barton L. Smith
Department of Mechanical and Aerospace Engineering
Utah State University
Logan, Utah

Douglas R. Smith
Air Force Office of Scientific Research
Arlington, Virginia

Yogen V. Utturkar
GE Global Research Center
Electronics Cooling Laboratory
Niskayuna, New York

David R. Williams
Fluid Dynamics Research Center
Department of Mechanical, Materials, and Aerospace Engineering
Illinois Institute of Technology
Chicago, Illinois

Nail K. Yamaleev
Department of Mathematics
North Carolina A&T State University
Greensboro, North Carolina

Shan Zhong
School of Mechanical, Aerospace and Civil Engineering
University of Manchester
Manchester, England

I

Fundamentals

1

Synthetic Jets: Basic Principles

Kamran Mohseni
University of Florida

Rajat Mittal
Johns Hopkins University

A jet flow is a class of fluid flow in which a stream of one fluid mixes with a surrounding fluid which might be at rest or in motion. Such flows occur in many natural and engineered applications. Flow properties for incompressible jets are often greatly dependent on the Reynolds number $Re \equiv UL/\nu$, where ρ is the fluid density, U is a characteristic velocity (e.g., jet exit velocity), L is a characteristic length scale (e.g., jet diameter), and ν is the kinematic viscosity. In case the flow shows compressible effects (typically for $M > 0.3$), jet flow characteristics are also affected by the Mach number defined as $M \equiv U/c$, where c is the speed of sound.

This monograph is dedicated to a class of incompressible jet flows where the jet is made up of the surrounding fluid, namely, a synthetic jet (SJ). SJs are typically formed by momentary and periodic ingestion and expulsion of fluid at the exit of a nozzle or across an orifice (Figure 1.1). Since there is no net flux of fluid during one full cycle of operation, such an SJ is also referred to as a zero-net mass-flux (ZNMF) jet. Given the inherent asymmetry of the flow conditions across the jet orifice (actuator cavity on one side and ambient flow on the other), for typical operating conditions, the net flux of momentum to the external flow is not zero. For certain operating conditions, a train of interacting vortices can be generated at the jet exit and this jet can penetrate far into the external flow and affect its dynamics. SJs are inherently pulsatile and may be periodic and their dynamics

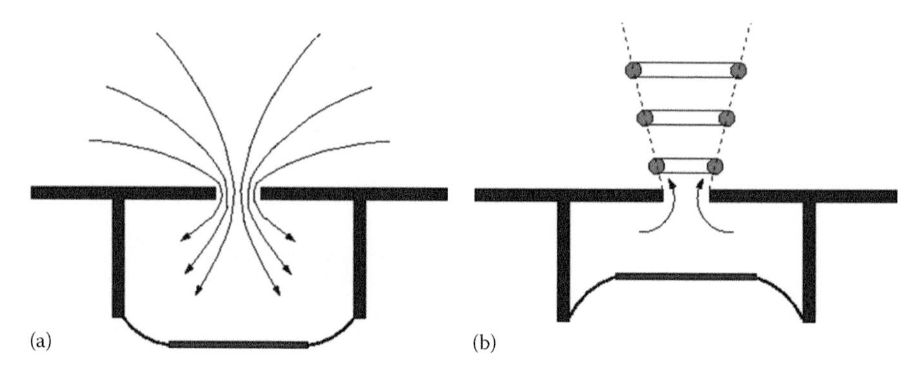

FIGURE 1.1 A typical SJ actuator. (a) Fluid ingestion phase; (b) fluid expulsion phase and vortex ring formation.

are fundamentally different from steady or pulsed jets. SJs can be produced in a number of ways through the use of piezoelectric, electromagnetic (e.g., solenoids), acoustic (e.g., speakers), or mechanical (piston) drivers.

SJs, similar to continuous jets (CJs), are studied by examining changes in the fluid velocity, density, temperature, and concentration of the component fluids both in the jet and in the ambient fluid. The readers are encouraged to consult a few available review articles for general experimental and simulation observations with SJs including the article by Glezer and Amitay (2002). In this chapter, we explore analytical models of SJs for two-dimensional (2D) plane SJs and axisymmetric SJs in cylindrical and spherical coordinates. The models start with the most simplified case for an SJ in a quiescent and infinitely large environment. We then gradually introduce more complex background flows (including coflow and crossflow) or interaction with a wall in the case of a wall SJ. The main results are based on a series of publications (Krishnan and Mohseni 2009a, 2009b, 2010; Xia and Mohseni 2010, 2011). We start by a brief discussion of the SJ actuators, the experimental setup, and a model of the actuator.

1.1 Preliminaries

1.1.1 SJ Actuator Devices

Several types of SJs have been developed and employed in the investigation of the physics and applications of SJs. Here, results of investigation based on three classes of SJs are presented (Figure 1.2). The first two types of actuators were used to generate round jets and rectangular jets, respectively. For the case of SJ in a coflow, one would like to have a minimal wake behind the actuator. The actuator in Figure 1.2c is an example of such a case where the orifice is located on the shorter side of the actuator device to minimize its projected area in the direction of the background flow.

The actuating element in all of these SJ actuators is designed to be a piezoelectric disk on a very thin brass shim. A typical actuator is comprised of a circular piezoelectric membrane sandwiched between two circular aluminum elements, which forms a cavity

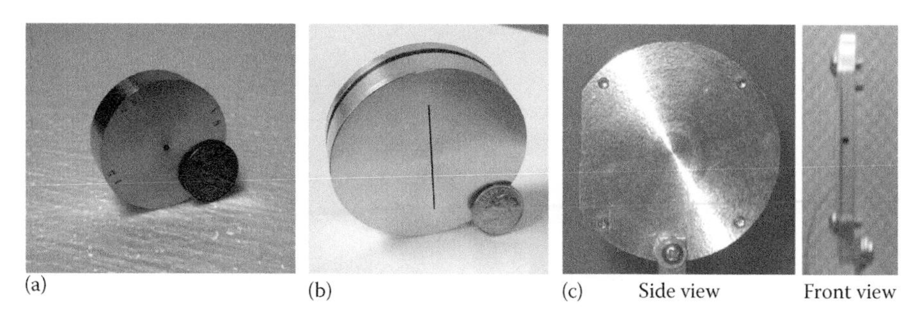

FIGURE 1.2 SJ actuators: (a) round with large frontal view; (b) rectangular with large frontal view; (c) round with a small frontal view. Case (c) is used for coflow SJ where minimal actuator wake is intended.

with an orifice either at the side or the base of the actuator device. The orifice could be round or rectangular depending on the application requirements. The piezoelectric membrane was driven by a sinusoidal input voltage, and the frequency and amplitude of the membrane deflection were controllable. The effects of other input voltage forms have also been investigated, but they are not reported here. The driving frequency (f) was usually of the order of 10^3 Hz, which was close to the resonant frequency of the system.* The driving voltage amplitude (V_d) could be varied to change the oscillation amplitude of the piezodisk and consequently increase the strength of an SJ.

1.1.2 Experimental Method

The mean velocity field measurements were conducted by using single and cross hotwire probes operating in constant-temperature anemometry mode. The hotwires were calibrated with an iterative procedure (Krishnan and Mohseni 2009a, 2009b) with a fourth-order polynomial curve used to convert voltage to velocity. Based on the accuracy in the calibration and scatter in the experiment, the uncertainty in the mean velocity measurements was estimated to be ±2% for velocities greater than 1 m/s and ±10% for velocities less than 1 m/s. As shown in Figure 1.3, the hotwire probe was affixed to a holder positioned on two computer-controlled stages capable of traversing the horizontal plane. To characterize the flowfield, the probe was moved in the horizontal plane of the orifice, in which the flow at each discrete location was sampled for about 10 s depending on the application. For the cross hotwire probe, there is typically an angle range of acceptance, which means that the velocity is valid only when the angle between the flow direction and the probe axis lies within a certain range. So to measure the velocity field correctly in all jet regions, the flow field was divided into several sections; then, the overall flow direction was estimated in

* Note that there are two relevant resonant frequencies: the moving membrane resonant frequency and the Helmholtz cavity resonant frequency. While one can design the cavity and the membrane disk to have the same natural frequency, this is not recommended since it could result in damaging the device.

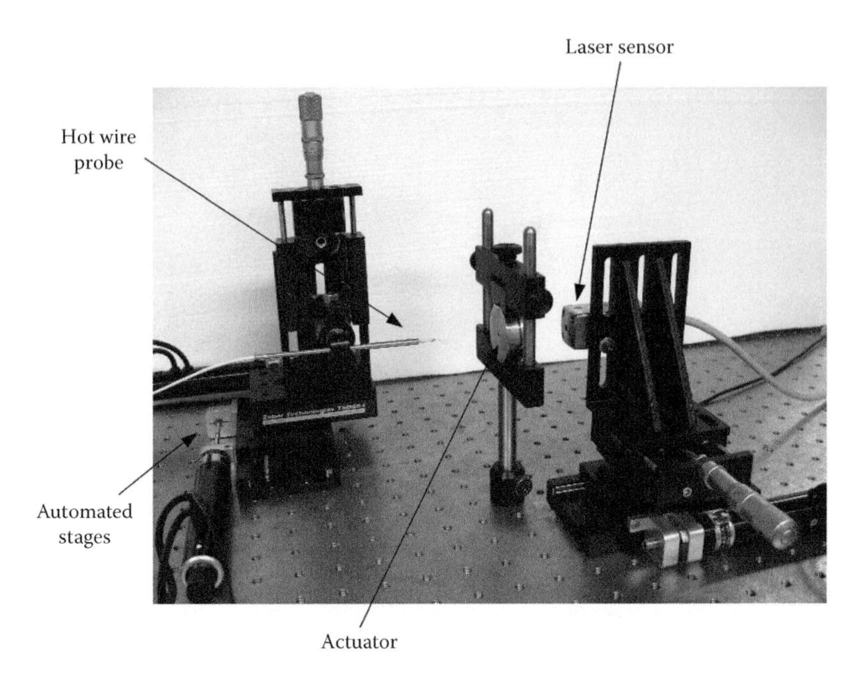

FIGURE 1.3 Experimental setup used to measure the velocity field and actuator diaphragm deflection.

each section. Finally, the direction of the probe in each section was adjusted according to this overall flow direction to make sure that the velocity fields were measured accurately using the cross hotwire probes.

The centerline deflection of the piezoelectric membrane is another key parameter in SJ studies. This membrane deflection is often not measured in SJ investigations, some exceptions being the works of Krishnan and Mohseni (2009a, 2009b, 2010) and Xia and Mohseni (2010, 2011). Here, the setup to measure the deflection consists of a laser nanosensor, a movable stage, and a small sliver of silicon affixed to the center of the piezoelectric membrane while it is housed in the actuator, as shown in Figure 1.3. The principle of the laser sensor is as follows: A laser beam generated by the sensor is incident upon a reflective surface (the piece of silicon serves this purpose) on the diaphragm. The reflected beam returns through the same sensor opening, whereupon it passes through an optical system and is projected on photodiodes. As the target moves back and forth, the position of the reflected beam translates on the photodiode surface from which this translation is correlated with the motion of the target through calibration. The calibration of the sensor is conducted as such. With the sensor attached to a movable automated stage and the diaphragm fixed in a particular location, the laser is moved in increments of 1 μm toward the diaphragm, with the signal response measured at each location. This nonlinear displacement-response curve then serves as the calibration curve. To make measurements, the laser is positioned at the location that allows for the largest sensitivity over the measurable range. To estimate the overall measurement accuracy of the laser, the uncertainties associated with stage position, sensor resolution, calibration, and experimental repeatability were taken into account. For a typical value, the total combined uncertainty was estimated to be ±1 μm by the root sum square.

1.1.3 Actuator Model

A nondimensional stroke ratio (L/d) and the Reynolds number (Re) have been identified as key actuator operational parameters that influence an SJ (Smith and Glezer 1998). For the cavity-diaphragm setup used in this experiment, they are obtained from an incompressible flow model, where it is assumed that the volume displaced by the membrane is equal to the volume ejected from the orifice (Figure 1.4). As a result, it is quite important to measure the piezodiaphragm displacement during the operation of the actuator to properly estimate the stroke ratio of the ejecting jet from the actuator. This has been achieved with the laser setup described earlier (see Krishnan and Mohseni, 2009a, for details).

To obtain the volume displaced by the membrane, the shape of the deflected membrane and the amplitude at the center of the membrane are required (Krishnan and Mohseni 2009a). The shape is obtained from the classical theory of plates (Timoshenko 1999), while the center amplitude is measured directly during the operation of the device using a laser sensor. With the ejected volume approximated as a cylindrical slug of fluid with the same cross section as the exit orifice, the conservation of volume is written as

$$\alpha \frac{\pi D^2}{4} \Delta = \frac{\pi d^2}{4} L \tag{1.1}$$

where:

α is the fraction of the volume displaced by an imaginary piston undergoing a peak to peak deflection of Δ and is expressed by

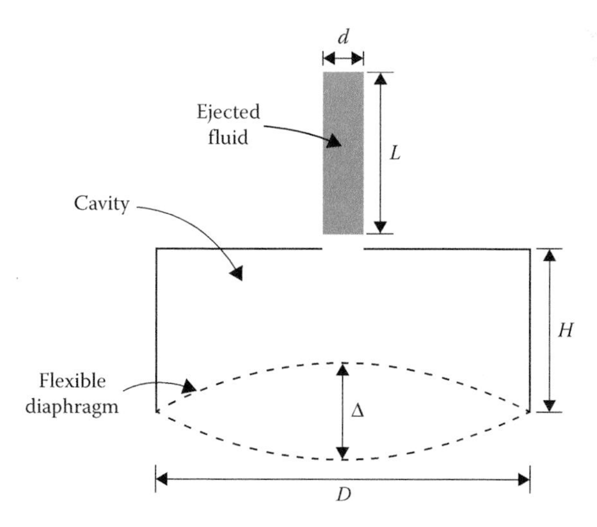

FIGURE 1.4 Schematic of the actuator model, where the volume of fluid displaced by the diaphragm is ejected through the orifice in the form of a slug.

$$\alpha = \frac{2\pi \int_0^{D/2} y(r) r dr}{\left(\pi D^2/4\right)\Delta} \tag{1.2}$$

where:

$y(r)$ is the deflection profile of the diaphragm

Assuming that the shape of the membrane is modeled by the static deflection of the circular membrane clamped on the edge subject to a uniform load, the deflection profile can be estimated by

$$y(r) = \frac{\Delta}{2}\left[1 - \frac{r^2}{R^2} + \frac{2r^2}{R^2}\ln\left(\frac{r}{R}\right)\right] \tag{1.3}$$

where:

r is the radial coordinate
R is the radius of the membrane

The above-mentioned deflection profile results in an α of 0.25. The nondimensional stroke ratio is then determined to be

$$\frac{L}{d} = \alpha\Delta\frac{D^2}{d^3} \tag{1.4}$$

It should be noted that the membrane displacement profile in Equation 1.3 is a reasonable estimate for axisymmetric vibration of the membrane. At frequencies close to the membrane resonant frequency or if there is any measurable asymmetry in the oscillating diaphragm, nonaxisymmetric modes of the membrane could be energized. This often results in a significant degradation of the strength of the produced jet since the volume changes by one side of the diaphragm could be cancelled by an opposite volume change of the other side of the actuator. Monitoring the vibration of the piezodisk during the operation of the actuators reveals any deviation from axisymmetric vibration of the membrane. For more discussion of this issue, the readers are referred to the works of Krishnan (2009) and Krishnan and Mohseni (2009a).

The periodic nature of SJs allows for the velocity scales to be defined based on either volume flux or momentum flux (Smith and Swift 2001). If defined based on volume flux, the velocity scale is given as $U_0 = \left(L/T\right) = fL$, and if defined based on momentum flux, it is given as $U_0 = \sqrt{2}\left(L/T\right) = \sqrt{2}fL$. In either case, it is seen that the average velocity of the SJ is directly proportional to frequency and the stroke length. Consequently, one might expect the momentum created by such a jet to be proportional to f^2 (Mohseni 2006). For $L/D = 4$, it is found that this momentum is of the order of $d^4 f^2$, which has been described in depth in Chapter 10 and experimentally verified in the works of Krieg and Mohseni (2008).

1.1.4 SJ Actuator Reynolds Number

To find the Reynolds number, it is more appropriate to use the velocity scale based on momentum flux because the models employed in these investigations define equivalent jets based on the same momentum flux and not mass flux. Consequently, the Reynolds number is defined as

$$Re = \frac{U_0 d}{\nu} = \frac{\sqrt{2} f L d}{\nu} = \frac{\sqrt{2} f \alpha \Delta D^2}{\nu d} \tag{1.5}$$

The Reynolds number is therefore explicitly seen to vary with both membrane driving frequency and amplitude, with the stroke ratio appearing to be independent of frequency. This independence of stroke ratio on frequency is not accurate as the use of a piezoelectric diaphragm as a driver gives rise to a coupling between the frequency and deflection, and consequently the stroke ratio. However, for purposes of calculating the jet parameters, the simple model described earlier serves the purpose. In summary, Equations 1.4 and 1.5 express the dependency of the critical actuator parameters on the diaphragm driving frequency and deflection amplitude.

The above-described flow and actuator models provide a framework to relate the input driving parameters (f, V_d) to the output jet parameters, decay and spreading rates (S_u, S_b), via the actuator variables $(L/d, Re)$. These will be the focus of Sections 1.1.5 and 1.1.6.

1.1.5 Incompressibility Check

We are now in the position to verify the validity of the incompressibility assumption used in the preceding calculations. In the following experiments, the actuation frequency is of the order of 1000 Hz and L is of an approximate magnitude of 0.01 m. Therefore, the velocity is of the order of 10 m/s. As a result, the operational Mach number for the actuators is quite below 0.3 that makes the flow virtually incompressible.

1.1.6 Eddy Viscosity Approach to Modeling SJs

The last century has seen a significant effort in the scientific community in developing models for the averaged flow properties of laminar and turbulent CJs. An elegant and strikingly simple solution for modeling a laminar continuos jet was offered by Schlichting (1933). He found that an axisymmetric laminar jet could have a self-similar solution. He predicted a linear spreading rate for the jet, where the spreading rate was controlled by the fluid viscosity. This result was later extended to axisymmetric turbulent jets by supplementing the kinematic viscosity with an eddy viscosity to compensate for the enhanced momentum mixing in a turbulent CJ (Schlichting 1979). The enhanced eddy viscosity could be easily calculated from the jet spreading rate.

In a similar fashion, it has been recently hypothesized (Krishnan and Mohseni 2009a, 2009b) that the enhanced mixing and spreading rate of an SJ above what has been observed for a related turbulent CJ could be attributed to an enhanced eddy viscosity due to the pulsatile nature of an SJ. Again, this enhanced eddy viscosity could be easily measured, as seen in Section 1.2, from the SJ spreading rate. This provides a rather simple technique for developing models for SJs while taking advantage of a century of model development for CJs. This approach has been employed here to develop models of SJs with increasingly complex background flow.

1.2 SJ Modeling

1.2.1 A Round SJ in Quiescent Environment

An SJ emanating from a round orifice in a quiescent environment could exhibit axial or spherical symmetry (Krishnan and Mohseni 2009b) depending on the properties of the jet and generating actuator. If the jet is coming out of a long cylindrical tube, one expects an axially symmetric flow, and if the jet is exiting a large cavity through a small and thin orifice, a spherical symmetry is expected. In this section, modeling of an SJ as an axisymmetric or a spherical jet is considered.

1.2.1.1 Axisymmetric SJs in Cylindrical Coordinates

Schlichting (1933) offered a similar solution to the axisymmetric boundary layer equations for the case of free incompressible CJ. This is the case for the far field of an axisymmetric jet emanating from a small orifice into a similar fluid at rest (Figure 1.5). At far distances from the orifice, the jet appears to issue from a point source of momentum in the jet direction. In this section, a summary of the Schlichting solution for a CJ, both laminar and turbulent, is offered and then it is extended to the case of SJs.

The far field of a CJ may be imagined to be generated by a continuous point source of momentum in an infinite incompressible fluid. It is admissible to describe the mean

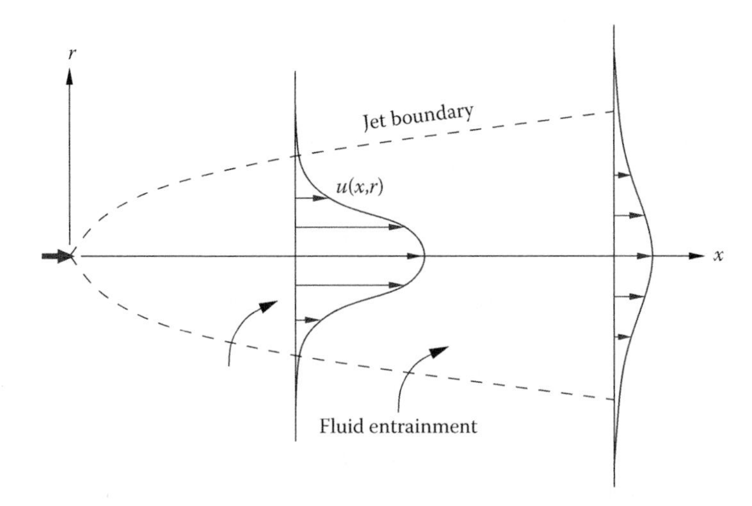

FIGURE 1.5 Schematics of an axisymmetric jet created by a source of momentum in x-direction in an infinite medium.

velocities in the CJ by boundary layer equations. In seeking a self-similar solution to the boundary layer equations, the streamwise pressure gradient is assumed to be zero, whereupon a closed form solution for a laminar jet exists (Schlichting 1933). Following along these lines, it is hypothesized here that the mean velocity field of an SJ may be modeled as a laminar free jet, along with the use of a virtual viscosity coefficient obtained empirically for an SJ. An overview of the similarity analysis is traced out next along with its pertinence to SJs.

In polar coordinates, the boundary layer equations with no pressure gradient are written as

$$\frac{\partial u}{\partial x}+\frac{\partial v}{\partial r}+\frac{v}{r}=0 \quad \text{and} \quad u\frac{\partial u}{\partial x}+v\frac{\partial u}{\partial r}=\frac{1}{\rho r}\frac{\partial (r\tau)}{\partial r} \tag{1.6}$$

where:

u and v are the streamwise and radial velocity components, respectively
τ is the total shear stress
ρ is the fluid density

The total shear stress may be related to the mean velocity using an eddy viscosity approximation

$$\tau=\rho\left(\nu+\varepsilon_{T}\right)\frac{\partial u}{\partial r}=\rho\varepsilon\frac{\partial u}{\partial r} \tag{1.7}$$

where:

ν is the laminar kinematic viscosity
ε_T is the turbulent eddy viscosity coefficient
ε is the total or effective eddy viscosity that takes into account both the laminar and turbulent contributions to the shear stress

While the eddy viscosity hypothesis assumes that momentum transfer in a turbulent flow is dominated by large-scale eddies, it does characterize the mixing due to turbulent fluctuations, which in turn is indicative of the rate of spreading of a free jet.

The eddy viscosity may be derived from the experimental data as follows: Assuming that the evolution of the jet is dependent only on local length and velocity scales and lacks memory of the orifice dimensions itself, the streamwise mean velocity profiles may be considered self-similar. From the conservation of streamwise momentum, the characteristic length (b) and velocity (u) of the jet scale may be represented as x and x^{-1}, respectively. The self-similar assumption then leads to a streamwise velocity profile of the form $u = x^{-1}f(r/x)$. The similarity variable written as $\eta = \sigma(r/x)$ is related to the virtual viscosity coefficient through a free constant σ. With the mixing length hypothesis showing that the virtual viscosity is constant over the entire jet, the boundary layer equations may then be reduced to an ordinary differential equation of the form $ff' = f' - \eta f''$. From the conservation of momentum and the assumed form of the velocity distribution, the streamwise velocity is solved to be

$$u = \frac{3K}{8\pi\varepsilon x} \frac{1}{\left[1 + (1/4)\,\eta^2\right]^2} \tag{1.8}$$

with the self-similarity variable given as

$$\eta = \frac{1}{4}\sqrt{\frac{3}{\pi}}\frac{\sqrt{K}}{\varepsilon}\frac{y}{r} \tag{1.9}$$

K, the kinematic momentum of the jet, is a measure of the strength of the jet and is obtained as $K = 2\pi \int_0^\infty u^2 r dr$. It is important to note here that the above analysis assumes a constant momentum flux in the streamwise direction. While this is applicable to the CJs, in the SJs it has been reported that the momentum flux at the orifice is higher than that in the far field (Smith and Glezer 1998; Spencer et al. 2005). The momentum flux was shown to decrease in the near field of the jet due to an adverse pressure gradient and then asymptote in the far field to some fraction of the exit momentum flux. It is this reduced asymptotic value of the momentum flux that should serve as the magnitude of the driving momentum flux in the above similarity analysis for an SJ and not the exit momentum flux at the orifice of the actuator. However, if we assume that the above asymptotic value applies to all SJs equally, then the use of the exit momentum flux is permissible, with the added benefit that it is obtainable from a simple actuator model.

The eddy viscosity (ε) is now obtained from the spreading and decay rates of the jet. At the centerline, the streamwise velocity may be expressed as

$$u_C = \frac{3K}{8\pi\varepsilon x} = \frac{1}{x\,S_u} \tag{1.10}$$

where:
$S_u = 8\pi\varepsilon/3K$ is a measure of the jet decay rate

The radial extent of the jet at a particular axial station may be characterized by a half width ($b_{1/2}$), defined as the radial distance from the centerline at which the streamwise velocity drops to half the centerline velocity. The linear streamwise variation of the half width may be written as

$$b_{1/2} = S_b x \tag{1.11}$$

where:
S_b is the spreading rate of the jet

From Equations 1.8 through 1.11, the free constant σ in the similarity variable is related to the spreading rate as

$$\sigma = \frac{2\sqrt{\sqrt{2}-1}}{S_b} \tag{1.12}$$

from which the eddy viscosity is related to the spreading rate and decay rate as

$$\varepsilon = \frac{1}{8(\sqrt{2}-1)}\frac{S_b^2}{S_u} \tag{1.13}$$

In summary, Equations 1.8 and 1.13 model the far field of the jet in cylindrical coordinates. Thus, in this study we employ the same technique as above where the eddy viscosity of an SJ is obtained from experimentally determined spreading and decay rates. Next, an alternative similarity model to an SJ is presented.

It should be noted that the independent variable x in the above equations is measured from the location of the virtual source of momentum ($x_{0,u}$) and not the location of the physical actuator. This virtual momentum source location is often obtained by extending the line of jet mean width at the far field to intersect the axis of symmetry of the jet.

1.2.1.2 Axisymmetric SJs in Spherical Coordinates

Landau (1944) and independently Squire (1951) have found an analytical solution to the Navier–Stokes equations in spherical coordinate for an axial and continuous source of momentum concentrated at the origin of the coordinate system. This solution to the Navier–Stokes equations should be contrasted with the Schlichting solution to the axisymmetric boundary layer equations for a similar axial source of momentum concentrated at the origin of the coordinate system. In this section, we present the Landau–Squire solution and describe its extension to SJs.

In the Schlichting solution, a self-similar solution to the boundary layer equations, the streamwise pressure gradient was required to be zero. However, in the Landau–Squire jet, a self-similar solution to the Navier–Stokes equations exists for a laminar jet with a type of axial pressure gradient in the axial direction. In replacing the viscosity coefficient of the laminar continuos jet with the virtual viscosity of a turbulent jet, the velocity distribution of the turbulent jet model was found to be in good agreement with experiments (Squire 1951). As the Schlichting model, our viscosity replacement hypothesis is extended to SJs here and the results of the Landau–Squire solution are briefly presented.

The analysis in the slender viscous region comprising the jet and in the limiting behavior of high Reynolds number yields the radial velocity and pressure distributions as (Sherman 1990)

$$u_r = \frac{4\epsilon}{RC}\frac{1}{(1+\xi^2)^2} \quad \text{and} \quad p - p_\infty = \frac{4\epsilon^2}{R^2C}\frac{\xi^2-1}{(1+\xi^2)^2} \tag{1.14}$$

where:

ϵ is the virtual kinematic viscosity of the jet

$\xi = \theta/\sqrt{2C}$ is the similarity variable

C is a constant that is shown to be inversely proportional to a Reynolds number, which is based on the centerline velocity of the jet and distance from the origin

In the limit as $Re \rightarrow \infty$, C is expressed in terms of the magnitude of the point force that drives the jet (F) as

$$\frac{F}{2\pi\rho\varepsilon^2} \rightarrow \frac{16}{3C} \tag{1.15}$$

The half-spreading angle ($\theta_{1/2}$) is defined as the angle at which the radial velocity along a constant radius is half the centerline radial velocity ($u_{r,c}$). Through the use of the definition of the half-spreading angle, the constant C is derived as

$$C = \frac{\theta_{1/2}^2}{2(\sqrt{2}-1)} \tag{1.16}$$

from which the virtual viscosity is shown to be

$$\varepsilon = \frac{1}{8(\sqrt{2}-1)} \frac{\theta_{1/2}^2}{S_u} \tag{1.17}$$

Notice the similarity of this relationship with the eddy viscosity of a cylindrically symmetric jet in Equation 1.13. The virtual kinematic viscosity is thus a product of the inverse of the centerline velocity decay rate (which is identical in both cylindrical and spherical coordinates) and the square of the spreading angle of the jet and is of the same form as that obtained from the Schlichting solution. Thus, Equations 1.14 and 1.17 model the free jet in a spherical coordinate system. To enable this solution for an SJ, the spreading angle and centerline decay rate need to be obtained from the experiment.

Equations 1.13 and 1.17 provide a relation between the spreading and decay rates of a jet and its apparent virtual viscosity. As argued in this section and in the works of Krishnan and Mohseni 2009a, the enhanced eddy viscosity of an SJ, beyond what has been observed for a turbulent CJ, could be directly measured from the decay and spreading rates of an SJ, these have been verified experimentally in this section.

1.2.1.3 Experimental Results for Round SJs

The experimental results in this section are based on a round SJ injected into a quiescent environment as reported in the works of Krishnan and Mohseni (2009a). Hotwire anemometry was employed to investigate the external flowfield of an SJ. The flow field could be divided into two distinct regions: a developing region near the orifice, named the near field, in which one expect to find signatures of coherent periodic vortex rings and their interactions, and a developed region further away from the orifice, named the far field, in which the vortical structures break down to turbulence and the jet exhibits characteristics of a round continuous turbulent jet. In the far field, the mean velocity profile exhibit a self-similar behavior, with the centerline velocity decaying as x^{-1} and the jet width increasing as x. This certainly resembles a continuous turbulent jet.

Three actuators were used in this study, the dimensions of which are stated in Table 1.1. Figure 1.6a shows the normalized velocity profiles of an SJ as compared with the Schlichting solution for a continuos jet with a similar jet momentum. The SJ velocity profiles show

TABLE 1.1 Geometric Dimensions of the Three Axisymmetric Actuators Tested in the Experiment

	Dimensions (mm)			
Actuator	d	h	D	H
1	1.5	0.5	24.8	1.7
2	2.5	4.2	24.8	1.7
3	2.8	0.6	40.0	3.4

self-similar behavior, while the width of the jet is significantly larger than a continuos turbulent jet with the same jet momentum. This self-similar behavior has led to the hypothesis that an SJ could be modeled similar to a continuos turbulent jet by replacing the eddy viscosity of a turbulent jet with the measured eddy viscosity of an SJ. It was further shown that, similar to a continuous turbulent jet, the eddy viscosity of an SJ could be obtained from the spreading and decay rates of the jet (Krishnan and Mohseni 2009a). The experiments on the flow field not only validated this hypothesis but also showed that the eddy viscosity of an SJ was larger than that of an equivalent (based on momentum flux) turbulent jet, as indicated in Figure 1.6. This enhanced eddy viscosity could be attributed to the additional mixing brought about by the initial introduction of the periodic vortical structures and their ensuing breakdown and transition to turbulence.

Figure 1.6b presents the variation of the calculated virtual viscosity from Equation 1.13 for the three actuators as the actuator stroke ratio is changed. The eddy viscosity of the equivalent CJ is also shown for comparison. SJ and CJ of the same exit diameter d are considered to be equivalent in this investigation, based on momentum flux; otherwise, if the steady bulk exit velocity of a CJ is equal to the mean velocity of an SJ, it is calculated as $U_0 = \sqrt{2}fL$. The eddy viscosity of the SJ is seen to far exceed that of CJs. Larger stroke ratios result in creating larger vortex rings and more energetic near-field region. This, in turn, results in more entrained flow in the jet and higher spreading rate and a higher eddy viscosity.

The Landau–Squire model, which is the solution to the Navier–Stokes equations in spherical polar coordinates, is also testified to be an appropriate model for SJs by experiment. Similar to Schlichting model, with the radial velocity and polar angle being normalized as suggested by Equation 1.14, the velocity profiles at different radial locations collapse onto a single curve (Figure 1.7). By using the measured decay rate and spreading angle, the eddy viscosity is obtained through Equation 1.17, following which the analytical velocity profile is obtained from Equation 1.14. As shown in Figure 1.7, the analytical solution agrees well with the experimental data, and the variation in the half-spreading angle with radial distance downstream of the orifice remains fairly constant. In addition, by comparing the width of the turbulent CJ with that of an equivalent SJ, it is concluded that SJs have larger spreadings than a CJ.

In summary, an SJ could be modeled using both the Schlichting solution to boundary layer equations in cylindrical polar coordinates and the Landau–Squire solution to the Navier–Stokes equations in spherical polar coordinates. Similar to a continuous turbulent jet, the eddy viscosity of an SJ can be obtained from the spreading and decay rates of

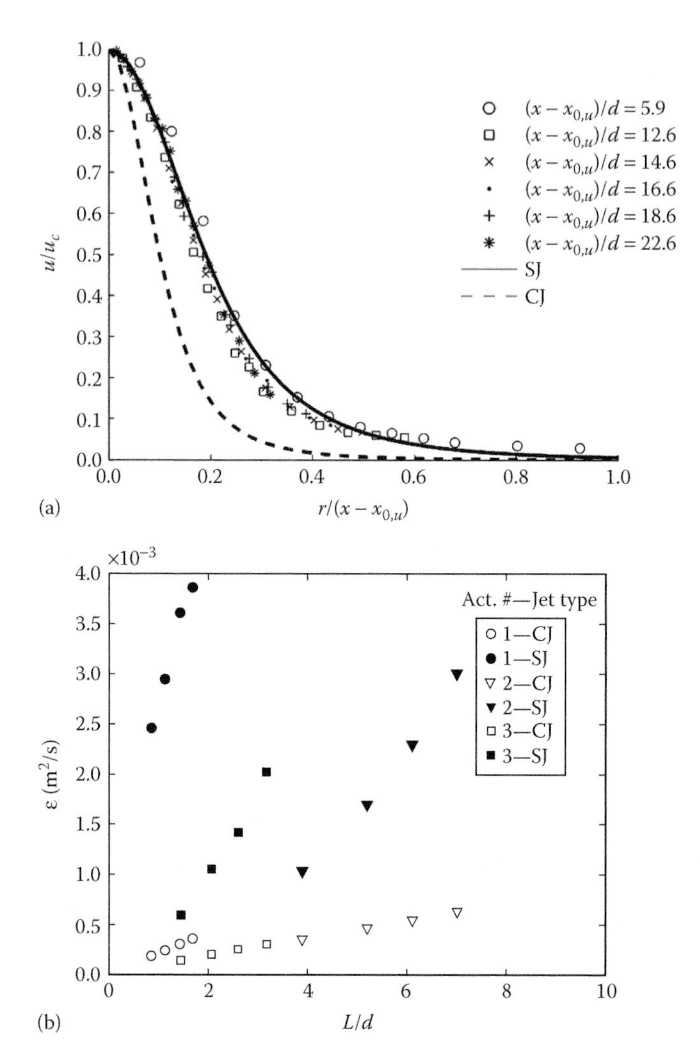

FIGURE 1.6 (a) Comparison of the analytical Schlichting model with experimental normalized velocity profiles for CJs and SJs. The solution for the SJ is shown as a solid line and the solution for a continuous turbulent jet is shown as a dashed line. (b) Dependence of eddy viscosity (ε) on stroke ratio for three actuators, with the eddy viscosity of equivalent turbulent CJs shown for comparison.

the jet. The experiments on the flow field validate this hypothesis, further showing that the eddy viscosity of an SJ is larger than an equivalent turbulent jet. This enhanced eddy viscosity is attributed to the additional mixing brought about by the initial introduction of the periodic vortical structures and their ensuing breakdown and transition to turbulence. Therefore, by using the adjusted value of the virtual viscosity, the theoretical models of a continuous turbulent jet may still be used to model a periodic SJ. The velocity decay rate and spreading rate of SJs are observed to increase with stroke ratio while being

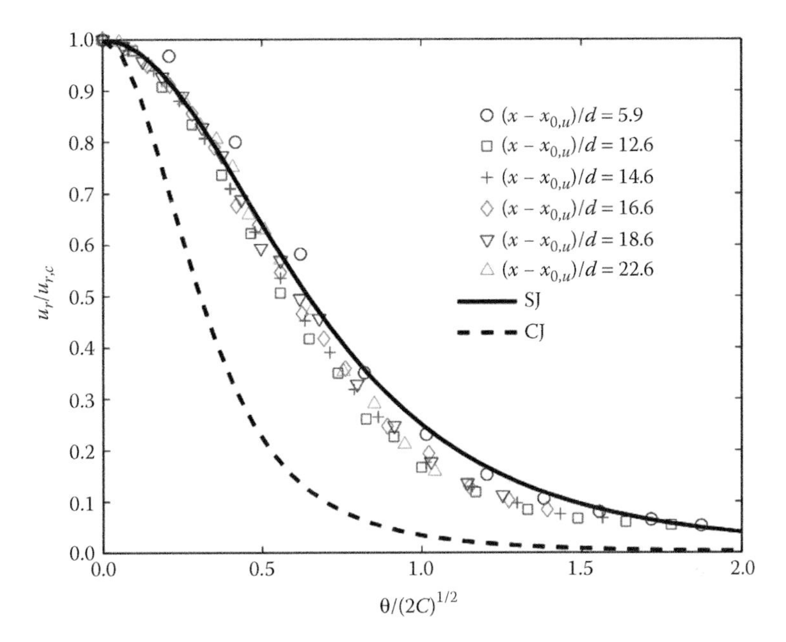

FIGURE 1.7 Comparison of the analytical Landau–Squire model with the experimental normalized velocity profiles. The solution for an SJ is shown as a solid line, and for comparison purposes, the solution for a continuous turbulent jet is shown as a dashed line.

independent of the Reynolds number within the limited range investigated. The geometry of the actuator is, however, seen to have an impact on the decay and spreading rates by means of influencing the initial conditions at the orifice. This dependency of spreading and decay rate on stroke ratio is accredited to the increased impulse, energy, and subsequent enhanced interactions of individual vortex rings emerging from the orifice as the stroke ratio increases. In summary, the semianalytical method proposed here for SJs connects the external flow field as characterized by the spreading (S_b) and velocity decay rates (S_u) to the actuator input driving function (V_d,f) via the actuator parameters ($L/d, Re$).

1.2.2 A Unified Model for Both CJs and SJs in Quiescent Environment

In this section, we offer a unified model for modeling jets in quiescent environment. This will cover laminar, turbulent, pulsatile, and synthetic jets. We will first outline the idea behind this unified modeling and then validate it based on the data from past published results and some recently conducted experiments. The material in this section is borrowed from Xia and Mohseni (2012, 2014, 2015) and interested readers should consult those references for a more detailed discussion.

In 1933, Schlichting (1933; 1979) offered the first analytical solution of the boundary layer equations for the far field of an axisymmetric laminar jet. The jet was generated from a point source of momentum at a virtual source location. He assumed a constant molecular viscosity coefficient for the entire domain and obtained a similarity solution to the axisymmetric boundary layer equations with no external pressure gradient. Although the solution was obtained for laminar jets, Schlichting observed that the spreading rate of this laminar jet is quite smaller than the observed rate from experimental data for turbulent jets. He attributed this mismatch to enhanced effective viscosity associated with turbulence. He then offered to measure this eddy viscosity by measuring the spreading rate of the jet and then estimating the effective viscosity of a turbulent jet by adding the molecular viscosity and this eddy viscosity due to turbulence. This resulted in a nice agreement between his similarity solution and data from turbulent jets. A similar similarity concept could be also extended to 2D plane jets (see Schichting, 1979 for details).

1.2.2.1 Unified Jet Modeling Concept in Quiescent Environment

Inspired by the Schlichting result, Krishnan and Mohseni (2009a) extended Schilchting's concept to obtain a general technique for modeling jets in quiescent environment. In this approach, any modification to a jet that results in enhanced mixing could be potentially modeled by an eddy viscosity and then directly measured from the spreading rate of the jet or from its velocity decay data. To this end, transition of a laminar to a turbulent flow under any perturbations could result in enhanced mixing and appearance of an added viscosity to the molecular viscosity governing the jet spreading. Krishnan and Mohseni (2009a) extended this model to synthetic jets by arguing that pulsation effect of a synthetic jet also contributes to mixing between the jet and its surrounding flow. This extension is further supported by the observation that the far field of synthetic jets displays self-similarity, which is a typical feature of continuous jets as well. Consequently, they replaced the turbulent eddy viscosity with an effective eddy viscosity that includes both turbulence and pulsation effects of a synthetic jet. In this way, the far field of synthetic jets can be modeled in the same manner as that of continuous jets, except that the synthetic jets have an enhanced eddy viscosity.

We believe other jet actuations such as jet pulsation as well as synthetic jets could be also modeled in the same fashion by an eddy viscosity measured directly from jet spreading rate. To this end, one can expect the jet flow properties to scale with the effective viscosity and result in a unified jet model valid for laminar, turbulent, continuous pulsatile, and synthetic jets. This will result in a unified modeling of axisymmetric jets with any axisymmetric actuation. In the rest of this section, we explore this idea and present supporting data for it. 2D plane jets could be modeled in the same fashion.

1.2.2.2 Scale Analysis and Unified Model

Lets apply dimensional analysis to the far field of a turbulent jet in order to identify key parameters relevant to the far-field flow features (Xia and Mohseni 2012). Considering that jet flows are prone to instabilities, the far field of continuous and synthetic jets are expected to be turbulent. To this end, one could imagine that the effects of the pulsation in a synthetic jet are implicitly captured by the turbulent characteristics of the far field. Therefore, pulsation frequency, f, and time are removed from the scaling parameters of the far field. As a result, the main parameters controlling the far field are the far-field eddy viscosity, ε, and the kinematic momentum flux of the jet, K, defined as

$$K = 2\pi \int_0^\infty u^2 r dr \tag{1.18}$$

Now, a straight forward application of the Buckingham Pi theorem to the jet centerline velocity, $u_c = f_1(K, \varepsilon, x)$, results in the following functional dependency:

$$\frac{u_c x}{\sqrt{K}} = \phi_1 \left(\frac{\varepsilon}{\sqrt{K}} \right) \tag{1.19}$$

A similar analysis could be conducted for the dependency of the far-field jet half width $b_{1/2} = f_2(K, \varepsilon, x)$ to obtain

$$\frac{b_{1/2}}{x} = \phi_2 \left(\frac{\varepsilon}{\sqrt{K}} \right) \tag{1.20}$$

Equations 1.19 and 1.20 constitute our basic dependency relationship among the parameter of the far field, namely, K, ε, the nondimensional centerline decay rate, and the spreading rate of the jet. At this point, theoretical relations between $b_{1/2}$, u_c, x, K, and ε can be obtained from Krishnan and Mohseni's (2009a) model of a round synthetic jet to be

$$\frac{1}{S_u \sqrt{K}} = \frac{3\sqrt{K}}{8\pi\varepsilon} \tag{1.21}$$

and

$$S_b = \sqrt{\frac{64\pi\left(\sqrt{2} - 1\right)}{3}} \frac{\varepsilon}{\sqrt{K}} \tag{1.22}$$

where S_u and S_b are the decay rate and the spreading rate of the jet, respectively. Using Equations 1.21 and 1.22, the functions ϕ_1 and ϕ_2 in Equations 1.19 and 1.20 can be found and recasted as

$$S_u \sqrt{K} = k_1 S_b \tag{1.23}$$

and

$$S_b = k_2 \frac{\varepsilon}{\sqrt{K}} \tag{1.24}$$

where:

$$k_1 = \sqrt{\frac{\pi}{3\left(\sqrt{2} - 1\right)}} = 1.590$$

$$k_2 = \sqrt{\frac{64\pi\left(\sqrt{2} - 1\right)}{3}} = 5.269$$

Note that, although synthetic jets and continuous jets are essentially differentiated by their estimated eddy viscosity, they should all satisfy the relationships proposed here by the unified model in Equations 1.23 and 1.24.

1.2.2.3 Experimental Validation

So far, we have proposed a unified model for the far field of any turbulent jet. With the relevant parameters estimated from experimental data, the model represented by Equations 1.23 and 1.24 is validated in this section. The results for both continuous and synthetic jet cases (Xia and Mohseni 2014, 2015) are shown in Figure 1.8. Xia and Mohseni (2014, 2015) reported average standard errors in the 5% range for these data. It can be

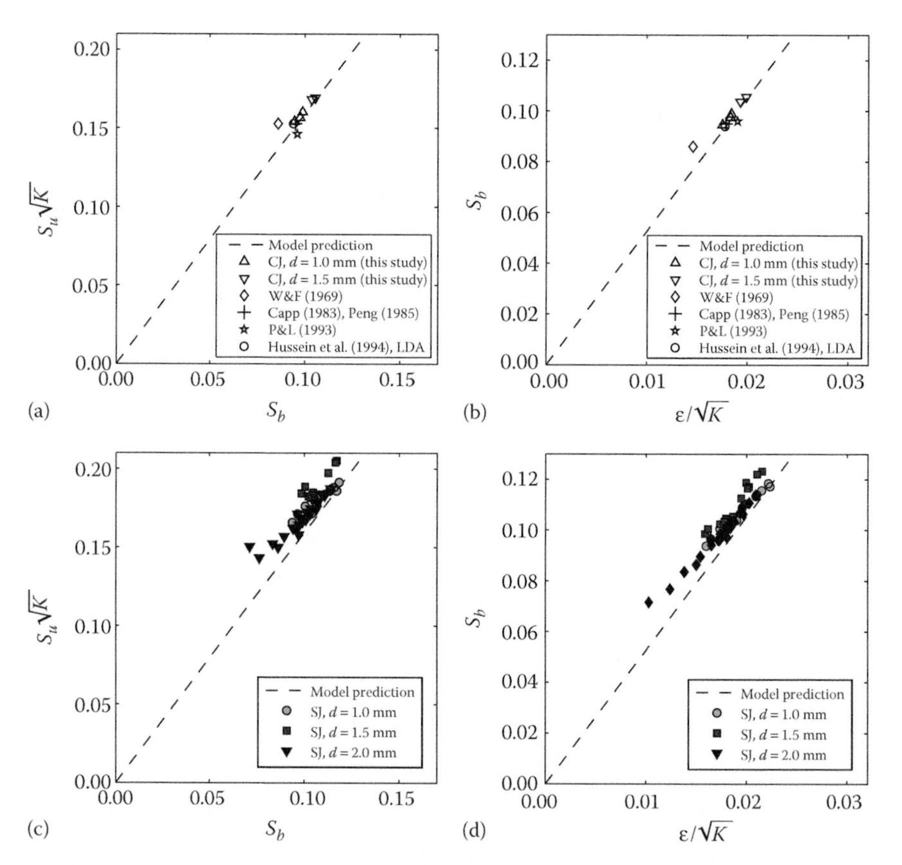

FIGURE 1.8 (a,b) Model validations for continuous jets and (c,d) synthetic jets. The experimental results are generated based on the 5 continuous jet cases and 50 synthetic jet cases (see Xia and Mohseni 2014, 2015 for more details). Results from previous continuous jet studies (Wygnanski and Fiedler 1969; Capp 1983; Peng 1985; Panchapakesan and Lumley 1993; Hussein et al. 1994) are also provided in (a) and (b) for comparison. Note that the model prediction line in (a) and (c) corresponds to Equation 1.23, whereas that in (b) and (d) correspond to Equation 1.24.

generally observed that the experimental data match reasonably well with the model prediction for both continuous jets and synthetic jets.

Although the agreement between experiment and model is promising, there is a slight upward shift in the synthetic jet data from the model. After examining the four relevant parameters of the model, Xia and Mohseni (2014, 2015) attributed this *upshift* to the inaccuracy of the far-field measurement, which treats the varying momentum flux as a constant. To demonstrate the effect of varying momentum flux, the two relations (Equations 1.23 and 1.24) are plotted in Figure 1.9 for a set of five selected test cases, with K (the saturated momentum flux) being substituted by K_x (the momentum flux corresponding to the axial location x). Clearly, it can be observed that higher momentum fluxes, corresponding to the locations closer to the transitional region of the jet, tend to have larger deviations from the model prediction. This suggests that the *upshift* might be due to a not fully saturated momentum flux. In other words, the better K_x approximates the far-field saturation value, K, the better the matching with the model. Therefore, the *upshift* behavior again indicates that the correct scaling is achieved by the far-field momentum K, usually much further downstream from the jet source than previously believed.

To further validate the model, we contrasted our model with several widely reported continuous jet experiments in the past from Wygnanski and Fiedler (1969), Capp (1983), Peng (1985), Panchapakesan and Lumley (1993), and Hussein et al. (1994) as well as our own continuous jet experimental data. These are shown in Figure 1.8a and b. Again, the model shows a good agreement with these previous experimental data. A slight deviation in the model is observed as it compares with the experiments from Wygnanski and Fiedler

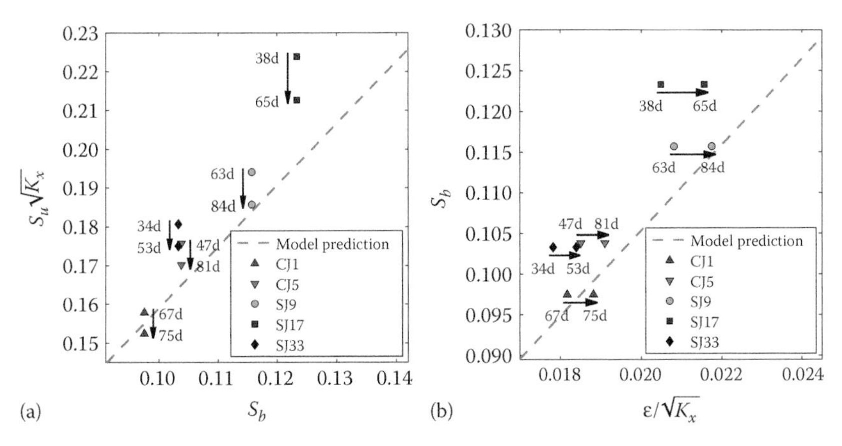

FIGURE 1.9 This figure demonstrates how the variation of momentum flux affects the accuracy of the unified turbulent jet model. For each jet, the two data points show the parameters calculated at two different axial locations, corresponding to the maximum and the minimum momentum flux in the far field, respectively. The coordinates of these locations are specified by the labels, and the directions from the maximum- to the minimum momentum-flux locations are shown by the arrows. Note that K_x instead of K is used in this calculation, whereas the other parameters remain constants in the far field. The dash lines in (a) and (b) represent the model predicted by Equations 1.23 and 1.24, respectively.

(1969). It should be pointed out that, in several previous experiments the jet momentum flux was measured at the jet exit and not at the far field. Considering the variation of the momentum flux with x, reported in Figure 1.9, this deviation between the model and data from Wygnanski and Fiedler (W&F) actually indicates a difference between the far-field momentum flux and the jet exit momentum flux. The momentum loss of the jet in W&F has also been confirmed by several other studies (Seif 1981; Panchapakesan and Lumley 1993; Hussein et al. 1994), whereas both Capp and Hussein et al. attributed the difference with the W&F's jet to a reverse flow caused by a confined spanwise dimension of the experimental setup. In contrast, the continuous jets studied by Capp and Hussein et al. were designed to conserve momentum flux. This is why their results match well with the current model. In summary, our analysis suggests that one must use the far-field momentum, and not the jet exit momentum or the momentum in the transition region, in order to properly scale the far field. Furthermore, our model could be used to identify a momentum loss in the jet at the far field due to a developing axial pressure gradient.

The relationship shown in Equation 1.23, and validated in Figure 1.8b and d, suggests that jet spreading and decay share a similar behavior if the momentum flux is preserved in the far field. On the other hand, the relationship shown in Equation 1.24 indicates that the spreading of a turbulent jet is essentially governed by a nondimensional effective eddy viscosity, $\varepsilon/\sqrt{K}$, regardless of the actuation technique. Combining these two relationships, we can conclude that $\varepsilon/\sqrt{K}$ governs the far field of turbulent jets. In summary, our model was able to reduce the number of required jet parameters from four parameters, S_b, S_u, K, and ε, to only two parameters, ε and $\sqrt{K}$. Further reduction in the number of parameters to predict the behaviors of the far field of a turbulent jet is reported in Xia and Mohseni (2014, 2015) where models for $\varepsilon/\sqrt{K}$ and K are presented.

1.2.3 Two-Dimensional Plane SJ in Quiescent Environment

As for a round SJ, a model for the external flow field of a rectangular SJ in terms of velocity profiles, spreading rates, and decay rates is introduced in this part. Consider the region of the SJ flow where the periodically formed vortical structures cease to be coherent and evolve into a turbulent jet directed downstream (Figure 1.10a). In this regime, the influence of the edge has not completely permeated the flow and the transverse velocity profiles may be thought to be independent of the spanwise coordinate. Therefore, the jet may be viewed as a 2D jet that is issued from a long, narrow slot (momentum source) into the external quiescent environment where it may be admissible to describe the time-averaged velocities in the jet by boundary layer equations. The basic premise in this model is that the observed far-field properties of an SJ allow one to follow the classic modeling approach for turbulent CJs (Schlichting 1933), with the key addition being the replacement of the eddy viscosity of the continuous turbulent jet with that of the SJ. Following the classic derivation of the jet (Schlichting 1933), the boundary layer equations with no pressure gradient may be expressed as

$$u\frac{\partial u}{\partial x} + v\frac{\partial u}{\partial y} = \frac{1}{\rho}\frac{\partial \tau}{\partial y} \text{ and } \frac{\partial u}{\partial x} + \frac{\partial v}{\partial y} = 0 \tag{1.25}$$

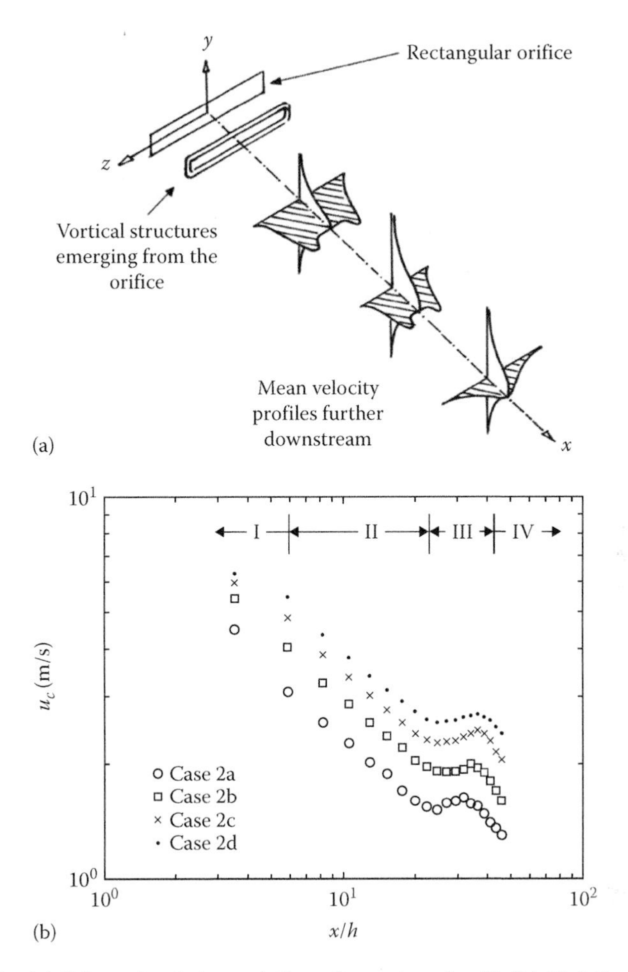

FIGURE 1.10 (a) Schematic of the evolution of a rectangular SJ. (b) Variation in the time-averaged centerline velocity with axial distance, displaying the following: I, developing; II, quasi-2D; III, transition; IV, axisymmetric regions. Cases in (b) are described in Table 1.3.

where:

τ is the turbulent shear stress
ρ is the fluid density

The turbulent shear stress may be related to the time-averaged velocity by an eddy viscosity approximation given as $\tau = \rho\varepsilon\left(\partial u/\partial y\right)$, where ε is a coefficient of eddy viscosity or virtual viscosity. The eddy viscosity hypothesis assumes that the momentum transfer in a turbulent flow is due to the eddies in turbulence in contrast to the laminar flow where molecular diffusion is responsible for momentum transport. With the eddy viscosity characterizing the time-averaged momentum transfer due to turbulent fluctuations, based on the similarity between synthetic rectangular jets and CJs, it is hypothesized here that SJs may be modeled as continuous planar jets with the replacement of the eddy viscosity of the continuous turbulent jet with the enhanced value associated with an SJ.

Assuming that the evolution of the jet is dependent only on local length and velocity scales and lacks memory of the slot dimensions itself, the streamwise time-averaged velocity profiles may be considered self-similar. From the mixing length hypothesis and conservation of streamwise momentum, the characteristic length and velocity of the jet scale may be represented as $b \propto x$ and $u \propto x^{1/2}$, respectively. The self-similar assumption then leads to a streamwise velocity profile of the form $u = x^{1/2} g\left(y/x\right)$. The similarity variable is written as $\eta = \sigma\left(y/x\right)$, where σ is a free constant. As a consequence of the form of self-similarity assumed, the eddy viscosity is proportional to the distance downstream of the momentum source and centerline velocity, $\varepsilon \propto x u_C$. This assumption implies that the eddy viscosity varies only in the streamwise direction and is constant in the lateral direction.

The governing equations along with the self-similar velocity profile assumption, the eddy viscosity model, and boundary conditions result in the reduction of the boundary layer equations to an ordinary differential equation, $\varphi^2 + \varphi' = 1$. This is the same equation obtained for a 2D laminar jet where $\varphi = \tanh(\eta)$ is a solution. From the conservation of momentum and the assumed form of the velocity distribution, the solutions for the velocity components are then written as

$$u = \frac{\sqrt{3}}{2}\sqrt{\frac{K\sigma}{x}}\left(1 - \tanh^2\eta\right) \text{ and } v = \frac{\sqrt{3}}{4}\sqrt{\frac{K}{x\sigma}}\left[2\eta\left(1 - \tanh^2\eta\right) - \tanh\eta\right] \tag{1.26}$$

where:

K is the kinematic momentum per unit length and is a measure of the strength of the jet

It may be obtained from $\rho K = J = \rho\int_{-\infty}^{+\infty} u^2 dy$. It is important to note that the above analysis assumes a constant momentum flux in the streamwise direction. While this is readily applicable to CJs, for SJs it has been reported that the momentum flux at the orifice is higher than that in the far field (Smith and Glezer 1998; Spencer et al. 2005). The momentum flux was shown to decrease in the near field of the jet due to an adverse pressure gradient and then asymptote in the far field to some fraction of the exit momentum flux. It is this reduced asymptotic value of the momentum flux that should serve as the magnitude of the driving momentum flux in the above similarity analysis for the SJ and not the exit momentum flux at the orifice of the actuator. However, if it is assumed that the asymptotic value applies to all SJs equally, then the use of the exit momentum flux as a scale is permissible, with the added benefit that it may be obtained from an actuator model.

The centerline streamwise velocity may be expressed as

$$U_C = \frac{\sqrt{3}}{2}\sqrt{\frac{K\sigma}{x}} = \frac{1}{\sqrt{x}\, S_u} \tag{1.27}$$

where:

$S_u = 2/\sqrt{3K\sigma}$ is the centerline velocity jet decay rate

In scaling this equation, the centerline velocity decay may be expressed as

$$\left(\frac{U_0}{U_c}\right)^2 = K_u\left(\frac{x - x_{0,u}}{h}\right) \tag{1.28}$$

where:

U_0 is an average velocity associated with the actuator
h is the width of the rectangular slot
K_u is a scaled decay rate of the jet
$x_{0,u}$ is the axial location of the virtual origin based on the centerline velocity

The location of the virtual origin is obtained from the intersection of the line asymptote of $1/U_c$ and the abscissa (x) and may be thought of as the self-similarity origin of the jet.

The width of the jet at a particular axial station may be characterized by a half width $b_{1/2}$, defined as the lateral distance from the centerline at which the streamwise velocity drops to half the centerline velocity. The axial variation of the jet width may be expressed either as $b_{1/2} = S_b x$ or identically as

$$\frac{b_{1/2}}{h} = K_b\left(\frac{x - x_{0,b}}{h}\right) \tag{1.29}$$

where:

S_b and K_b are the spreading rates
$x_{0,b}$ is the axial location of the virtual origin based on the jet width

From Equations 1.26 through 1.28, the free constant σ in the similarity variable is related to the spreading rate as $\sigma \approx 0.88/S_b$. From Prandtl's second hypothesis for eddy viscosity (Schlichting 1979), the eddy viscosity may be expressed as

$$\varepsilon \approx \frac{0.28 U_c b_{1/2}}{\sigma} \approx 0.32 S_u S_b^2 \sqrt{x} \approx \frac{0.32 U_0 \sqrt{hx} K_b^2}{\sqrt{K_u}} \tag{1.30}$$

From Equation 1.30, it is seen that the eddy viscosity (1) varies in the axial direction as $\sqrt{x}$, (2) depends on both the dimensions of the orifice and the exit velocity of the jet, and (3) increases with the square of the spreading rate.

In summary, Equations 1.26, 1.28, and 1.29 describe the velocity profiles, jet decay, and spreading rate, respectively, and are used to characterize the external flow field of an SJ. The inputs required are U_0, h, K_u, and K_b that are obtained from experimental measurements.

1.2.3.1 Experimental Results for 2D Plane SJs

First, the results of the external flow field are presented and different regions that comprise the SJ are identified. Figure 1.10b shows the axial progression of the streamwise time-averaged velocity (U_c) along the centerline axis of the jet. The flow field may comprise of four regions. The first region, referred to as a developing region ($0 < x/h < 3$), is not captured in this experiment due to the limitations associated with the hotwire. However, it has been shown that in this region the velocity increases, reaches a maximum, and then starts to decrease (Smith and Glezer 1998). The second region, referred to as a quasi-2D region ($3 < x/h < 25$), is where the velocity decays like a planar jet or as $x^{1/2}$. The third region, referred to as a transition region ($25 < x/h < 35$), is where the jet once again deviates from planar jet behavior, while further downstream the fourth region is where the jet starts to exhibit axisymmetric behavior.

Two actuators were tested for 2D plane jets in this section. The dimensions common to both included nozzle length $w = 38.6$ mm, nozzle depth $t = 0.5$ mm, cavity diameter $D = 44.5$ mm, and cavity height $H = 2$ mm, with the difference being the slot width $h = 0.5$ and 0.85 mm. These were then translated into aspect ratios (w/h) of 77:1 (Actuator 1) and 45:1 (Actuator 2). It was observed that at aspect ratios less than 20:1, a jet with a measurable quasi-planar region was not present and hence not considered in this study. The following table shows the test matrix for each case (Tables 1.2 and 1.3).

Due to a nonperfect 2D actuator and the resulting flow, the flowfield of a 2D rectangular jet shows some three-dimensional (3D) features. To this end, as shown in Figure 1.10b, the flow field of a rectangular SJ could be divided into four regions: (1) an initial developing region where the coherent periodic vortical structures exist and interact, (2) a quasi-2D region where the jet exhibits characteristics similar to a planar CJ, (3) a transition region where the jet once again deviates from planar jet behavior on account of the spreading influence of the narrow edges, and (4) an axisymmetric region where the jet starts to decay like a round jet. In the quasi-2D region, the time-averaged velocity profiles in the lateral direction exhibit self-similar behavior. The centerline velocity decay and jet width growth show trends similar to a continuous turbulent jet, and it is this region that could be modeled as a CJ with a replacement of the eddy viscosity.

TABLE 1.2 Test Matrix for 2D Actuator 1 with $w/h = 77$

Case Number	f (Hz)	V_d (V)	L/h	Re
1a	560	3	7.5	141
1b	560	3.5	9.3	173
1c	560	4	10.9	203
1d	560	4.5	12.3	230
1e	560	5	13.6	254

TABLE 1.3 Test Matrix for 2D Actuator 2 with $w/h = 45$

Case Number	f (Hz)	V_d (V)	L/h	Re
2a	560	1.5	2.5	137
2b	560	2	4.2	228
2c	560	2.5	5.1	275
2d	560	3	5.8	311

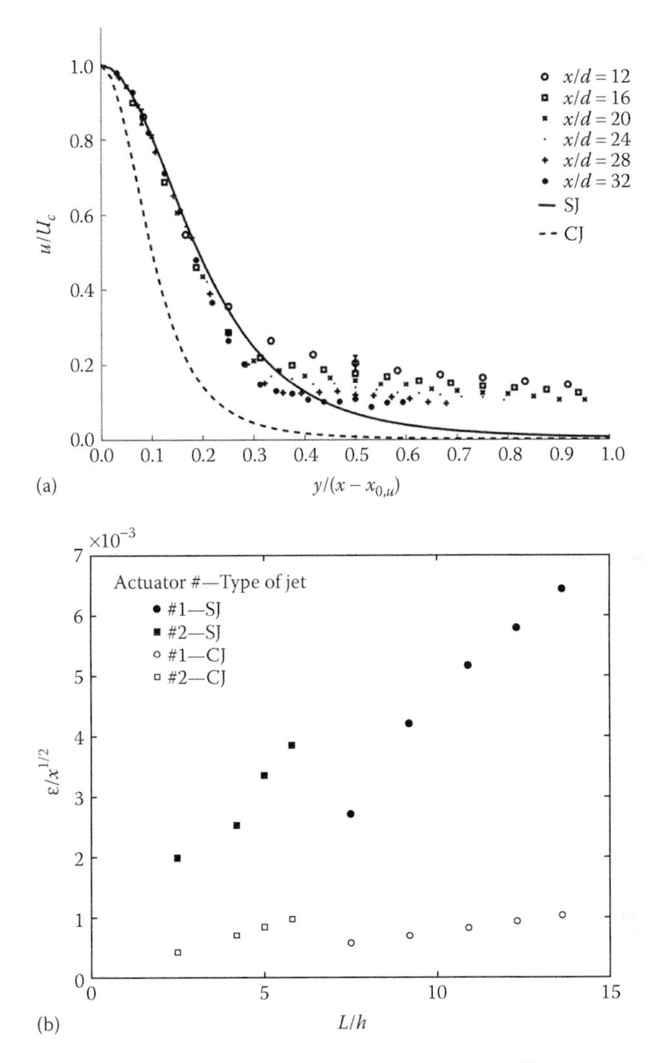

FIGURE 1.11 (a) Normalized time-averaged streamwise velocity profile and the comparison with a CJ. (b) Dependence of eddy viscosity (ε) on stroke ratio for the two SJ actuators. The eddy viscosity of equivalent turbulent CJs is shown for comparison.

Figure 1.11a shows the transverse time-averaged velocity profiles in the quasi-2D region at different streamwise stations. The streamwise velocity and lateral distance are normalized by the centerline velocity and streamwise distance from the velocity virtual origin, respectively. From Figure 1.10b, it is observed that as the jet moves downstream, it slows down, and from Figure 1.11a, it may be deduced that the jet widens due to the entrainment of the surrounding fluid. The profiles collapse reasonably well in the central region and appear to be self-similar and Gaussian in nature. Figure 1.11a also compares the analytical time-averaged velocity profiles of the jet obtained from Equation 1.26 with the experimental data. A profile of an equivalent continuous turbulent jet with a spreading rate of 0.1 is presented for comparison. The enhanced mixing present in the SJ in comparison with the CJ is clearly seen. The model approximates data well toward the center of the jet, thereby validating the eddy viscosity replacement hypothesis. However, progressing away from the center results in some deviations where the model under-predicts the flow.

The variation of the eddy viscosity with stroke ratio for two actuators is shown in Figure 1.11b. The eddy viscosity of an equivalent CJ is also shown for comparison. Again, the eddy viscosity of an SJ is seen to far exceed that of the equivalent CJ. With the eddy viscosity encompassing the capacity to transfer momentum to the surrounding fluid, it appears that the periodic nature of an SJ greatly enhances the momentum transfer in comparison with the CJs. This higher eddy viscosity of an SJ makes it appropriate for applications where changes in the surrounding fluid are desired as in fluid mixing or flow control.

1.2.4 Wall Jet Formed by Round SJ Impingement

The model and experiments reported in this section are based on the works of Krishnan and Mohseni (2010). Figure 1.12a presents the evolution of the flow of an SJ impinging normally upon a wall where the flow may be divided into four regions. The region in the near field of the orifice is typified by the presence of discrete coherent vortex rings and fully periodic flow. Further downstream, the vortex rings begin to interact and break down, reducing their coherence as they transition toward a free turbulent jet directed normal to the wall. The synthetic free jet exhibits similarities to continuous turbulent free jets in their self-similar behavior, while demonstrating an enhanced spreading rate (Smith and Glezer 1998). As the jet approaches the wall, the axial velocity decreases rapidly with the static pressure increasing on account of the presence of a stagnation point at the wall. The jet then is directed radially outward along the wall where the flow is temporarily accelerated on account of the local pressure gradient. Away from the immediate region of impingement, the flow develops into a fully developed wall jet where both a free and wall boundary exists. Here, three established models for the mean velocity of a continuous steady wall jet are outlined and their possible applicability toward synthetic wall jets is discussed.

1.2.4.1 Laminar Model

The classic solution for a laminar wall jet (Tetervin 1948; Glauert 1956) is briefly introduced. For a radial, steady, incompressible, laminar wall jet, it is possible to describe the flow using boundary layer approximations. The governing equations may be expressed as

$$u\frac{\partial u}{\partial r} + v\frac{\partial u}{\partial z} = \nu\frac{\partial^2 u}{\partial z^2} \quad \text{and} \quad \frac{\partial (ru)}{\partial r} + \frac{\partial (rv)}{\partial z} = 0 \tag{1.31}$$

with boundary conditions

$$atz = 0;\; u = v = 0 \;\text{ and }\; z \to \infty;\; u \to 0 \tag{1.32}$$

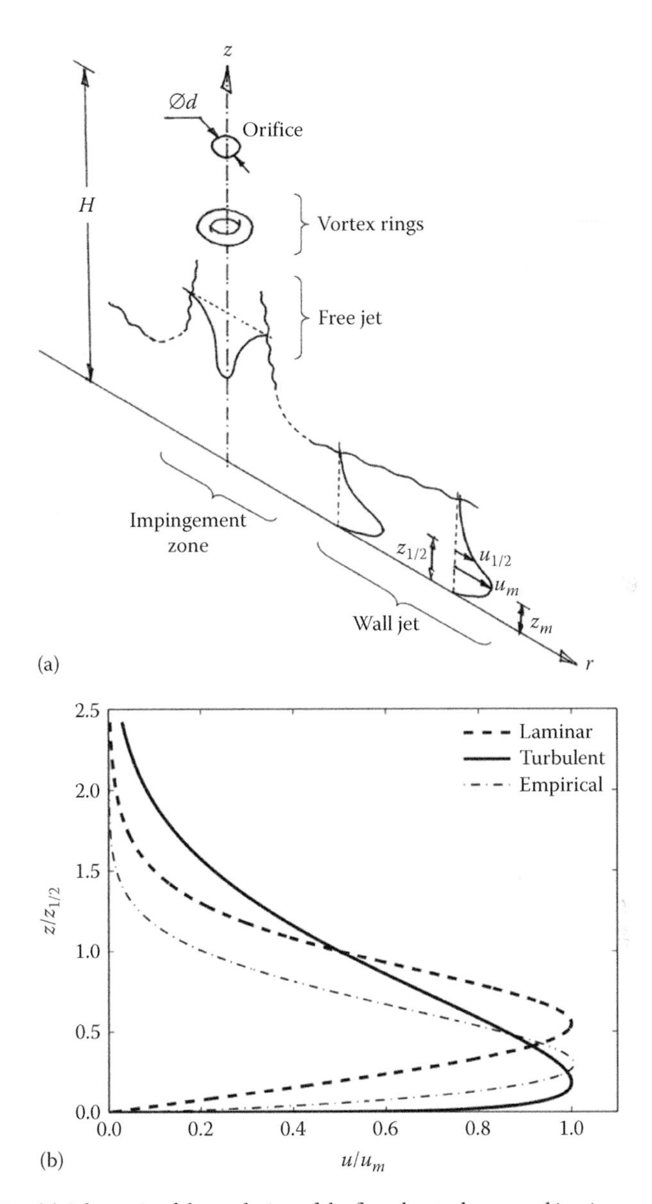

FIGURE 1.12 (a) Schematic of the evolution of the flow due to the normal impingement of a round SJ upon a wall. Shown are vortex rings in the near field of the orifice and mean velocity profiles of both free and wall jets in the far field. (b) Representative velocity profiles of the closed form laminar, semiempirical turbulent, and empirical wall jet solutions. The dashed line represents the classic laminar wall jet solution, the solid line shows a turbulent jet solution for $\alpha = 1.4$, while the dash–dot line shows an intermediate profile that represents the synthetic wall jet for the parameters employed.

where:

r and z are the radial and axial coordinates, respectively
u and v are the radial and axial velocity components, respectively
ν is the kinematic viscosity

Considering U and δ as characteristic velocity and length scales, a similarity solution of the form $u/U = f'(r/\delta)$ is sought. The similarity variable is expressed as $\eta = \eta(r/\delta)$, and the characteristic velocity and length scales are proposed to vary as $U \propto r^a$ and $\delta \propto r^b$, respectively. Through the use of (1) a stream function formulation, (2) the conservation of exterior momentum flux, and (3) the given boundary conditions, the self-similar equation is written as

$$f''' + ff'' + 2f'^2 = 0 \tag{1.33}$$

with boundary conditions $f(0) = f'(0) = 0$ and $f''(\infty) = 0$. An analytical closed form solution to the above exists and is displayed in Figure 1.12b.

1.2.4.2 Turbulent Semiempirical Model

For a fully turbulent steady wall jet, a semiempirical model of the mean velocity profiles exists (Glauert 1956). Boundary layer equations (Equation 1.31) form the basis of the analysis along with an eddy viscosity approximation used in replacing the kinematic viscosity. The flow is divided into two regions. In the inner layer, the eddy viscosity is assumed to vary as $\varepsilon \propto z^{6/7}$, much like a Blasius boundary layer in a pipe, while in the outer layer, it remains constant $\varepsilon = \varepsilon_0$, in the manner of a free turbulent jet. In addition to the first two assumptions stated above for the laminar case, the use of (1) the assumed variation in the eddy viscosity within the respective regions and (2) a method of matching solutions of the inner and outer layer (see Glauert, 1956, for a complete derivation) results in a self-similar equation in the outer layer (f_0) written as

$$f_0''' + f_0 f_0'' + \alpha f_0'^2 = 0 \tag{1.34}$$

and the inner layer (f_i) as

$$\frac{\partial}{\partial \eta}(A f_i'^6 f_i'') + f_i f_i'' + \alpha f_i'^2 = 0 \tag{1.35}$$

with boundary conditions

$$\eta \to 0; f_i'\eta^{-1/7} \to C, f_0(\infty) = 0, \eta = \eta_m; f_0 = f_i, f_0' = f_i', f_0'' = f_i'' \tag{1.36}$$

The above equations are numerically integrated to obtain the inner and outer solutions, where A and α are constants that emerge from the analysis.

Previous work (Wygnanski et al. 1992) has shown that the jet Reynolds number does affect the velocity decay and spreading rates in a turbulent wall jet. However, beyond a certain threshold jet Reynolds number, the self-similar collapse of the mean velocity profiles do not differ and are approximated well with an analytical solution corresponding to $\alpha = 1.3$–1.4. Figure 1.12b shows the turbulent solution for $\alpha = 1.4$. In moving from the laminar to turbulent solution, an increase in the velocity gradient in the inner layer is seen, as is the reduction of the location of the local maximum velocity. This may be attributed to the change in momentum transfer mechanism from viscous diffusion to turbulent mixing. The outer layer tends to spread away from the wall much like a free shear layer, and the inner layer develops a steeper gradient akin to the way a boundary layer reacts to the transition from laminar to turbulent flow.

1.2.4.3 Empirical Model

Under the conditions that either (1) the purely laminar or fully turbulent assumptions do not apply, or (2) the unsteady nature of the jet results in significantly influencing the scaled mean velocity profiles, the above laminar and turbulent formulations may not be pertinent. An alternative approach is based on fitting a description to the velocity profile that is derived informally from the individual forms of the velocity profiles in boundary layer and shear layer. One such function commonly used is given by (Wood et al. 2001)

$$\frac{u}{U_m} = A\left(\frac{z}{z_{1/2}}\right)^{1/n}\left\{1 - erf\left[B\left(\frac{z}{z_{1/2}}\right)\right]\right\} \tag{1.37}$$

where:

A, B, and n are constants to be determined by a fitting procedure, respectively

The wall jet possesses some characteristics similar to a free jet in the outer layer and a boundary layer in the inner layer. Thus, Equation 1.37 may be considered as the product of a power law and error function, where the power law dominates near the wall and the error function dominates in the outer layer. While the function has no rigorous theoretical basis, it provides a simple alternative to modeling the velocity profiles, since at both intermediate Reynolds number flow and externally excited flows, no fundamental theory exists to describe the wall jet. A representative profile intermediate between a steady laminar wall jet and fully turbulent wall jet is shown in Figure 1.12b.

1.2.4.4 Experimental Results

Figure 1.13a presents the normalized mean velocity profiles at different radial locations downstream of the stagnation point. The velocity is scaled by the maximum velocity, while the axial distance from the wall is scaled by the outer half width. The mean velocity profiles are observed to collapse onto a single curve indicating that the velocity and length scales employed in the CJ-induced wall jets are appropriate for a synthetic wall jet as well. While the inner region does appear to collapse, it is known that this region does not scale precisely with the outer variables (Launder and Rodi 1983). Due to the low local Reynolds number of the flow, measurements may have to be made closer to the wall to reveal this deviation.

Both the laminar and turbulent solutions are incapable of faithfully representing the scaled velocity profiles of the synthetic wall jet for the actuator conditions considered. With the turbulent nature of wall jet being evident, the laminar solution quite expectedly did not agree well with the experimental data. The deviation of the turbulent solution

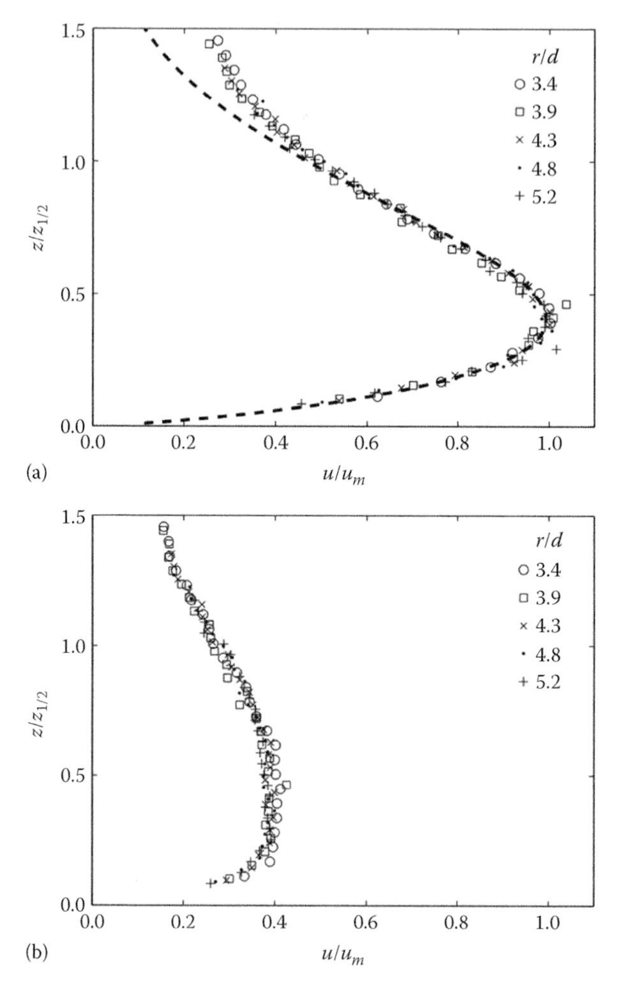

FIGURE 1.13 (a) Normalized radial mean velocity profiles exhibiting self-similar behavior. The analytical solution is represented by the dashed line. (b) Normalized radial rms velocity profiles.

may be attributed to either the low local wall jet Reynolds number or the effect of the high amplitude oscillations driving the flow. With the laminar and turbulent solutions failing to model the synthetic wall jet, Equation 1.37 is observed to agree well with the experimental profile when there is a fit to the data. The deviation toward the outer layer may be attributed to a secondary flow that develops due to both entrainment effects of the free jet and confinement due to the surface of the actuator as is also observed in impinging continuous wall jets (Fairweather and Hargrave 2002).

The profiles of the radial root mean square (rms) velocity when scaled appear to collapse onto a single curve (Figure 1.13b). The profiles display a local maximum, but due to the lack of measurements closer to the wall a secondary maximum associated with the high shear stress near the wall is not captured (Launder and Rodi 1983). However, the collapse of both the mean and rms profiles suggests that at the radial locations considered the jet has achieved self-similarity.

In summary, with the flow field measured at the wall jet region, the synthetic wall jet shows similarities to continuous wall jets in (1) the linear growth of the outer layer, (2) the inverse manner of decay of the maximum velocity, and (3) the collapse of the mean and radial rms velocity profiles when normalized by the outer layer scaling. The spreading and decay rates, however, differ considerably from that of a high Reynolds number continuous wall jet. In contrast to continuous radial wall jets, the flow field of the synthetic wall jet was dominated by vortical structures associated with the actuator driving frequency and harmonics connected with the interaction of the vortex structures. The effect of an increase in actuator driving amplitude at a fixed frequency was found to (1) increase the growth rate of the outer layer and (2) decrease the decay rate of the local velocity maximum. For the actuator conditions investigated, neither the classical laminar nor the fully turbulent analytical solutions to continuous wall jets were able to model the synthetic wall jets. However, a fully empirical form did indeed accommodate the transitional and unsteady nature of the synthetic wall jet. Interested readers are encouraged to consult the works by Krishnan and Mohseni (2010).

1.2.5 Round SJ in a Coflow Wake

The next case of modeling of an SJ is an SJ in a coflow wake, that is, in a mean background flow aligned with the direction of the jet itself. Such a case occurs in applications when an SJ is positioned downstream of a thin obstacle in a background flow. The operation of the SJ is then expected to energize the wake and close the wake region. Considering that an SJ actuator itself could create a wake in a background flow, we also propose modeling procedures for SJs in a coflow wake. Most of the presented results are based on an investigation reported in the works of Xia and Mohseni (2011).

Figure 1.14a shows a schematic of an SJ in an ideal coflow. The near field of the SJ is typified by the presence of discrete coherent vortex rings and fully pulsatile flow. The effect of coflow on the formation of vortex rings has been investigated and shown to reduce the strength and vorticity of the starting vortex rings and to delay the vortex pinch-off process (Krueger et al. 2006). Further downstream, the flow spreads away from the orifice and breaks into a turbulent jet with many similarities to a turbulent CJ.

1.2.5.1 SJ in an Ideal Coflow

Consider the formation of a turbulent jet from a round orifice of diameter d injected parallel to a free stream with a uniform velocity U_a. In this case, ideal coflow here refers to a coflow with no effect of actuator wake, that is, $U_a = u_\infty$. For modeling purposes, the far field of the jet in a coflow may, in principle, be modeled as generated by a continuous point source of momentum in a flow parallel to the jet. The velocity in excess of the coflow velocity at any position is then given by

$$u_e(x,r) = u(x,r) - U_a \tag{1.38}$$

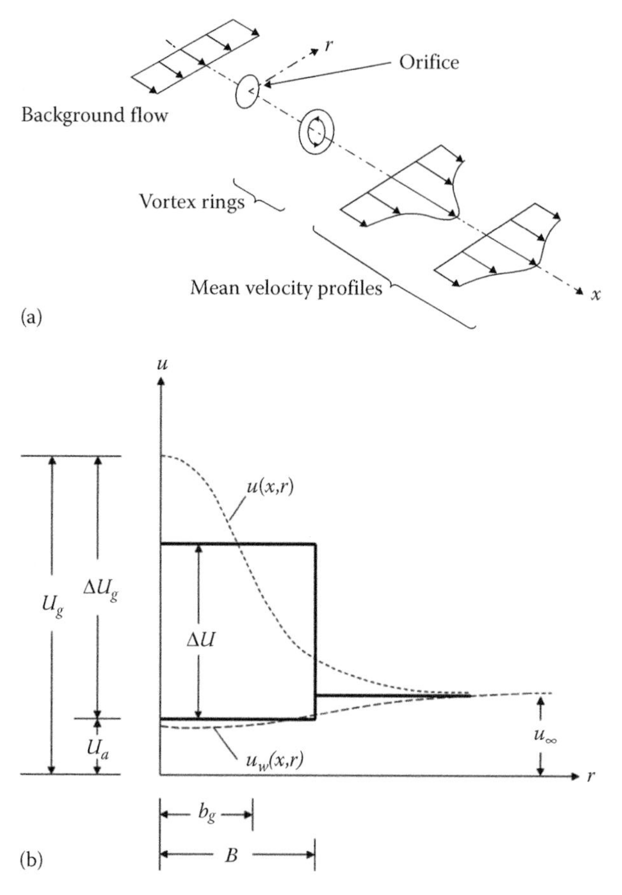

FIGURE 1.14 (a) Schematic of the evolution of a round SJ in a coflow, showing a vortex ring in the near-field and mean velocity profiles of the jet in the far field. (b) Schematics of the equivalent wake velocity profile and the top-hat jet velocity profile. Note that u_∞ is different from U_a in this case. The coflow wake is simplified to a constant velocity of $U_a(x)$ with the same width (B) as the top hat.

The excess velocity profile in the far field of a continuous turbulent jet is shown to exhibit a self-similar behavior (Antonia and Bilger 1973; Nickels and Perry 1996) and to be well approximated by a Gaussian profile given by

$$\frac{u_e}{\Delta U_g} = e^{-r^2/b_g^2} \tag{1.39}$$

where:

ΔU_g is the centerline excess velocity

b_g is the half width of the jet, defined as the radial distance where $u_e = \left(1/e\right)\Delta U_g$

Using boundary layer approximations, it can be shown that the excess kinematic momentum flux (K_e) is conserved in the streamwise direction for a CJ in a coflow (Kotsovinos and Angelidis 1991), defined as

$$K_e = 2\pi \int_0^{\infty} u u_e r dr \tag{1.40}$$

With the behavior of the jet being dependent primarily on K_e and U_a, an excess momentum length scale may be defined by

$$l^* = \frac{\sqrt{K_e}}{U_a} \tag{1.41}$$

In this study, we make the assumption that the width and the mean strength of the excess velocity profile are the most relevant parameters for modeling a jet in a coflow. While a Gaussian velocity profile approximates well the real excess velocity distribution in the far field of a CJ, it carries more information than needed for our modeling purposes. To avoid unnecessary complications, in this study we limit ourselves to a top-hat excess velocity profile with only two main characteristic scales: the width B and the amplitude ΔU. As seen shortly, this implies that the changes in the velocity profile may be rather simply characterized by two parameters of an equivalent top-hat velocity profile (shown in Figure 1.14b) that can be described as

$$u_e = \begin{cases} \Delta U, & r < B \\ 0, & r > B \end{cases}$$

The use of this top-hat profile can be motivated by the need to identify proper velocity and width scales for the jet so that the spreading and decaying behaviors are modeled straightforwardly. Therefore, these characteristic scales are determined from a conservation prospective of the excess mass and excess momentum fluxes so that the large-scale eddies with a length scale on the order of B are responsible for jet spreading and they are advected with respect to the ambient coflow at a velocity of ΔU. The relationship between the top-hat and Gaussian profiles could be derived by equating their excess mass and excess momentum fluxes. These conditions are given by

$$\pi B^2 \Delta U = 2\pi \int_0^{\infty} u_e r dr \tag{1.42}$$

and

$$\pi B^2 (\Delta U + U_a) \Delta U = 2\pi \int_0^{\infty} u u_e r dr \tag{1.43}$$

Solving the above equations for the velocity profile given in Equation 1.39 yields

$$\Delta U = \frac{\Delta U_g}{2} \tag{1.44}$$

and

$$B = \sqrt{2} b_g \tag{1.45}$$

To relate the jet width to the jet velocity, it is assumed that the change in the width of the shear layer moving with the eddies is proportional to the relative velocity between the jet and the surroundings. That is,

$$\frac{dB}{dt} = |U_a + \Delta U|\frac{dB}{dx} = \beta|\Delta U| \tag{1.46}$$

where:
 β is the proportionality factor

The characteristic velocity and length scales are obtained from the top-hat velocity profile. Assuming that β remains constant for a given jet, the above relation should hold for a CJ in a quiescent environment in which case $U_a = 0$. Here, the spreading rate for this free CJ is designated as β_g, which is calculated from the characteristic jet width b_g. Therefore,

$$\beta = \sqrt{2}\beta_g \tag{1.47}$$

The conservation of excess momentum flux defined in Equation 1.40 for the top-hat velocity profile yields

$$K_e = \pi B^2 U \Delta U = \pi B^2 (\Delta U + U_a)\Delta U \tag{1.48}$$

which, using Equation 1.41, can be rearranged to

$$\left(\frac{\Delta U}{U_a}\right)^2 + \left(\frac{\Delta U}{U_a}\right) - \left(\frac{l^{*2}}{\pi B^2}\right) = 0 \tag{1.49}$$

The excess momentum Equation 1.49 and the spreading hypothesis in Equation 1.46 can be nondimensionalized to

$$U^{*2} + U^* - \frac{1}{\pi B^{*2}} = 0 \tag{1.50}$$

and

$$\frac{dB^*}{dx^*} = \sqrt{2}\beta_g \frac{U^*}{1 + U^*} \tag{1.51}$$

where:

$U^* = \Delta U / U_a$

$B^* = B / l^*$

$x^* = x / l^*$

The initial conditions for this set of equations can be written as $U_0^* = \Delta U_0 / U_a$ and $B_0^* = B_0 / l^*$ where U_0 and B_0 are the measured values at some convenient point in the turbulent far-field region. Equations 1.50 and 1.51 are coupled differential equations that predict the streamwise evolution of the velocity magnitude and its spreading. This approach has been successfully employed in modeling CJs in a coflow (Chu et al. 1999).

1.2.5.2 SJ in a Real Coflow Wake

To set up a simplified model for an SJ in a coflow wake, there are several issues that need to be taken into account. The first is the enhanced spreading of an SJ in comparison with a counterpart continuous turbulent jet. While a nominal spreading rate for a continuous turbulent jet is about 0.11 (Schlichting 1979; Pope 2000), the value for an SJ may vary greatly depending on the forcing conditions and upon a nondimensional operational parameter called the stroke ratio (L/d) (Smith and Glezer 1998; Smith and Swift 2001; Shuster and Smith 2007; Krishnan and Mohseni 2009a).

The other is the existence of the coflow wake instead of a uniform coflow. In this study, the coflow velocities vary in both spanwise and streamwise directions inside the wake region. As discussed in Section 1.2.5.1, the coflow velocity U_a is an important parameter effecting the characteristic length and velocity scales. Therefore, it is natural to define the characteristic coflow velocities for the wake with respect to the wake velocity at different streamwise locations from the actuator. To simplify the characterization of the wake velocity field, the same approach as the top-hat profile for the jet velocity field, described in Section 1.2.5.1, is adopted here, and the wake velocity profiles are then averaged in a momentum balanced sense as shown in Figure 1.14b. The local equivalent coflow velocities for the wake region could then be represented by

$$U_a = \frac{2\pi \int_0^B u_w^2 r dr}{2\pi \int_0^B u_w r dr} \tag{1.52}$$

where:

u_w is the actuator wake velocity profile measured without the SJ actuator being actuated

It should be noticed that the choice of the integration region for the wake is the same as the jet top-hat width because this region is where an SJ is mostly affected by the coflow wake.

1.2.5.3 Experimental Results

The introduction of a coflow wake changes the mean properties of the SJ as the coflow serves to both contain and advect the jet. To account for the wake after the jet actuator, the velocity field of the coflow wake is measured first. Then, the scaled collapse of the excess velocity profiles is shown in Figure 1.15a after subtracting the wake velocities from the

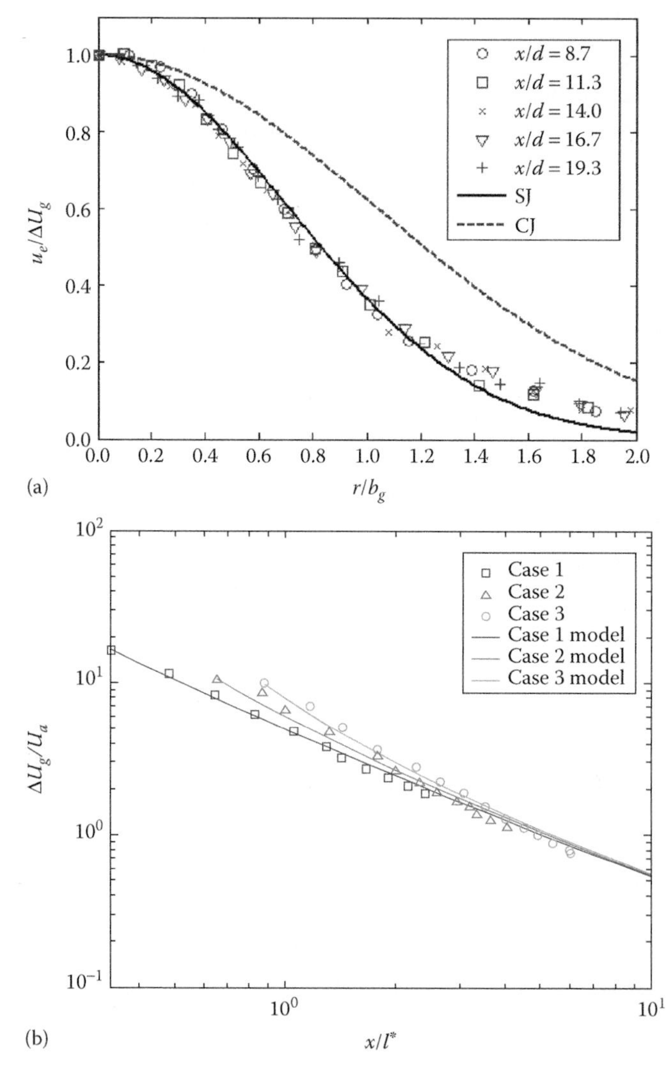

FIGURE 1.15 (a) Normalized excess mean velocity profiles of a n SJ operating in a coflow wake with $U_t = 1.5$ m/s and $L/d = 10.4$ m/s. Note that for comparison, the velocity profile for a CJ in quiescent environment is also plotted. b_g in Equation 1.39 is computed from the Gaussian profile that fits the experimental data. (b) Normalized centerline velocity decay ($\Delta U_g/U_a$) as a function of the scaled streamwise distance (x/l^*) for cases 1, 2, and 3. Analytical solution for SJs solved from our model in Equations 1.50 and 1.51 is also plotted for comparison.

coflow jet velocities. As in Equation 1.39, the excess velocity is scaled by the centerline excess velocity, while the radial distance is scaled by the jet half width. Once again, much like a jet in a quiescent environment, a Gaussian profile approximates the data well except the tail of the profile. This verifies the self-similarity behavior of an SJ in a coflow wake and validates the previous assumptions made in this theory.

The experiments are performed in an open circuit wind tunnel with a cross section of 0.34×0.34 m and a length of 0.9 m. The wind tunnel speed is varied from 1.5 to 2.5 m/s for cases 1, 2, and 3 (specified in Table 1.4) to study the effect of different coflow wakes. The actuator is 45 mm in diameter and 5 mm in thickness. The dimensions of the orifice diameter (d) and depth are 1.5 and 0.5 mm, respectively, while the cavity diameter and depth are 35 and 2 mm, respectively. A sinusoidal input voltage drove the piezoelectric membrane, the frequency and amplitude of which are controllable. The driving frequency (f) is fixed at 2100 Hz, which is close to the resonant frequency of the piezoelectric membrane. The driving amplitude (V_d) is varied from 20 to 30 V for cases 1, 4, and 5 to study the influence of the jet strength on the overall flow. The test matrix along with the relevant nondimensional parameters is summarized in Table 1.4.

Close to the orifice, the jet velocity will be the dominant velocity scale and the effect of the coflow and wake could be neglected. However, further downstream, where the jet decays and the coflow wake velocity is more pronounced, the excess jet velocity would be comparable to the coflow wake velocity. At this region, the jet exhibits some deviation from the behavior of a jet in a quiescent environment. After experimentally identifying the starting point (x_0^*) of the far-field region (see Krishnan and Mohseni, 2009a; 2009b, for calculating the virtual origion of the jet), $U^*(x^*)$ could be obtained in this region by solving Equations 1.50 and 1.51 numerically. The result is compared with the experimental data in Figure 1.15b. As seen in Figure 1.15b, the centerline excess velocity, scaled by the coflow characteristic velocity, is plotted against the axial distance from the jet as scaled by the characteristic momentum length scale. It can be observed that the experimentally scaled velocity decay shows good agreement with the theoretical model in all cases. This indicates that our jet spreading hypothesis provides an accurate model of an SJ in a coflow environment. Furthermore, the wake compensation strategy appears to be an effective technique for defining the relevant coflow velocity and length scales. Moreover, the velocity

TABLE 1.4 Actuator Parameters for an SJ in Coflow Wake.
Test Matrix Displaying the Driving Frequency (f), Voltage (V_d), Corresponding Nondimensional Actuator Parameters (L/d, Re_{U_j}), and Wind Tunnel Velocity (U_t)

Case Number	f (Hz)	V_d (V)	L/d	Re_{U_j}	U_t (m/s)
1	2100	30	10.4	4650	1.5
2	2100	30	10.4	4650	2.0
3	2100	30	10.4	4650	2.5
4	2100	25	9.0	4000	1.5
5	2100	20	7.3	3200	1.5

decay results indicate that the nondimensional velocity decay of the jet is determined only by the spreading rate of the same jet in a quiescent environment. The different initial points of these curves show that different coflows affect the initial momentum ratios between the jet and the ambient flow which is the result of the interaction between the SJ and the coflow in the near field. A weaker coflow velocity could result in a higher initial jet to coflow momentum ratio. However, further downstream, the collapse of the three curves shows the properly scaled velocities decay similarly albeit the changes in the coflow velocity.

In summary, an integral model much like that employed to model CJs in a coflow was shown to be applicable, with some adjustments, to SJs in ideal coflow. However, in most applications of SJs in coflow, the actuator wake significantly affects the mean background coflow in the near-actuator region. In such cases, it has been shown that the same ideal coflow model is valid if the local value of the coflow wake (measured separately without the actuator operating) is employed in the model instead of uniform coflow velocity at far. The excess velocity profiles of the SJ in a coflow demonstrates self-similar behavior when normalized appropriately. The models for the velocity decay showed good agreement with experimental data for different coflow velocities and jet strengths. The presented model indicates that the decay nature of the SJ in a coflow is not affected directly by the coflow but only by the SJ itself. A change in the spreading rate of the jet indicates that the existence of the coflow would prevent the spreading of the SJ and, moreover, increasing the jet strength caused an enhancement in jet spreading.

1.2.6 Round SJ in Crossflow

In many applications, such as flow control problems, modeling of SJs in a crossflow is quite important. The model presented here is based on the works of Xia and Mohseni (2010). For a continuous transverse jet, similarity analysis (Paul 1996; Mungal and Hasselbrink 2001) was used for modeling the flow field. This section considers extending the observed self-similarity in the jet near field to the far field of an SJ in crossflow. With momentum balance applied to the control volume that contains the jet region and ignoring the pressure terms in the equations, we are seeking for the similarity relations for an SJ in crossflow by analogy to a CJ in crossflow with necessary corrections.

1.2.6.1 Theoretical Modeling

Our modeling approach is to extend the techniques developed in Mungal and Hasselbrink (2001) for CJs in crossflow to an SJ. We apply conservation laws to a control volume of the jet in crossflow, upon which similarity relations for jet trajectory and velocities can be derived.

To extend similarity analysis from CJ to SJ in crossflow, we assume the velocity ratio of jet to crossflow is large enough so that the SJ reaches its self-similar region, usually the far field of an SJ, before starting to deflect under the influence of the crossflow. Combining the features of conventional transverse jet and SJ in quiescent environment, one could divide this jet flow into three main regions: (I) A synthetic region, in the first few diameters away from the jet orifice, is the region where strong vortex structures are formed, interact, and coalesce. This region is actually similar to the near field of an SJ in quiescent flow, which could be further divided into two subregions: a region that is very close to the orifice where momentum transfer from the actuator to the jet flow take place; another region further away where the flow structures move away from the orifice and start to mix and interact; (II) A near-field region, just beyond the synthetic region, is the region where the flow becomes turbulent but has not been deflected much. This region is already the far field of an SJ in quiescent environment; and (III) A far-field region is the region where the flow is fully turbulent and has been deflected significantly under the influence of the crossflow.

Figure 1.16a shows a schematic of the control volume including the synthetic transverse jet. To simplify the problem, it is reasonable to assume that the jet region is confined to a characteristic area A_s, and therefore, we have $u \to u_\infty$ and $w \to 0$ outside the jet region. With the above approximations, we can define A_s to be the corresponding cross-sectional area that is perpendicular to the centerline velocity. As a result, the mass flux $\dot{m}$, z-momentum flux $\dot{M}_z$, and x-momentum deficit $\Delta\dot{M}_x$ through the cross section at a given distance s from the jet orifice could be expressed as

$$\dot{m}(s) = \int_{A_s} \rho\sqrt{(u^2 + w^2)}\mathrm{d}A_s \tag{1.53}$$

$$\dot{M}_z(s) = \int_{A_s} \rho w\sqrt{(u^2 + w^2)}\mathrm{d}A_s \tag{1.54}$$

$$\Delta\dot{M}_x(s) = \int_{A_s} \rho(u_\infty - u)\sqrt{(u^2 + w^2)}\mathrm{d}A_s \tag{1.55}$$

where:

u is the x-component of the velocity vector
w is the z-component of the velocity vector
u_∞ is the crossflow velocity where the jet has reached the far field completely

For simplicity, several assumptions are made. These include neglecting the net pressure integral over the surface volume and ignoring any mass or momentum transfer outside A_s. Consequently, from the mass and momentum conservation laws, it is implied that there are two flow invariants that characterize this jet flow: $\dot{M}_z$ and $\Delta\dot{M}_x$. It is then possible to do similarity analysis based on one of the invariants. It should be pointed out that different from continuous transverse jet in which case $\dot{m}_j$ comes from other source flow, $\dot{m}_j$ of synthetic transverse jet comes essentially from the crossflow, and it is clearly seen from the momentum balance equations that the actuator is actually absorbing the x-momentum of $\dot{m}_j u_\infty$ from the crossflow and converting that to z-momentum of $\dot{m}_j w_j$ that synthesizes the jet.

Assuming that self-similar jet regions exist in the near and far fields, and ρ, w_c, and $u_\infty - u_c$ be normalized by a characteristic width of $\delta(z_c)$ in the local n-direction, we derive similarity relations for the near and far fields, respectively.

Near field: the velocity is dominated by w_c. It is natural to assume $\sqrt{u_c^2 + w_c^2} \approx w_c$ and $\dot{M}_z$ as the invariant. Then, the related parameters are $w_c, \dot{M}_z, z, \rho$, and ν. A nondimensional analysis for the conservation of $\dot{m}, \dot{M}_z$, and $\Delta\dot{M}_x$ results in

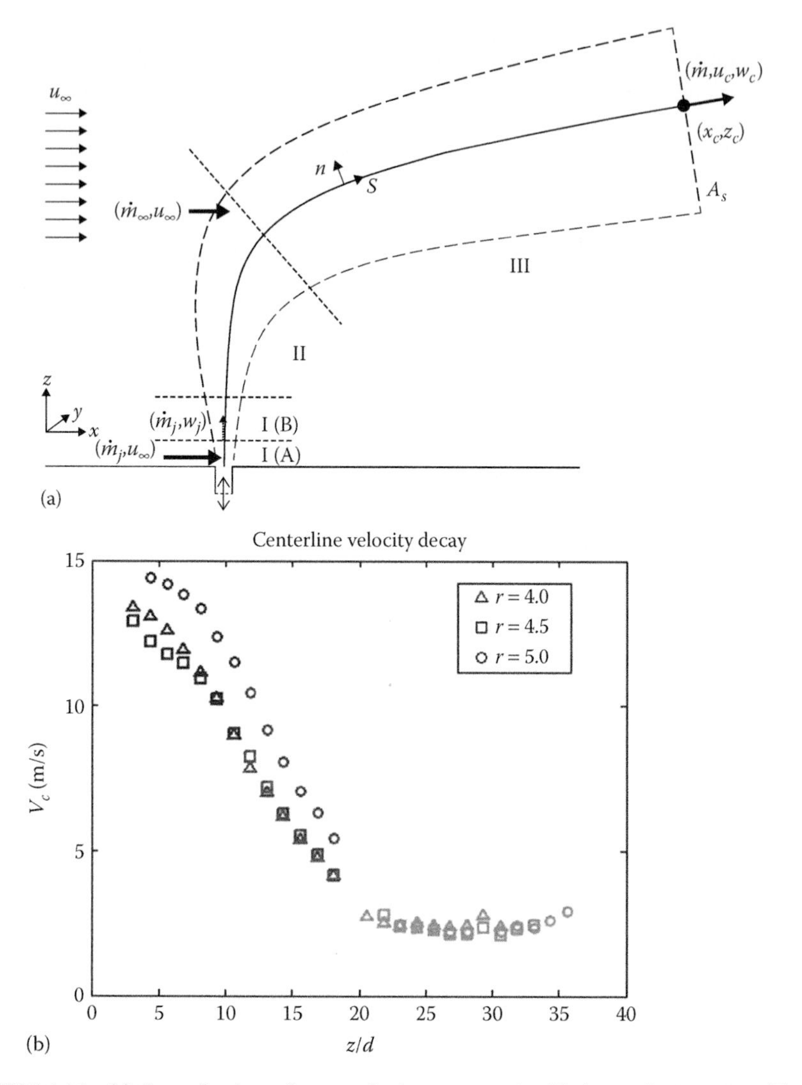

FIGURE 1.16 (a) Control volume for a synthetic transverse jet: (I) the synthetic region, (II) the near-field region, and (III) the far-field region. (b) Centerline velocity decay from experimental data: the left, middle, and right set of data point represent regions (I), (II), and (III), respectively. The triangle, square, and circle points represent the three test cases.

$$w_c \propto \dot{M}_z^{1/2}\rho^{-1/2}z^{-1} \tag{1.56}$$

$$\delta \propto z \tag{1.57}$$

$$u_\infty - u_c \propto \Delta\dot{M}_x\dot{M}_z^{-1/2}\rho^{-1/2}z^{-1} \tag{1.58}$$

Far field: In this region, the jet is significantly deflected in the crossflow direction. It is therefore assumed that $\sqrt{u_c^2 + w_c^2} \approx u_\infty$, and $u_\infty - u_c$ is the objective parameter we are interested in. u_c is ignored in this case, so the related parameters are $u_c, u_\infty, \dot{M}_z, z, \rho$, and ν. In this case, the conservation equations for $\dot{m}$, $\dot{M}_z$, and $\Delta\dot{M}_x$ could be represented by

$$u_\infty - u_c \propto \dot{M}_z \rho^{-1} u_\infty^{-1} z^{-2} \tag{1.59}$$

$$\delta \propto z \tag{1.60}$$

$$w_c \propto \dot{M}_z \rho^{-1} u_\infty^{-1} z^{-2} \tag{1.61}$$

With these scaling relations, similarity solutions for the jet trajectories can be obtained in the near-field and far-field regions. The jet trajectory is defined as the jet center streamline such that $dz/dx = w_c/u_c$, where the Equations 1.56 and 1.61 provide w_c in the near field and far field, respectively, and the assumption that $u_c \rightarrow u_\infty$ in this region. These are, respectively,

$$\frac{z_c}{rd} = A_n \left(\frac{x_c}{rd}\right)^{1/2} \tag{1.62}$$

and

$$\frac{z_c}{rd} = A_f \left(\frac{x_c}{rd}\right)^{1/3} \tag{1.63}$$

Here, r is the velocity ratio that is defined as w_j/u_∞, and d is the diameter of the jet orifice. With z_c/d as the length metric for near field, combined with Equation 1.62, we can write the scaling relations of velocities (Equations 1.56 and 1.58) to be

$$\frac{w_j}{w_c} = B_n \frac{z_c}{d} \tag{1.64}$$

$$\frac{u_\infty}{u_\infty - u_c} = C_n \frac{z_c}{d} \tag{1.65}$$

In the far field, with x_c/rd as the length metric, combined with Equation 1.63, we can write the scaling relations of velocities (Equations 1.59 and 1.61) to be

$$\frac{w_j}{w_c} = B_f r \left(\frac{x_c}{rd}\right)^{2/3} \tag{1.66}$$

$$\frac{u_\infty}{u_\infty - u_c} = C_f r \left(\frac{x_c}{rd}\right)^{2/3} \tag{1.67}$$

The constants of proportionality, A_n, A_f, B_n, and B_f, are expected to be determined experimentally.

1.2.6.2 Experimental Results

The flow field of an SJ in crossflow is characterized by the three regions, the differences between which are essentially reflected in their different characteristics of the velocity field. Basically, in this study, the three regions are determined by the centerline velocity decay as shown in Figure 1.16b. In the synthetic region (I), the centerline velocities are observed to be decaying in a manner similar to that of an SJ in quiescent flow. Further experimental data have shown that the velocity was increased in the first couple of diameters away from the jet orifice due to the pulsatile nature of SJs. In the near-field region (II), the decay of the jet tends to be stable. In the far-field region (III), the jet has deflected significantly and the u velocity starts to dominate the jet flow, so the centerline velocities are observed to decay remarkably slower than the near-field region and then start to increase as it spreads further into the crossflow.

After the near-field and far-field regions are determined, from previous similarity modeling for the trajectories (Equations 1.62 and 1.63), it is possible to fit experimental data with those relations and find the coefficients A_n and A_f for different cases. From the previous study, the length of the potential core is set to be $4d$ (Mungal and Hasselbrink 2001) and the horizontal dimension of the near field is computed to be approximately $4d$ (Paul 1996). The coefficients A_n and A_f are set to be 2.5 and 1.6, respectively. The results for these comparisons are shown in Figure 1.17a and b for the near and far fields.

The linear trends reflected from the coefficients validate the previous assumptions and derivations for this similarity modeling approach. Comparing our SJs to the CJ of velocity ratio equal to 5.0, we found that the coefficients for the SJs are significantly larger than those of the CJ (2.5 and 1.6 for the near and far fields, respectively). This means that the transverse SJ has a much less entrainment in the near and far fields. Moreover, as we compare the coefficients A_n and A_f between the different cases, A_n for the near field has a relatively good consistency that verifies the self-similar behavior for the SJ in the near field. However, in the far field, A_n tends to increase slightly as velocity ratio is increased. This could be explained by the influence of the boundary layer growth along the channel's wall. In the case with a larger velocity ratio, we have a lower crossflow velocity and the jet bends less in the far field than a jet with lower r. So, the crossflow velocity changed more in the z vertical direction for a larger r, and it is reasonable that in this case, the r is overestimated more than the case with smaller velocity ratio because of a more scattered crossflow velocity distribution in vertical direction. And it could be predicted that, for an SJ that interact with crossflow boundary layer, the coefficient A_f would be approximately 1.6 as r decreases in a certain range (this theory do not apply for very small r). So with the trajectory law testified, this self-similarity approach should provide appropriate model for SJs in crossflow.

In summary, similarity analysis in CJs could be effectively extended to model SJs in a crossflow. To model the flow field of a transverse SJ, three regions—the synthetic region, the near-field region, and the far-field region—are proposed. The near field could be modeled similar to the far field of an SJ in quiescent flow, while the far field could be treated as turbulent far field of a continuous transverse jet. In the synthetic region, we found that the velocity magnitude of jet was affected by the crossflow, while the length of the region remains the same and it was not affected by the crossflow. In the near-field region, by comparing the trajectory coefficients with different velocity ratios, it is found that the

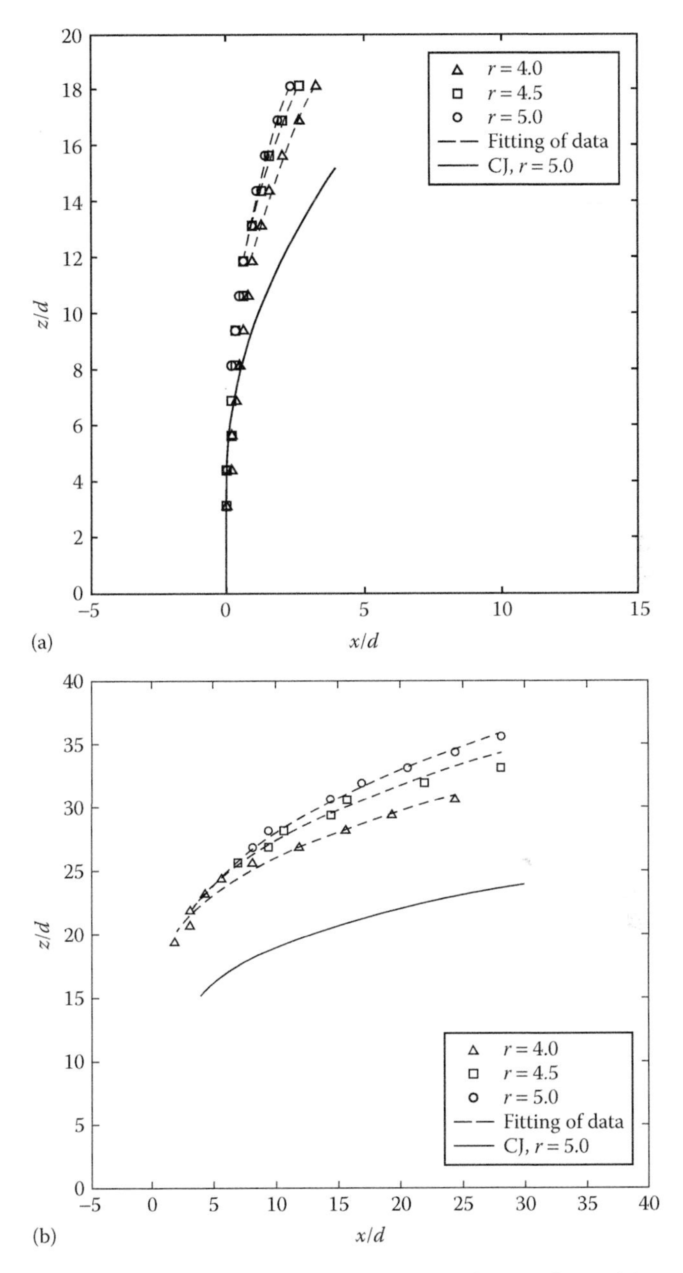

FIGURE 1.17 Comparison of trajectories between cases of SJs and CJs: (a) near field and (b) far field.

entrainment increases with stronger crosswind. Moreover, the decay rates and spreading rates of transverse SJs are larger than that of a free SJ. In the far-field region, we found that the entrainment is not only related to the velocity ratio but also affected by the velocity profile of the crossflow boundary layer.

References

Antonia, R. A. and R. W. Bilger. 1973. Experimental investigation of an axisymmetric jet in a co-flowing air stream. *Journal of Fluid Mechanics*, 61: 805–822.

Capp, S. P. 1983. Experimental investigation of the turbulent axisymmetric jet. PhD thesis, University at Buffalo, SUNY, Buffalo, NY.

Chu, P. C. K., J. H. W. Lee, and V. H. Chu. 1999. Spreading of turbulent round jet in coflow. *Journal of Hydraulic Engineering*, 125: 193–204.

Fairweather, M. and G. K. Hargrave. 2002. Experimental investigation of an axisymmetric, impinging turbulent jet. 1. Velocity field. *Experiments in Fluids*, 33: 464–471.

Fugal, S. R., B. L. Smith, and R. E. Spall. 2005. Displacement amplitude scaling of a two-dimensional synthetic jet. *Physics of Fluids*, 17(4): 045103-1–10.

Glauert, M. B. 1956. The wall jet. *Journal of Fluid Mechanics*, 1: 625–643.

Glezer, A. and Amitay, M. 2002. Synthetic jets. *Annual Review of Fluid Mechanics*, 34: 503–529.

Hussein, H. J., S. P. Capp, and W. K. George. 1994. Velocity measurements in a high Reynolds number, momentum conserving, axisymmetric, turbulent jet. *Journal of Fluid Mechanics*, 258: 31–76.

Kotsovinos, N. E. and P. B. Angelidis. 1991. The momentum flux in turbulent submerged jets. *Journal of Fluid Mechanics*, 229: 453–470.

Krieg, M. and K. Mohseni. 2008. Thrust characterization of pulsatile vortex ring generators for locomotion of underwater robots. *IEEE Journal of Oceanic Engineering*, 33(2): 123–132.

Krishnan, G. 2009. Experimental and theoretical investigation of synthetic jets. PhD thesis, University of Colorado, Boulder, CO.

Krishnan, G. and K. Mohseni. 2009a. Axisymmetric synthetic jets: An experimental and theoretical examination. *AIAA Journal*, 47(10): 2273–2283.

Krishnan, G. and K. Mohseni. 2009b. An experimental and analytical investigation of rectangular synthetic jets. *Journal of Fluids Engineering*, 131(12). doi:10.1115/1.4000422.

Krishnan, G. and K. Mohseni. 2010. An experimental study of a radial wall jet formed by the normal impingement of a round synthetic jet. *European Journal of Mechanics B/Fluids*, 29: 269–277.

Krueger, P., J. Dabiri, and M. Gharib. 2006. The formation number of vortex rings formed in a uniform background co-flow. *Journal of Fluid Mechanics*, 556(1): 147–166.

Landau, L. 1944. A new exact solution of Navier-Stokes equations. Akademija Nauk SSSR (Moscow): Doklady Akademie Nauk SSSR (Moscow), Vol. 43, pp. 286–288.

Launder, B. E. and W. Rodi. 1983. The turbulent wall jet—Measurements and modeling. *Annual Review of Fluid Mechanics*, 15: 429–459.

Mohseni, K. 2006. Pulsatile vortex generators for low-speed maneuvering of small underwater vehicles. *Ocean Engineering*, 33(16): 2209–2223.
Mungal, M. G. and E. F. Hasselbrink. 2001. Transverse jets and jet flames. Part 1. Scaling laws for strong transverse jets. *Journal of Fluid Mechanics*, 443: 1–25.
Nickels, T. B. and A. E. Perry. 1996. An experimental and theoretical study of the turbulent coflowing jet. *Journal of Fluid Mechanics*, 309: 157–182.
Panchapakesan, N. R. and J. L. Lumley. 1993. Turbulence measurements in axisymmetric jets of air and helium. Part 1. Air jet. *Journal of Fluid Mechanics*, 246: 197–223.
Paul, C. C. K. 1996. Mixing of turbulent advected line puffs. PhD thesis, The University of Hong Kong, Pokfulam, Hong Kong.
Peng, D. 1985. Hot wire measurements in a momentum conserving axisymmetric jet. Master's thesis, University at Buffalo, SUNY, Buffalo, NY.
Pope, S. B. 2000. *Turbulent Flows*. New York: Cambridge University Press.
Schlichting, H. 1933. Laminare strahlausbreitung. *Journal of Applied Mathematics and Mechanics (ZAMM)*, 13: 260–263.
Schlichting, H. 1979. *Boundary-Layer Theory*. New York: McGraw-Hill.
Seif, A. A. 1981. Higher order closure model for turbulent jets. PhD thesis, University at Buffalo, SUNY, Buffalo, NY.
Sherman, F. S. 1990. *Viscous Flow*. New York: McGraw-Hill.
Shuster, J. M. and D. R. Smith. 2007. Experimental study of the formation and scaling of a round synthetic jet. *Physics of Fluids*, 19(4): 45109-1-21.
Smith, B. L. and A. Glezer. 1998. The formation and evolution of synthetic jets. *Physics of Fluids*, 10(9): 2281–2297.
Smith, B. L. and G. W. Swift. 2001. Synthetic jets at larger Reynolds number and comparison to continuous jets. *The 31st AIAA Fluid Dynamics Conference and Exhibit*, Anaheim, CA, June 11–14.
Squire, H. B. 1951. The round laminar jet. *The Quarterly Journal of Mechanics & Applied Mathematics*, 4(3): 321–329.
Tetervin, N. 1948. *Laminar Flow of a Slightly Viscous Incompressible Fluid that Issues from a Slit and Passes over a Flat Plate*. Washington, DC: National Advisory Committee for Aeronautics, p. 40.
Timoshenko, S. P. 1999. *Theory of Plates and Shells*. New York: McGraw-Hill.
Wood, G. S., K. C. S. Kwok, N. A. Motteram, and D. F. Fletcher. 2001. Physical and numerical modelling of thunderstorm downbursts. *Journal of Wind Engineering & Industrial Aerodynamics*, 89: 535–552.
Wygnanski, I. and H. E. Fiedler. 1969. Some measurements in the self-preserving jet. *Journal of Fluid Mechancis*, 38: 577–612.
Wygnanski, I., Y. Katz, and E. Horev. 1992. On the applicability of various scaling laws to the turbulent wall jet. *Journal of Fluid Mechanics*, 234: 669–690.
Xia, X. and K. Mohseni. 2010. Modeling and experimental investigation of synthetic jets in cross-flow. *The 48th AIAA Aerospace Sciences Meeting Including the New Horizons Forum and Aerospace Exposition*, Orlando, FL, January 4–7.
Xia, X. and K. Mohseni. 2011. An experimental and modeling investigation of synthetic jets in a coflow wake. *International Journal of Flow Control*, 3(1): 19–35.

Xia, X. and K. Mohseni. 2012. Axisymmetric synthetic jets: A momentum-based modeling approach. In *Proceedings of the 50th AIAA Aerospace Sciences Meeting including the New Horizons Forum and Aerospace Exposition*, AIAA Paper 2012–1246, Nashville, TN, January 9–12.

Xia, X. and K. Mohseni. 2014. Far field modeling of high-frequency synthetic jets. *Submitted for publications.*

Xia, X. and K. Mohseni. 2015. Far field of high-frequency synthetic jets: Modeling and momentum flux. *Submitted to the 53rd Aerospace Sciences Meeting*, Kissimmee, FL, January 5–9.

2

Design of Synthetic Jets

Louis N. Cattafesta
Florida State University

Matias Oyarzun
University of Florida

2.1 Introduction

Zero-net mass-flux (ZNMF) devices, otherwise known as synthetic jet actuators, have garnered much attention over the past several decades from the active flow control community. ZNMF devices have been used in a wide range of active flow applications, which include separation control (Amitay et al. 2001; Amitay and Glezer 2006), jet vectoring (Smith and Glezer 2002, 2005), mixing enhancement (Chen et al. 1999; Chiekh et al. 2003), thermal management (Mahalingam and Glezer 2005; Pavlova and Amitay 2006), aero-optics (Vukasinovic et al. 2009, 2010), and cavity oscillations (Cattafesta et al. 2008). Glezer and Amitay (2002) provide a detailed review of synthetic jets and their applications. While usage of ZNMF devices is widespread, the systematic design of such devices remains somewhat of a mystery.

Even with the wealth of information that exists with regard to ZNMF actuators in the literature, it can be a daunting and time-consuming task to properly design a ZNMF actuator for a specific application. The performance requirements of any actuator for active flow control applications may include the following: level of control authority, bandwidth, robustness, power consumption, efficiency, size, weight, and cost, to name a few. In addition to these requirements, the geometry of the driver, cavity, and orifice/slot has to be considered as well as the materials that are used. There are many variables that have to be accounted for when designing ZNMF actuators, and trial-and-error methodologies are often prohibitive due to financial cost and time constraints.

The purpose of this chapter is to provide an overview of the design of synthetic jet actuators, in particular performance metrics and several implementation issues that often arise when designing these devices. Lumped element modeling (LEM) of synthetic jet actuators is introduced as a design tool, including several illustrative examples of how LEM is used for piezoelectric and electrodynamic synthetic jet actuators. The limitations are addressed, as well as the design trade-offs, optimization, and possible future trends.

2.2 Actuator Components

Before discussing the specific details of the design of synthetic jet actuators, a brief discussion of each actuator component is warranted. In general, a typical synthetic jet actuator consists of the following components: a driver (e.g., piezoelectric and electrodynamic), a cavity, and an orifice/slot, as illustrated in Figure 2.1.

Figure 2.1 represents the most generic design of a synthetic jet actuator, which consists of multiple drivers, cavities, and orifices/slots. The primary purpose of the driver in a synthetic jet actuator is to displace large volumes of fluid at desired frequencies, which then forces the fluid into and out of the cavity through the orifice or slot. The two most commonly employed drivers, which are discussed in detail in this chapter, are a

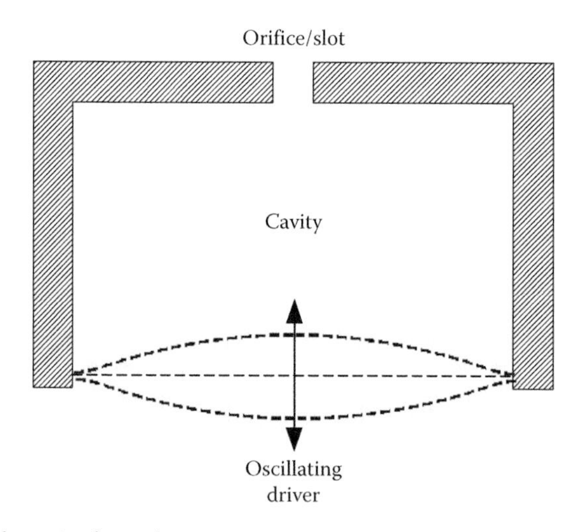

FIGURE 2.1 Schematic of a synthetic jet actuator.

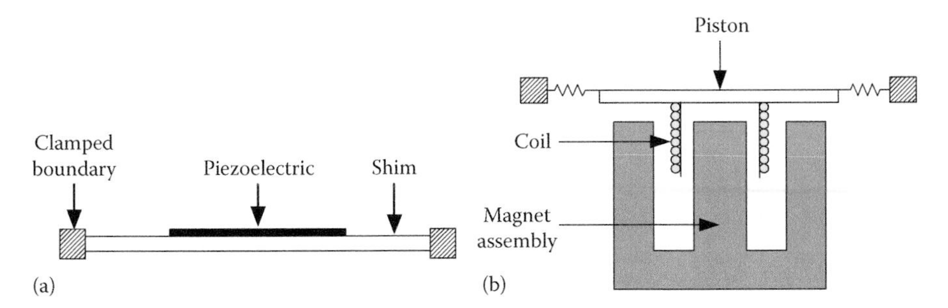

FIGURE 2.2 (a) Typical piezoelectric composite diaphragm configuration (unimorph) and (b) typical electrodynamic piston configuration.

piezoelectric composite diaphragm and an electrodynamic (e.g., voice-coil) piston or diaphragm, illustrated in Figure 2.2.

The piezoelectric composite diaphragm driver type typically consists of a metallic shim (usually brass) that has either a single piezoceramic (e.g., PZT-5A) patch bonded to its surface (unimorph) or two symmetrically bonded piezoceramic patches (bimorph). The piezoceramic patches have thin surface-coated electrodes (typically silver) that allow for electrical connections. The metallic shim is then clamped, as illustrated in Figure 2.2a, or pinned (Chen et al. 2000), depending on the application and/or design constraints. In general, for a given diaphragm, these two boundary conditions will lead to markedly different deflection profiles and will have an impact on the resonant frequency of the diaphragm, as well as its mechanical properties (e.g., stiffness and mass). The different boundary conditions are discussed in greater detail in Section 2.4. When an alternating current (ac) sinusoidal voltage is applied across the piezoelectric patch, an electric field is created, and via the indirect piezoelectric effect, the piezoelectric material experiences strain. This strain causes a bending moment in the piezoelectric diaphragm, which in turn causes the piezoelectric diaphragm to deflect, leading to a volume displacement (ΔVol) and volume velocity ($Q = j\omega\Delta\text{Vol}$).

The electrodynamic piston driver type, illustrated in Figure 2.2b, typically consists of a piston, a coil, and a magnetic assembly (McCormick 2000; Agashe et al. 2009). The magnetic assembly is responsible for creating a static, permanent magnetic field with a magnetic flux density (B). When an ac current (I) is passed through the coil with coil length ℓ_{coil}, the interaction results in an electromagnetic force, or Lorentz body force ($F = B\ell_{\text{coil}}I$), which causes the piston to oscillate once again leading to a volume displacement (ΔVol). The volume displacement is related to the volume velocity ($Q = j\omega\Delta\text{Vol}$) and is related to the magnetic properties and coil properties by $V_{\text{ac}} = B\ell_{\text{coil}}Q/S_{\text{eff}}$, where V_{ac} is the ac voltage and S_{eff} is the effective area of the piston or diaphragm.

The cavity of a synthetic jet actuator plays an important role on the overall device performance and it is where the fluid is periodically compressed and expanded. In general, the cavity can be of any shape (e.g., cylindrical and rectangular) provided the lumped assumption is valid, with the important parameter being the overall cavity volume (V_C). The cavity also plays an important role in determining whether or not the flow is compressible or incompressible, which, from a computational standpoint, is of utmost importance. The issue of compressibility versus incompressibility is discussed in subsequent sections.

The final component of a synthetic jet actuator is the orifice/slot. The shape and size of the orifice/slot design will greatly impact the actuator response and the nature of the flow inside of the orifice/slot, which are important to understand to accurately model the actuators. Several different orifice/slot configurations have typically been utilized, including axisymmetric orifices (Smith and Glezer 1998; Cater and Soria 2002; Holman et al. 2005; Shuster and Smith 2007; Krishnan and Mohseni 2009), two-dimensional (2D) slots (large aspect ratios) (Chen et al. 2000; Bera et al. 2001; Holman et al. 2003, 2005), three-dimensional (3D) slots (small aspect ratios) (Yao et al. 2004; Amitay and Cannelle 2006), and beveled and/or rounded (Cater and Soria 2002; Shuster and Smith 2004; and Holman et al. 2005), each of which has advantages and disadvantages depending on the application of interest.

2.3 Performance Metrics

Tailoring a synthetic jet actuator, or any actuator for that matter, for a particular application can often be a difficult task, and determining a suitable objective function is often left to the creativity of the designer. There are no clear choices or figures of merit that are universal for all applications (Seifert 2007); however, the purpose of this section is to provide some insight into the desirable characteristics and to provide some potential objective functions that can be used to optimize synthetic jet actuators for different applications.

2.3.1 Desirable Actuator Characteristics

The obvious objective, or goal, of any actuator is that it should possess sufficient control authority to achieve some desired control objective. Control authority can be defined as the output from the actuator or the disturbance/perturbation that is introduced to the flow. The disturbance that is introduced can be velocity, mass flow, momentum, and vorticity. There is no clear definition of what "sufficient control authority" actually is, which is in most cases application specific (Joslin and Miller 2009; Cattafesta and Sheplak 2008, 2011).

Aside from control authority, bandwidth is also an important actuator characteristic. Bandwidth refers to the frequency range over which the actuator can operate or the range over which it provides sufficient control authority. Depending on the application, large control authority may be required at a discrete frequency, such that the actuator output can be modulated, or over a wide frequency range. In addition to being able to operate over a large frequency range, it may be desirable for the actuator to produce a broadband output, meaning that it should be able to produce multiple frequencies at any given instant in time. Furthermore, the actuator should have a fast time response to adjust its output in accordance with rapid changes in the flowfield.

In addition to bandwidth and broadband output, the actuator should be robust, meaning that it is reliable and can operate over a wide range of conditions without failure. Low cost is also an important issue. If the cost of implementing the actuator outweighs the potential benefit, then perhaps another approach is warranted. Along the same lines, size, weight, and energy efficiency are also desirable traits of an actuator.

If the actuator is to be used in conjunction with linear feedback control, there are additional characteristics such as broadband output (discussed earlier in this section), flat frequency response, and a linear response. A relatively flat frequency response is a desirable actuator trait because the controller design becomes much more complex if the actuator has a resonant frequency response (Rowley and Batten 2008). Linearity is an important feature by which the actuator can be modeled in the controller. In this instance, linearity simply means that the output is linearly proportional to the input. A final desirable characteristic of an actuator is that an accurate model exists for design purposes.

2.3.2 Potential Objective Functions

As was mentioned earlier, the choice of objective function will depend greatly on the purpose of the actuator as well as its intended application. For instance, if the actuator is primarily used to leverage multiple natural instabilities of a particular system, such as a shear layer instability or a global wake instability indicative of post-stall flow over an airfoil, to delay separation, then perhaps large control authority is not as important as being able to operate at several frequencies (Tian et al. 2006). If the global modification of an aerodynamic body (i.e., virtual aeroshaping or circulation control) is the goal, then the jet velocity must be comparable or greater than the external freestream velocity (Mittal and Rampunggoon 2002). As a final example, some feedback control applications may require large control authority over a wide frequency range, in which case a high bandwidth actuator is necessary (Cattafesta et al. 2008).

In practice, since these devices are more often than not used in the presence of an external crossflow, the actuators are characterized in terms of momentum flux or in terms of the dimensionless momentum coefficient. The dimensionless momentum coefficient represents the ratio of the momentum added by the actuator to that of the freestream flow

$$C_\mu = \frac{\rho_j U_0^2 S_n}{1/2 \rho_\infty U_\infty^2 S_{ref}} \tag{2.1}$$

where:

ρ_j is the density of the jet
S_n is the cross-sectional area of the orifice/slot
ρ_∞ is the density of the crossflow
U_∞ is the crossflow velocity
S_{ref} is some reference area (i.e., based on chord length and boundary layer thickness)
U_0 is the velocity scale

The velocity scale, as defined by Cater and Soria (2002), is given by

$$U_0 = \left[\frac{1}{A}\frac{1}{T}\int_A \int_0^T u(A,t)^2 \, dt dA\right]^{1/2} \tag{2.2}$$

where:

T is the period of oscillation
A is the area of the orifice/slot
$u(A, t)$ is the spatial- and time-varying streamwise velocity component at the exit plane

In the limiting case of a "slug" velocity profile, U_0 represents the root mean square (rms) velocity. This occurs at large Stokes numbers ($S = \sqrt{\omega a^2/\nu}$), where ω is the frequency, a the orifice radius or slot "half width," and ν is the kinematic viscosity of the fluid. When the velocity profile is spatially varying, U_0 does not represent the rms velocity. As is illustrated by Equation 2.2, the momentum coefficient is directly proportional to the square of the momentum velocity scale U_0, which is a function of $u(A, t)$. The lumped element model computes the volumetric flow rate, which can readily be used to compute the average velocity of the orifice/slot and not the time- and spatially varying velocity directly. However, the exact solution of fully developed oscillatory flow through an orifice/slot (White 2006) can be used to compute the shape of the velocity profile, and assuming that the velocity profile varies sinusoidally with time, the jet momentum coefficient can be estimated given the external freestream parameters. Therefore, the momentum coefficient is a potential objective function.

Another parameter that can be used to characterize a ZNMF actuator is the vorticity flux (Didden 1979)

$$\Omega_u = \int_0^{T/2} \int_0^{d/2} \xi_z(t, x)\, V(t, x)\, \mathrm{d}x\mathrm{d}t \quad (2.3)$$

where:

Ω_u is the strength of each shed vortex
ξ_z is the spanwise vorticity component at the exit for a 2D slot

By using the definition of ξ_z and assuming that the velocity varies sinusoidally within the slot, it can be shown that the vorticity flux scales with the square of the jet centerline velocity (Holman et al. 2005), which the lumped element model computes. Vorticity flux is an important parameter when discussing synthetic jet actuators because shed vorticity is the mechanism by which the flow control is achieved in many applications. Therefore, vorticity flux and peak output velocity are the other candidate objective functions.

2.4 Design/Implementation Issues

During the course of designing a synthetic jet actuator, several issues may arise including where to obtain the necessary components, how to properly assemble the actuator, and what the potential failure mechanisms and device limitations are. The cavity and orifice/slot designs are primarily machined out of metal or can be rapidly prototyped using one of many rapid prototyping processes (SLA, SLS, 3DP, etc.). The most difficult component to acquire is the driver.

It is possible to obtain the raw components to assemble a piezoelectric composite diaphragm (e.g., piezoceramic patch, metallic shim, and adhesive) and a custom piezoelectric composite diaphragm, as outlined by Chen et al. (2000). However, there are several companies that sell a wide range of off-the-shelf, commercially available piezoelectric composite diaphragms. These off-the-shelf diaphragms are often referred to as "disc benders" and are primarily intended for use in buzzer applications as well as in alarms and sirens. A schematic of a typical commercially available diaphragm is illustrated in Figure 2.3, which is a bimorph-type piezoelectric composite diaphragm that consists of a metallic shim (brass) that is sandwiched between two piezoceramic cylindrical patches (PZT-5A).

Table 2.1 lists several companies that supply commercially available, off-the-shelf diaphragms similar to that shown in Figure 2.3. These commercially available diaphragms typically range between 10 and 50 mm in diameter with shim and piezoceramic thicknesses of $O(100\ \mu m)$. The typical resonant frequency for these off-the-shelf diaphragms is between a few hundred hertz and approximately 10 kHz. The recommended input voltage for these devices varies from 20 to 30 V_{PP}, and these types of devices are not recommended for use with direct current (dc) voltages. For implementation in synthetic jet actuators, these off-the-shelf diaphragms are driven in excess of 100 V_{PP}, before mechanical failure

FIGURE 2.3 Typical piezoelectric composite diaphragm ($\phi \sim 41$ mm) (APC International, Ltd.)

TABLE 2.1 Vendors of Commercially Available, Off-the-Shelf Piezoelectric Composite Diaphragms

APC International, Inc.	Mouser Electronics
Piezo Systems, Inc.	Murata Electronics N.A., Inc.
Sinoceramics	PI Ceramics
Omega Piezo	PUI Audio, Inc.
Morgan Electro Ceramics	Digi-Key
Kyocera	

and fatigue become the issues, which are discussed in Section 2.4.2. Typical displacement for these devices ranges from tens of μm at dc to hundreds of μm at resonance.

The advantage of these off-the-shelf composite diaphragms is that they are relatively inexpensive (~$1) and they are easy to obtain. One of the disadvantages is the lack of options in the design. A particular diaphragm may have appropriate dimensions. However, the resonant frequency may not match the specific requirement. Another drawback is the durability of these types of diaphragms for synthetic jet actuators, especially at the voltages that are typically needed to generate the desired output.

Another option is to have custom diaphragms manufactured by companies such as the ones listed in Table 2.1. Custom-made diaphragms are typically made with higher quality piezoceramics and exhibit better performance, especially since they can be tailored to meet a specific need (e.g., displacement and resonant frequency). However, these diaphragms typically cost an order of magnitude more than off-the-shelf diaphragms (~$50–$100) and generally require a minimum order.

In the case of electrodynamic drivers, they typically consist of a magnetic assembly comprised of permanent and soft magnets, a multiturn solenoidal coil, and a composite diaphragm. The coil is rigidly attached to the diaphragm, and in the presence of the fixed, uniform magnetic field established by the magnetic assembly, the coil/diaphragm structure is driven into motion. Voice-coil-type drivers are commercially available in the form of speakers, which were used by McCormick (2000) and Agashe et al. (2009). Aside from traditional audio speakers, custom electrodynamic drivers have been developed by Agashe et al. (2009), the design details and optimization of which can be found in the work by Agashe (2009).

2.4.1 Boundary Conditions

The boundary condition of the diaphragm is an important aspect of a synthetic jet actuator design and can have a large impact on the performance of the device. Figure 2.4 shows the typical boundary conditions and how they are achieved in practice.

The first boundary condition is the clamped boundary condition, illustrated in Figure 2.4a. At the clamped boundary ($r = R$), there is no transverse or radial displacement ($w = 0$ and $u = 0$), zero slope ($dw/dr = 0$), and a finite radial moment (M_r = finite). A clamped boundary condition is typically achieved by machining a plate, usually a hard metal (e.g., steel), with a recessed portion that is approximately the thickness of the shim. The piezoelectric composite diaphragm is then seated in the recessed portion and a clamp plate is then flush mounted, creating a fixed boundary.

In a pinned boundary condition, illustrated in Figure 2.4b, there is no transverse or radial displacement ($w = 0$ and $u = 0$), a nonzero slope ($dw/dr \neq 0$), and zero radial moment ($M_r = 0$) at the pinned boundary ($r = R$). A pinned boundary condition is more difficult to achieve in practice. Typically, the piezoelectric diaphragm is fixed in place by two metal O-rings, one on either side of the diaphragm, as illustrated in Figure 2.4b. The O-ring does not allow the diaphragm to displace, but does allow rotation. Another implementation is to use a rubber or compliant gasket around the perimeter of the diaphragm. This boundary condition lies somewhere between a clamped and pinned boundary condition.

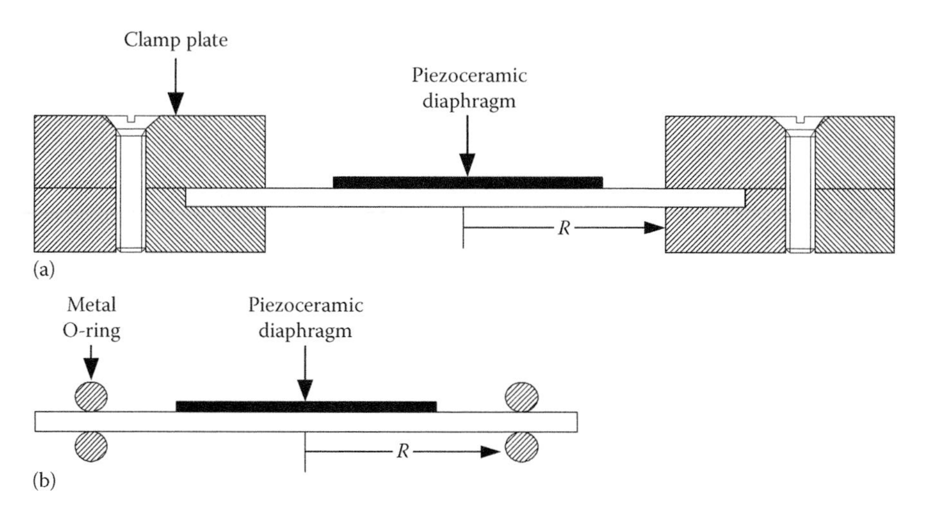

FIGURE 2.4 Typical illustration of (a) a clamped and (b) a pinned boundary condition.

The qualitative differences between a clamped and pinned boundary condition are best exemplified by Figure 2.5. There is a marked difference near the outer boundaries, where there is a clear zero slope for the clamped boundary condition and a finite slope for the pinned boundary condition. In terms of performance, a pinned boundary condition will generate more displacement per unit applied voltage than a clamped boundary, but at the consequence of reduced bandwidth, where bandwidth is defined from the dc to the first resonant frequency of the diaphragm.

2.4.2 Failure

Due to the large voltages and large displacements that are typically generated when operating synthetic jet actuators, failure of the piezoelectric composite diaphragm can become a significant issue that must be addressed.

One potential failure mode is depolarization of the piezoelectric material. Many ferroelectric materials, such as lead zirconate titanate (PZT), do not exhibit piezoelectricity on the macroscale, due to the random orientation of poles within the material (Solecki and Conant 2003; Henderson 2004). Therefore, they must be made piezoelectrically active on a macroscale through a process termed "poling." During the poling process, the polarization direction is obtained by applying heat and a strong external electric field. The external electric field at which this reorientation occurs is called the polarization field of the piezoelectric material. Once the heat and electric fields are removed, the new polarization direction remains. After polarization, exposure to strong electric fields of polarity opposite to that of the poling process will lead to depolarization of the piezoelectric material. This essentially causes the material to become piezoelectrically inactive. The so-called coercive field (E_C) for PZT-5A is approximately 1200 V/mm.

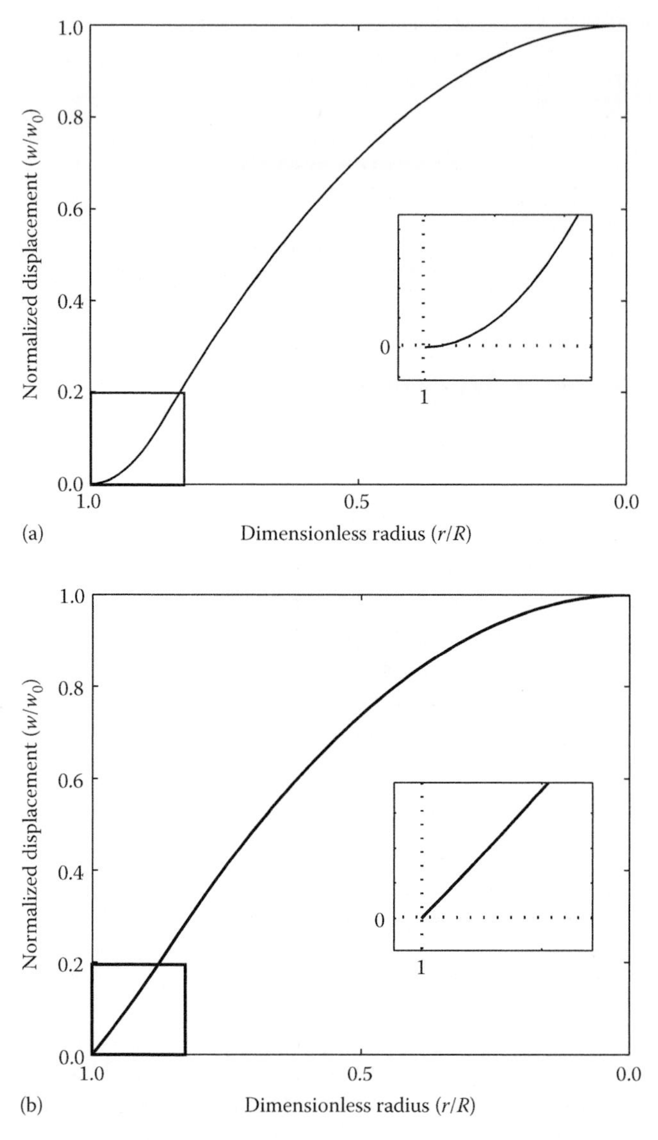

FIGURE 2.5 Different mode shapes of (a) clamped and (b) pinned boundary conditions for a typical piezoelectric diaphragm.

Piezoelectric ceramics are brittle in nature and are particularly susceptible to cracking or fracture. Fracture can occur due to various defects, such as domain walls, grain boundaries, and impurities (Zhang and Gao 2004), which in turn cause geometric, electric, and mechanical discontinuities that induce high stress concentrations. These stress concentrations can induce crack initiation and lead to crack growth and subsequent fracture and failure. Figure 2.6 shows a piezoelectric composite diaphragm that has undergone failure where

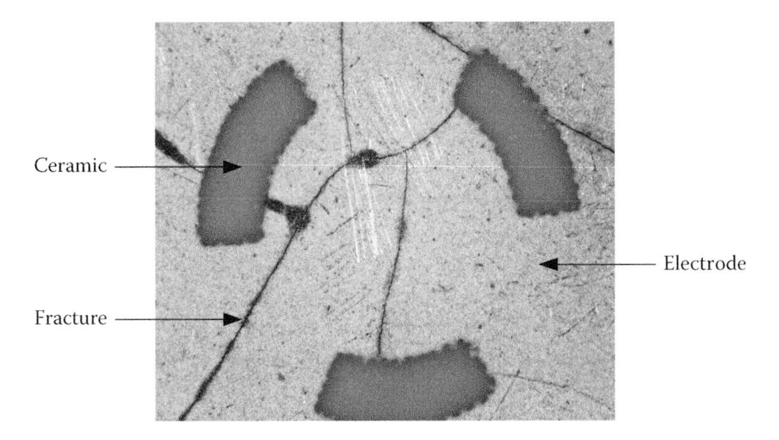

FIGURE 2.6 Failure in piezoelectric composite diaphragm: Fracture.

several cracks in the electrode and/or in the ceramic are clearly visible. This particular diaphragm was driven at approximately 150 V_{PP} and achieved center displacements in excess of 300 μm. The electrodes used are typically made of silver and are a few μm in thickness. If the electrode cracks, this can lead to an electrical short that will cause the diaphragm to no longer operate. Stress concentrations can also cause the diaphragm to fail. The clamped boundary condition causes a stress concentration near the outer boundary, and if the piezoelectric ceramic is clamped, the stress concentration can cause it to fail.

2.5 Design Tool

To effectively design and optimize synthetic jet actuators, low-order models of sufficient fidelity, capable of predicting the dynamic response of synthetic jet actuators, are necessary. One such low-order model is LEM. LEM provides a simple method for predicting the low-frequency dynamic response of synthetic jet actuators with reasonable fidelity and accuracy. The lumped element model is also well suited for incorporation into well-known optimization approaches, such as sequential quadratic programming (SQP). The goal of this section is to provide some background on the LEM and its implementation in the modeling of synthetic jet actuators. Several sample designs are illustrated with comparison between the model and experimental results. Finally, the limitations of LEM are discussed, as the well as the design trade-offs and optimization.

2.5.1 Lumped Element Modeling

LEM is a design tool that has been used for more than a half century in the modeling of electromechanical-acoustical systems (Fischer 1955; Merhaut and Gerber 1981; Rossi and Roe 1988). It is a low-dimensional, low-frequency approximation that can be used to decouple the spatial and temporal variations of complicated coupled, nonlinear systems. McCormick (2000) used this modeling technique to accurately model an electrodynamic synthetic jet actuator. Agashe et al. (2009) have subsequently extended the electrodynamic model developed by McCormick (2000). LEM was applied to piezoceramic synthetic jet actuators by Gallas et al. (2003a). The lumped element model developed and validated by Gallas et al. (2003a) was for a single piezoceramic diaphragm, a single cavity, and a single orifice/slot. Arunajatesan et al. (2009) have since extended the model to the case in which a single diaphragm separates two cavities and orifice/slot pairs. Experiments were performed to validate the model, and good agreement was shown.

The fundamental underlying assumption of LEM is that the characteristic length scale of the physical phenomenon under consideration is much greater than the characteristic size of the device in question. This assumption, in a general sense, is illustrated pictorially in Figure 2.7 for an acoustic transducer, where the wavelength is the length scale. If this assumption is valid, the spatial and temporal variations in the governing partial differential equations (PDEs) can be decoupled, and the temporal and spatial variations can be considered independently, with the latter being lumped into discrete or spatially averaged variables. Consider the pressure wave equation, with frequency ω and characteristic length scale L. A simple order-of-magnitude scaling analysis yields

$$\frac{1}{c^2}\frac{\partial^2 p}{\partial t^2} = \nabla^2 p \rightarrow \frac{p\omega^2}{c^2} = \frac{p}{L^2} \rightarrow (kL)^2 = 1 \tag{2.4}$$

where:

$k = \omega/c$

If $kL \ll 1$, then $\nabla^2 p \sim 0$, and the spatial variations of pressure are negligible.

For this example, if the size of the cavity is small compared to the acoustic wavelength of the pressure fluctuations inside the cavity, the distribution of pressure inside the cavity is essentially uniform. There is no spatial variation in the cavity pressure. Instead, the pressure inside the cavity is some "bulk" average pressure. The pressure continues to vary with time. However, these changes are experienced everywhere inside the cavity simultaneously.

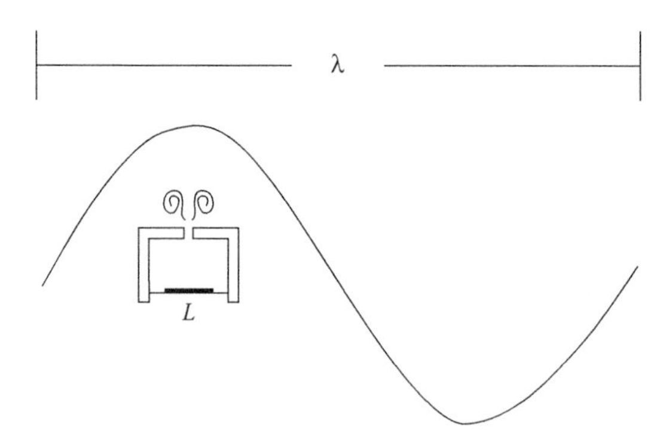

FIGURE 2.7 Schematic of lumped assumption.

This means that the distributed acoustic system, which consists of a linear PDE, can be lumped into an ordinary differential equation.

If the lumped assumption is valid ($\lambda \gg L$), the distributed system can be transformed into an equivalent mass–spring–damper system in which the total kinetic energy, potential energy, and dissipation of the distributed system are lumped into equivalent discrete energy storage elements called "lumped elements." The total kinetic energy is represented by lumped mass, the total potential energy is represented by a lumped spring or compliance ($1/k$), and the total dissipation is represented by a lumped damper. The lumped elements are found by equating the distributed energy of the true system to that of the ideal lumped system. The equivalent single-degree-of-freedom second-order system illustrated in Figure 2.8 is then represented as an equivalent circuit with the lumped mass being equivalent to an inductor, the compliance being equivalent to a capacitor, and the damper being equivalent to a resistor.

To ensure that the energy exchange between the true system and equivalent system is conserved, an ideal transformer (piezoelectric case) or an ideal gyrator (electrodynamic case) is used. Power is defined as power = effort × flow. The effort and flow variables for the different energy domains are summarized in Table 2.2.

The coupling between energy domains (e.g., electrical to mechanical/acoustical domain) is achieved by using equivalent two-port models of the physical system. The resulting equivalent circuit model provides an efficient and reasonably accurate model of low-frequency dynamic behavior (e.g., output volume flow rate per unit applied voltage or current) of multi-energy domain systems that is valid up to and just beyond the first resonant frequency of the system (Merhaut and Gerber 1981).

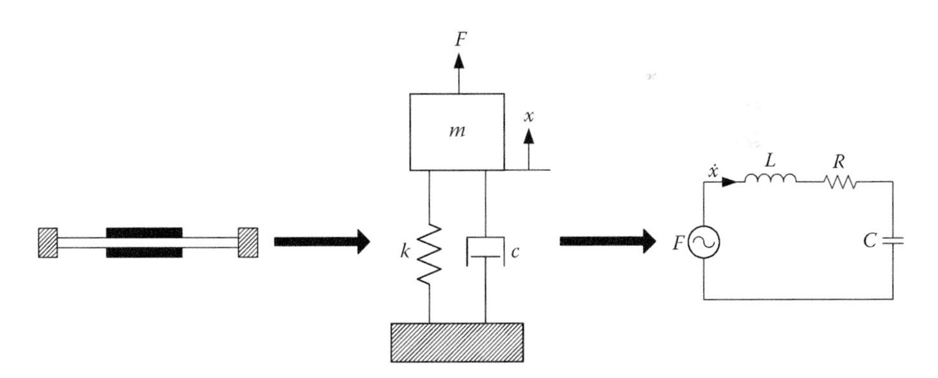

FIGURE 2.8 Representation of distributed system as equivalent discrete system.

TABLE 2.2 Effort and Flow Variables in Various Energy Domains

Energy domain	Effort (E)	Flow (f)
Electrical	Voltage (V)	Current (I)
Mechanical	Force (F)	Velocity (U)
Acoustic/fluidic	Pressure (P)	Volume velocity (Q)

2.5.2 Transduction Models

Transduction refers to the conversion of one form of energy to another. In the case of synthetic jet actuators, the conversion of energy is from the electrical domain to the acoustic/fluidic domain. In LEM, the conversion of energy between the electrical and acoustic domain is realized by using two-port models. An impedance analogy is utilized in which the elements that share a common flow (I or Q) are connected in series and the elements that share a common effort (V or P) are connected in parallel. The equivalent two-port model for piezoelectric and electrodynamic synthetic jet actuators is illustrated in Figure 2.9.

As is demonstrated by Figure 2.9, the energy conversion for a piezoelectric synthetic jet actuator is realized through an ideal transformer, while an ideal gyrator is used for an electrodynamic synthetic jet actuator. The piezoelectric coupling between the electrical and acoustic domain is characterized by the electroacoustic transduction coefficient (ϕ_a)

$$\phi_a = -\frac{d_a}{C_{aD}} \tag{2.5}$$

where:

d_a is the effective acoustic piezoelectric coefficient
C_{aD} is the acoustic compliance of the piezoelectric diaphragm

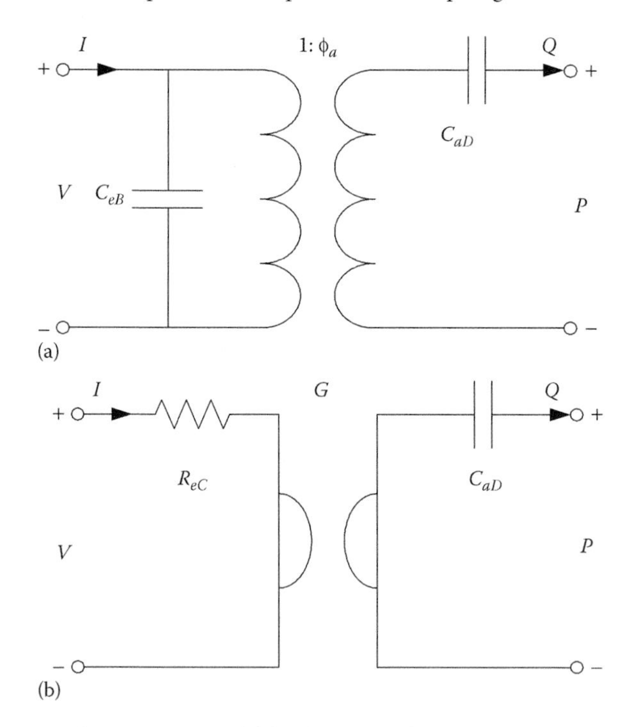

FIGURE 2.9 Equivalent two-port model for (a) a piezoelectric synthetic jet actuator and (b) an electrodynamic synthetic jet actuator.

Similarly, the electroacoustic transduction coefficient (G) for electrodynamic coupling is given by

$$G = \frac{B\ell_{\text{coil}}}{S_{\text{eff}}} \tag{2.6}$$

where:

B is the magnetic flux density
ℓ_{coil} is the length of the coil that carries the current
S_{eff} is the effective area of the diaphragm (or piston) (Agashe et al. 2009)

2.5.3 Driver Model

In general, the driver, whether piezoelectric or electrodynamic in nature, is modeled as a second-order system, the properties of which are lumped into an acoustic mass (M_{aD}), an acoustic compliance (C_{aD}), and an acoustic resistance (R_{aD}). In each case, a static model that computes the displacement of the driver for a given input (e.g., voltage, current, and pressure) is required to estimate these lumped parameters. The piezoelectric driver model is first described, followed by the electrodynamic driver model.

Piezoelectric materials are those that demonstrate coupling between strain and electric field. An electric field induced by mechanical strain is referred to as the direct piezoelectric effect, with the converse referred to as the indirect piezoelectric effect. For actuators, the indirect piezoelectric effect is directly applicable. The one-dimensional (1D), time harmonic piezoelectric coupling two-port equations are given by (Prasad et al. 2006)

$$\begin{Bmatrix} Q \\ I \end{Bmatrix} = \begin{bmatrix} j\omega C_{aD} & j\omega d_a \\ j\omega d_a & j\omega C_{ef} \end{bmatrix} \begin{Bmatrix} P \\ V \end{Bmatrix} \tag{2.7}$$

where:

I is the current
Q is the volume velocity ($Q = j\omega\Delta\text{Vol}$)
ΔVol is the volume displaced by the diaphragm
P is the differential pressure across the diaphragm
V is the applied voltage

The matrix of equations presented in Equation 2.7 shows that the mechanical response of the diaphragm (e.g., displacement) is a function of a linear combination of pressure and voltage given by

$$Q = j\omega C_{aD}P + j\omega d_a V \tag{2.8}$$

From the above equation, the short-circuit acoustic compliance (C_{aD}) and the effective acoustic piezoelectric coupling coefficient (d_a) are given as

$$C_{aD} = \left.\frac{\Delta\text{Vol}}{P}\right|_{V=0} \tag{2.9}$$

and

$$d_a = \left.\frac{\Delta \text{Vol}}{V}\right|_{P=0} \tag{2.10}$$

In addition to the mechanical response, Equation 2.7 demonstrates the electrical response of the diaphragm (e.g., current) to a linear combination of pressure and voltage. Here, note that due to the reciprocal nature of the piezoelectric diaphragm, $d_a = I/j\omega P$ when $V = 0$ (short circuit) or if $I = 0$, then $d_a = -C_{ef}V/P$. The free electrical capacitance is defined as $C_{ef} = \varepsilon A_e/t_p$, where ε is the permittivity of the piezoelectric ceramic, A_e is the area of the exposed electrode, which corresponds to the area of the piezoelectric ceramic, and t_p is the thickness of the piezoelectric ceramic. The blocked electrical capacitance is related to C_{ef} through the electromechanical coupling factor (k), which is defined as $k^2 = d_a^2/C_{ef}\,C_{aD}$. The blocked electrical capacitance, shown in Figure 2.9a, is therefore

$$C_{eB} = \left(1 - k^2\right) C_{ef} \tag{2.11}$$

The acoustic mass (M_{aD}) is found by equating the kinetic energy of the distributed system to that of the lumped acoustic mass

$$\frac{1}{2} M_{aD} Q^2 = \frac{1}{2} \int_0^{2\pi} \int_0^R \rho_A \left[j\omega w(r)\right]^2 r dr d\theta \tag{2.12}$$

where:
$Q = j\omega \Delta \text{Vol}|_{V=0}$

Solving for M_{aD} in Equation 2.12 yields

$$M_{aD} = \frac{2\pi}{C_{aD}^2} \int_0^R \rho_A \left[\frac{w(r)|_{V=0}}{p}\right]^2 r dr \tag{2.13}$$

where

$$\rho_A = \int_{z_b}^{z_t} \rho(r, z)\, dz \tag{2.14}$$

z_b and z_t define the top and bottom locations of each layer of the composite plate, respectively.

The acoustic resistance (R_{aD}) is found by using the definition of the damping ratio (Bendat and Piersol 2011) given by

$$R_{aD} = 2\zeta \sqrt{\frac{M_{aD}}{C_{aD}}} \tag{2.15}$$

where:

ζ is the diaphragm damping ratio and is typically an empirical value that is determined via experiment by either a second-order fit (Bendat and Piersol 2011) or the log-decrement method (Meirovitch 2001)

As is evident from Equations 2.9, 2.10, and 2.12, knowledge of the electromechanical behavior of the piezoelectric diaphragm is needed. The piezoelectric diaphragm is modeled using linear classical laminated plate theory (Prasad et al. 2006). Small deflection theory of thin plates, also called classical plate theory (or Kirchhoff theory), assumes Kirchhoff's hypothesis* and in the case of laminated plates it assumes perfect bonding between layers (Reddy 1997). The details of the model can be found in the work by Prasad et al. (2006). From the model, the transverse displacement $[w(r)]$ due to both pressure and voltage loading is found, from which the lumped element parameters of the diaphragm can be derived.

Given the transverse displacement, the volume displaced by the diaphragm, due to either pressure or voltage loading, is

$$\Delta\text{Vol} = \int_0^{2\pi}\int_0^R w(r)rdrd\theta = 2\pi\int_0^R w(r)\,rdr \tag{2.16}$$

where:

R is the outer radius of the diaphragm

The vibration of the exposed side of the diaphragm and the ambient air results in a radiation impedance and is represented by an acoustic radiation mass ($M_{aD,\text{rad}}$) and a radiation resistance ($R_{aD,\text{rad}}$). The low frequency approximation for radiation from a baffled piston (Blackstock 2000) is used to approximate these two quantities by

$$M_{aD,\text{rad}} = \frac{8\rho}{3\pi^2 R} \tag{2.17}$$

and

$$R_{aD,\text{rad}} = \frac{\rho\omega^2}{2\pi c} \tag{2.18}$$

where:

R is the radius of the diaphragm

ρ is the density of the medium

* Thickness of the plate is small compared to the outer radius, middle plane of the plate does not undergo in-plane deformation, transverse displacement of midsurface of the plate is small compared to the thickness, influence of shear deformation is negligible, and transverse normal strain is negligible (Rao 2007).

ω is the frequency
c is the speed of sound of the medium

This definition has to be modified, however, because the solution assumes a piston of uniform velocity. Therefore, an effective diaphragm radius must be found. The effective diaphragm radius is found by equating the volume displaced by the diaphragm to some effective area ($\Delta\text{Vol} = w_0 A_{\text{eff}} = w_0 \pi R_{\text{eff}}^2$). Solving for the effective radius (R_{eff}) yields

$$R_{\text{eff}} = \sqrt{\frac{\Delta\text{Vol}}{w_0 \pi}} \tag{2.19}$$

where:
w_0 is the center displacement of the diaphragm

In the electrodynamic case, the key transduction equations are the Lorentz force law ($F = B\ell_{\text{coil}} I = PA$), which relates the pressure (P) acting across the diaphragm/piston to the current (I), and Lenz's law ($V = B\ell_{\text{coil}} U = (B\ell_{\text{coil}}/S_{\text{eff}})Q = GQ$), which relates the volume velocity of the diaphragm/piston (Q) to the voltage induced on the coil winding (V). These transduction equations are represented using an ideal gyrator. The ideal gyrator equations can be written as

$$\begin{Bmatrix} P \\ Q \end{Bmatrix} = \begin{bmatrix} 0 & G \\ 1/G & 0 \end{bmatrix} \begin{Bmatrix} V \\ I \end{Bmatrix} \tag{2.20}$$

The definitions for C_{aD}, M_{aD}, R_{aD}, and R_{eff} are the same for both the electrodynamic and piezoelectric cases, in that they are all a function of the deflection mode shape. Having said that, knowledge of the deflection mode shape is required, which requires a separate static mechanical model to compute. Agashe et al. (2009) developed an analytical model for an electrodynamic-driven composite diaphragm, illustrated in Figure 2.10.

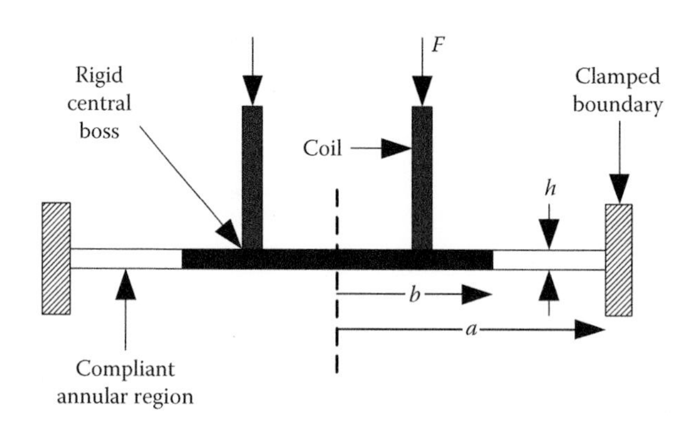

FIGURE 2.10 Schematic of an electrodynamic composite diaphragm.

The composite diaphragm was modeled as a transversely isotropic, linearly elastic, axisymmetric plate (Agashe et al. 2009) with two regions (rigid central boss and compliant annular regions). Assuming that the central boss region is perfectly rigid and that the diaphragm is uniformly loaded with pressure P, the displacement profile for regions 1 and 2 is expressed as

$$w(r)\big|_{0<r<b} = \frac{3P\left(1-\nu_2^{\,2}\right)a^4}{16E_2h^3}\left[1+4\left(\frac{b}{a}\right)^2 \ln\left(\frac{b}{a}\right)-\left(\frac{b}{a}\right)^4\right]$$

$$w(r)\big|_{b<r<a} = \frac{3P\left(1-\nu_2^{\,2}\right)a^4}{16E_2h^3}\left[1+4\left(\frac{b}{a}\right)^2 \ln\left(\frac{r}{a}\right)-2\left(\frac{r}{a}\right)^2+2\left(\frac{b}{a}\right)^2 -2\frac{b^2r^2}{a^4}+\left(\frac{r}{a}\right)^4\right] \tag{2.21}$$

where:

a and b are the radius of the compliant annular region and rigid central boss region, respectively

h is the composite plate thickness

ν_2 and E_2 are the Poisson's ratio and the Young's modulus of the compliant annular region

Given Equation 2.21, the lumped parameters can be computed for the composite diaphragm.

2.5.4 Cavity Model

The acoustic compliance of the cavity is found by solving the classic acoustics problem of a harmonic wave field in an acoustically compact duct with a sound hard termination (Blackstock 2000). Figure 2.11 shows a sound source (e.g., piston) at one end of the duct of length ℓ and height H and the other end of the duct terminated by a sound hard boundary.

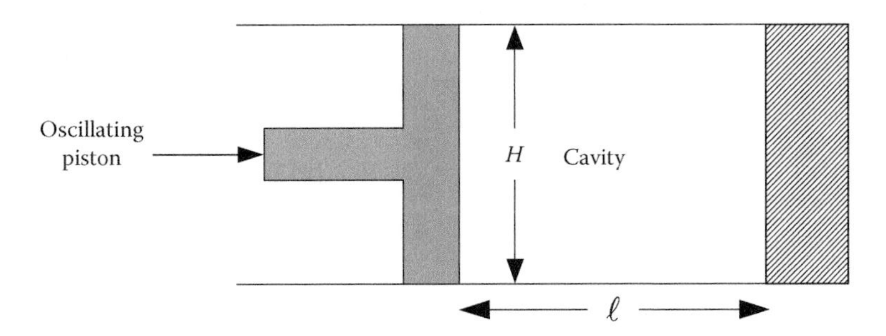

FIGURE 2.11 Schematic of a piston mounted in a closed tube.

The acoustic impedance for this case is given by

$$Z_{aC} = \frac{\Delta P}{Q} = -\frac{j\rho c \cot(k\ell)}{S_{\text{cav}}} \tag{2.22}$$

where:

S_{cav} is the cross-sectional area of the cavity

For compact cavities ($k\ell \ll 1$ and $kH \ll 1$), the truncated Taylor series expansion reduces to

$$Z_{aC} = \frac{1}{j\omega C_{aC}} \tag{2.23}$$

where:

$C_{aC} = V_c/\rho c^2$ and V_c is the cavity volume

2.5.5 Orifice/Slot Model

The orifice/slot, illustrated in Figure 2.12, is modeled as a series of acoustic resistances and mass terms. The relevant orifice/slot dimensions are the orifice/slot radius or "half width" (a), the orifice/slot height (h), and in the case of a slot, the slot width in the long dimension (w). The linear resistance (R_{aN}) and mass (M_{aN}) are derived from the solution for incompressible, fully developed, laminar flow in a pipe or channel (e.g., Poiseuille or Hagen–Poiseuille flow) (White 2006).

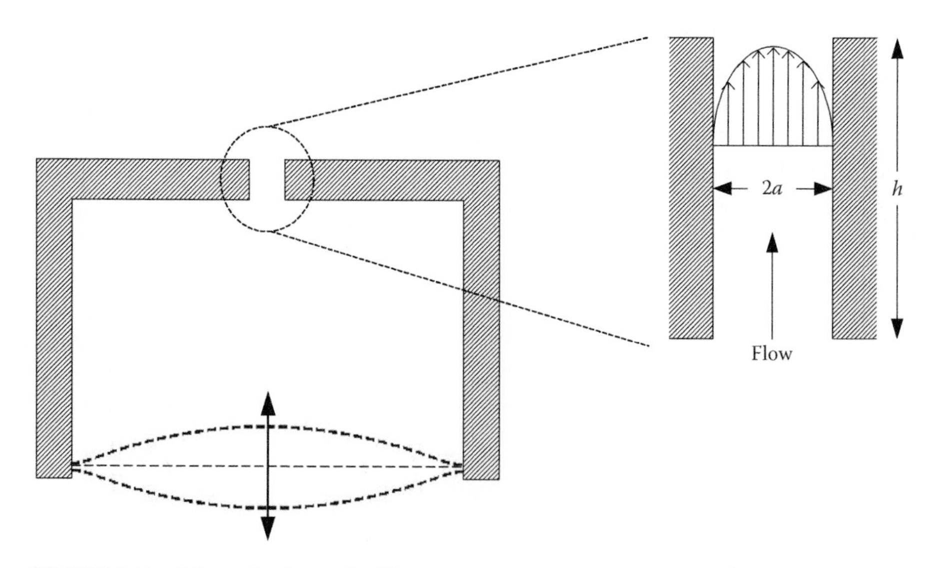

FIGURE 2.12 Schematic of an orifice/slot.

The resistance is found by relating the volume flow rate (Q) in the duct to the associated pressure drop (ΔP), and the mass is found by equating the kinetic energy of the distributed pipe flow to the lumped kinetic energy at the center of the pipe. The acoustic resistance and mass for an axisymmetric orifice are given as (Gallas et al. 2003a)

$$R_{aN} = \frac{8\mu h}{\pi a^4} \tag{2.24}$$

and

$$M_{aN} = \frac{4\rho h}{3\pi a^2} \tag{2.25}$$

where:

μ is the dynamic viscosity of the fluid
h is the height or the length of the pipe
a is the pipe radius

A similar analysis was performed for flow in a 2D channel results in the following equations (Gallas 2005):

$$R_{aN} = \frac{3\mu h}{2wa^3} \tag{2.26}$$

and

$$M_{aN} = \frac{3\rho h}{5wa} \tag{2.27}$$

where:

a is the slot half width
w is the slot length in the long dimension
h is the slot height

The full derivation for the linear acoustic resistance and mass can be found in the work by Gallas (2005).

The linear acoustic resistance and mass terms given in Equations 2.24 through 2.27 are only valid at low frequencies when the velocity profile in the orifice/slot is parabolic. At higher frequencies, the velocity profile becomes more "plug like," or uniform. Therefore, in general, the linear acoustic resistance and mass will be a function of frequency as well as geometry.

The frequency-dependent linear acoustic resistance and mass are derived from the exact solution for oscillatory pressure-driven flow through a constant area pipe or channel (White 2006). The steady-state solution for an axisymmetric pipe is

$$u(r,t) = \frac{\Delta P}{j\rho\omega h} e^{j\omega t} \left[1 - \frac{J_0\left(r\sqrt{-j\omega/\nu}\right)}{J_0\left(a\sqrt{-j\omega/\nu}\right)} \right] \tag{2.28}$$

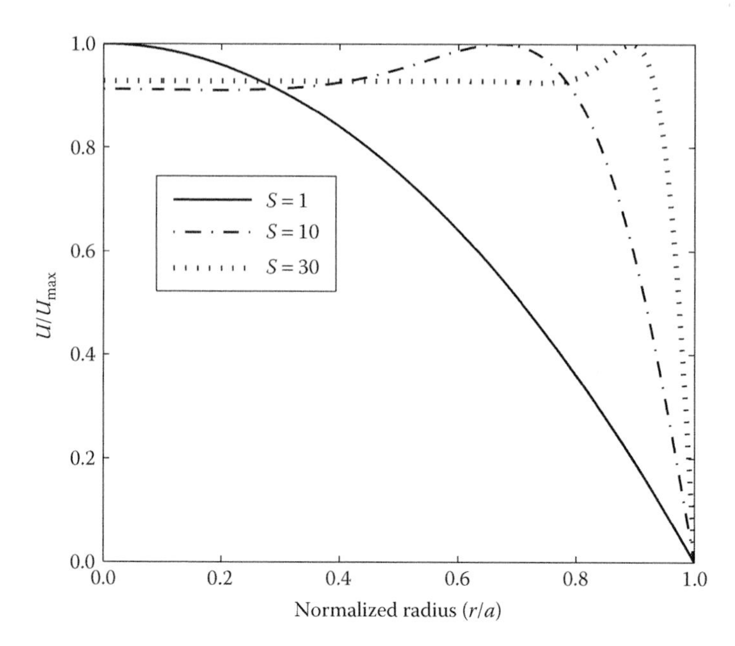

FIGURE 2.13 Velocity profile at various Stokes numbers for oscillatory pipe flow.

where:

J_0 is a zeroth-order Bessel function of the first kind
ν is the kinematic viscosity
r is the radial direction in the pipe
a is the pipe radius

The magnitude of the velocity profile is plotted versus the Stokes number $S = \sqrt{\omega a^2/\nu}$ in Figure 2.13.

In the limit $S \to 0$, the velocity profile asymptotes to the Haigen–Poisseuille flow steady-state solution. As S increases, the velocity profile becomes more "plug like," and the maximum velocity is no longer at the center of the pipe. The overshoot near the pipe walls is known as Richardson's annular effect (White 2006). The flow near the center of the pipe is out of phase with the pressure gradient by approximately 90°. Under certain conditions, the viscous forces are in phase with the pressure gradient, causing a constructive interference in which the velocity overshoots near the pipe walls.

The ratio of the average velocity (U_{avg}) to the center velocity (U_0) versus Stokes number is plotted in Figure 2.14. Once again, as $S \to 0$, the solution reduces to the Hagen–Poiseuille solution. For Hagen–Poiseuille flow, the ratio of U_{avg}/U_0 equals 0.5 (Fox et al. 1998), as shown in Figure 2.14. As $S \to \infty$, this ratio approaches unity, which makes physical sense since the velocity profile becomes more uniform with the increasing S.

The acoustic resistance and mass are computed from the velocity profile given in Equation 2.28 by computing the complex acoustic impedance

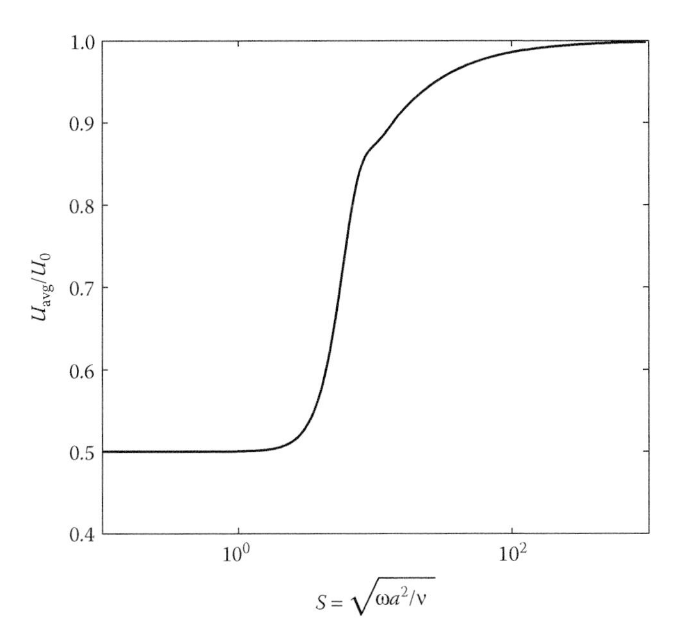

FIGURE 2.14 Ratio of average velocity to center velocity versus Stokes number for oscillatory pipe flow.

$$Z_{aO} = \frac{\Delta P}{Q} = R_{aN} + jX_{aN} = R_{aN} + j\omega M_{aN} \tag{2.29}$$

in which the real part (R_{aN}) represents the acoustic resistance and the imaginary part (M_{aN}) is the acoustic mass. The normalized resistance and mass as a function of Stokes number (S) are plotted in Figure 2.15. The resistance and mass are normalized by their respective low-frequency estimates given by Equations 2.24 and 2.25. Once again, at low frequencies, the mass and resistance terms asymptote to the steady values. At higher frequencies, the acoustic resistance increases substantially due to the Richardson effect, while the acoustic mass decreases slightly.

The nonlinear orifice resistance ($R_{a,nl}$) represents the losses due to nonlinear entrance and exit region effects. In the entrance and exit regions, the pressure is proportional to the velocity squared ($\Delta P \sim u^2$). The nonlinear orifice resistance is modeled as a Bernoulli flow meter (McCormick 2000) given by

$$R_{a,nl} = \frac{0.5\rho K_d |Q_{out}|}{S_n^2} \tag{2.30}$$

where:

K_d is a dimensional dump loss coefficient
Q_{out} is the volume flow rate exiting the orifice/slot
S_n is the orifice/slot cross-sectional area

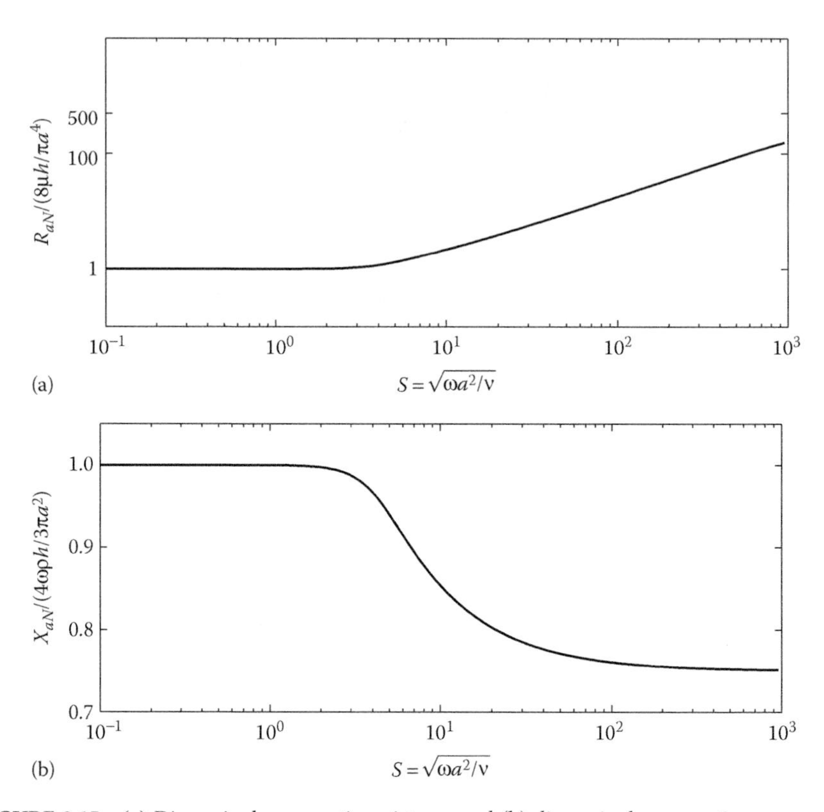

FIGURE 2.15 (a) Dimensionless acoustic resistance and (b) dimensionless acoustic mass versus Stokes number for oscillatory pipe flow.

Traditionally, a value of $K_d = 1$ is used in the lumped element model. However, by assuming a steady flow that transitions from a uniform plug flow at the inlet to fully developed parabolic profile at the exit, the minor loss coefficient is $K_d = 2/3$ for an axisymmetric pipe and $K_d = 2/5$ for a 2D slot (White 2006; Oyarzun 2013). Since $R_{a,nl}$ represents the nonlinear losses due to entrance and exit region effects, it is expected to dominate when the ratio of the particle displacement, or stroke length ($\bar{L}$), to the orifice/slot height (h) is much greater than unity ($\bar{L}/h \gg 1$). The linear orifice resistance (R_{aN}) dominates when $\bar{L}/h \ll 1$. The stroke length ($\bar{L}$) is defined as

$$\bar{L} = \int_0^{\tau} \bar{u}(t)\, dt \tag{2.31}$$

where:

$\bar{u}(t)$ is the spatially averaged streamwise velocity in the orifice/slot
$0 < t < \tau$ is the expulsion portion of the cycle

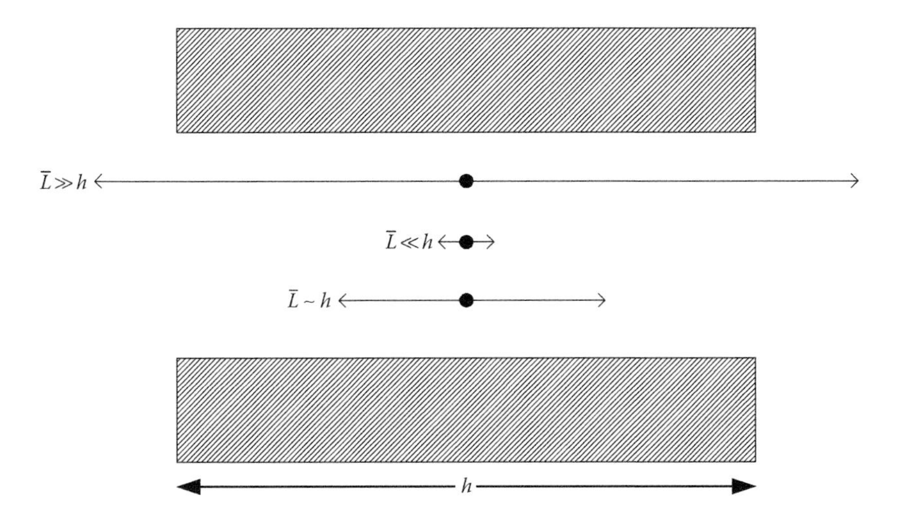

FIGURE 2.16 Schematic of different stroke length regimes.

Physically, the stroke length is a measure of the average distance that a fluid particle travels during the expulsion portion of the cycle. Figure 2.16 illustrates the different stroke length regimes previously mentioned. When the particle displacement is small compared to the orifice/slot height ($\bar{L}/h \ll 1$), the flow looks fully developed and the linear resistance losses dominate. When the particle displacement is large compared to the orifice/slot height ($\bar{L}/h \gg 1$), the flow looks like quasi-steady bidirectional flow (first in one direction and then the other), in which case the nonlinear (i.e., minor) entrance and exit losses dominate. When $\bar{L}/h \sim 1$, linear and nonlinear losses are both expected to be important.

The acoustic radiation resistance and mass of the orifice are estimated from the low-frequency approximation for radiation from a baffled piston (Blackstock 2000) and are defined as

$$M_{a,\mathrm{rad}} = \frac{8\rho}{3\pi^2 a} \tag{2.32}$$

and

$$R_{a,\mathrm{rad}} = \frac{\rho\omega^2}{2\pi c} \tag{2.33}$$

Just as was the case for the diaphragm, this definition must be modified since the solution from which this quantity is derived assumes a piston of uniform velocity. The effective orifice radius is found by equating the volume velocity through the orifice to some effective area. Solving for the effective radius yields

$$a_{\mathrm{eff}} = \sqrt{\frac{Q}{\pi u_0}} \tag{2.34}$$

where:

Q is the volume velocity
u_0 is the jet centerline velocity

The equivalent expressions for a 2D slot (Meissner 1987) are given by

$$M_{a,\text{rad}} = \frac{\rho}{\text{w}}\left\{\frac{1}{\pi \ln(\text{w}/a)} + \frac{1}{2\pi\left[1 - \omega^2\text{w}^2/6c^2\right]}\right\} \tag{2.35}$$

and

$$R_{a,\text{rad}} = \frac{\rho\omega^2}{2\pi c}\left(1 - \frac{\omega^2\text{w}^2}{36c^2}\right) \tag{2.36}$$

where:

w is the slot width in the long dimension
a is the slot radius or "half width"
ω is the frequency
ρ is the density
c is the isentropic speed of sound

2.6 Sample Designs

Section 2.5 introduced the concept of LEM and how to compute the lumped element parameters for the various actuator components. In this section, LEM is applied to several illustrative examples.

2.6.1 Piezoelectric-Driven Synthetic Jet Actuator

A schematic of a traditional piezoelectric synthetic jet actuator is shown in Figure 2.17. It consists of a single uniform piezoelectric diaphragm, a single cylindrical cavity, and an axisymmetric orifice. In this particular case, the diaphragm is directly below the orifice, as opposed to side mounted, as is the case in a subsequent example.

The equivalent circuit representation for the actuator shown in Figure 2.17 is provided in Figure 2.18. By traditional circuit analysis, the transfer function of volume velocity through the orifice (Q_{out}) to the input voltage (V_{ac}) can be derived and is given as

$$\frac{Q_{\text{out}}}{V_{\text{ac}}} = \phi_a \frac{Z_{aC}}{Z_{aC}Z_{aD} + Z_{aD}Z_{aO} + Z_{aC}Z_{aO}} \tag{2.37}$$

where

$$Z_{aD} = R_{aD} + R_{aD,\text{rad}} + j\omega\left(M_{aD} + M_{aD,\text{rad}}\right) + \frac{1}{j\omega C_{aD}}$$

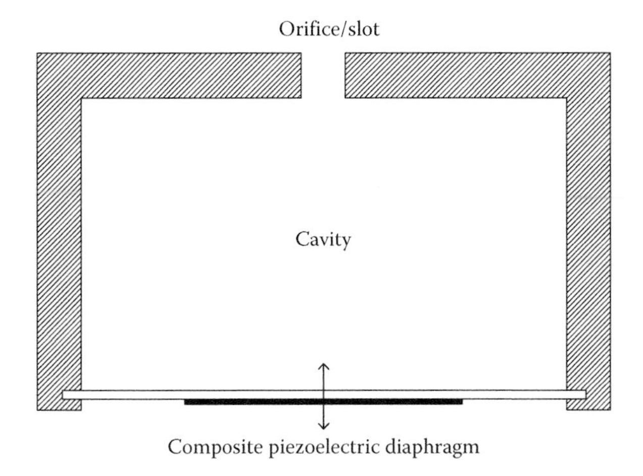

FIGURE 2.17 Schematic of a piezoelectric synthetic jet actuator.

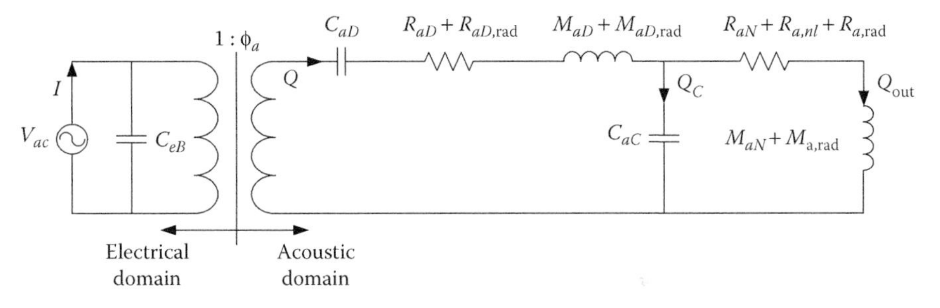

FIGURE 2.18 Equivalent circuit representation for a piezoelectric synthetic jet actuator.

$$Z_{aC}=\frac{1}{j\omega C_{aC}} \tag{2.38}$$
$$Z_{aO}=R_{aN}+R_{a,nl}+R_{a,\text{rad}}+j\omega\left(M_{aN}+M_{a,\text{rad}}\right)$$

For comparison purposes, two illustrative examples from Gallas et al. (2003a) are presented, Case 1 and Case 2, in Figure 2.19. For both cases, unimorph piezoelectric diaphragms were used as well as axisymmetric orifices. An input voltage amplitude (V_{ac}) of 25 V was used. The details of the actuator dimensions can be found in the work by Gallas et al. (2003a). As can be seen, there is a good agreement in terms of both amplitude and frequency for both cases. The experimental data were found using laser Doppler velocimetry (LDV). For Case 1, illustrated in Figure 2.19a, the diaphragm natural frequency is f_D = 2114 Hz and the Helmholtz frequency is f_H = 941 Hz. These two frequencies are relatively far apart from one another, and as a result, two distinct resonant frequencies appear, f_1 = 970 Hz and f_2 = 2120 Hz, which are related to but not equivalent to f_D and f_H. For Case 2,

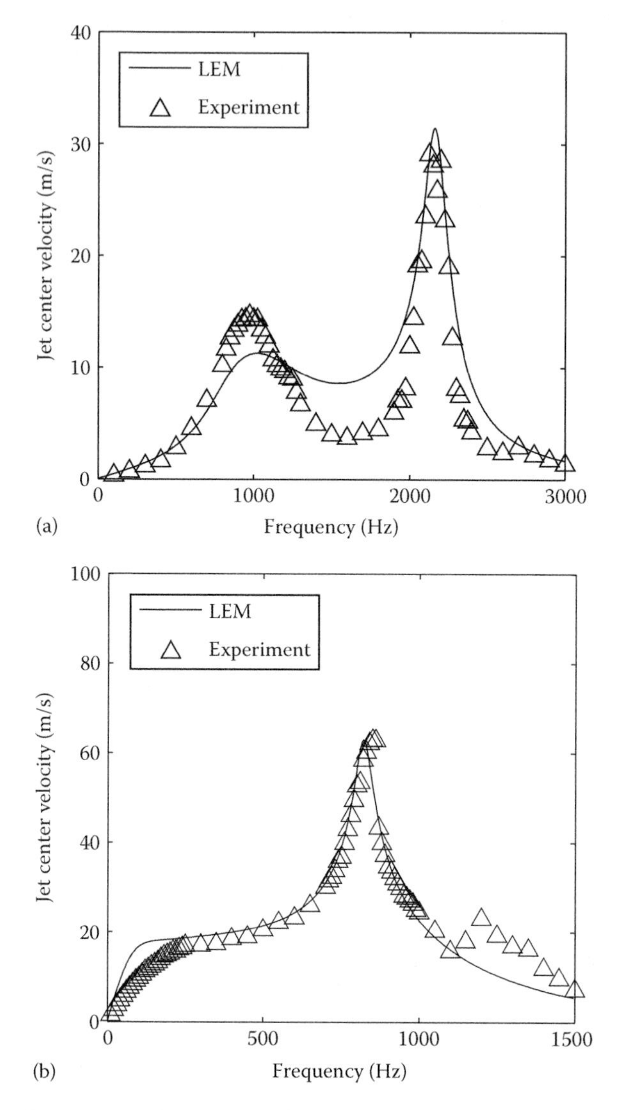

FIGURE 2.19 Comparison of jet center velocity versus frequency between model and experiment for (a) Case 1 ($K_d = 1$ and $\zeta = 0.03$) and (b) Case 2 ($K_d = 1$ and $\zeta = 0.02$). Symbols represent experimental values and solid lines represent the model prediction. (From Gallas, Q. et al., *AIAA J.*, 41, 240–247, 2003a.)

$f_D = 632$ Hz and $f_H = 452$ Hz. These two characteristic frequencies are close to one another, as opposed to Case 1, and as a result, there is a single dominant resonant peak, as illustrated in Figure 2.19b. The issue of a single peak versus two peaks is discussed in Section 2.8.

2.6.2 Electrodynamic-Driven Synthetic Jet Actuator

A schematic of the electrodynamic synthetic jet actuator that was used by Agashe et al. (2009) and Sawant et al. (2012) is illustrated in Figure 2.20. This particular actuator consists of a magnetic assembly, a coil, a composite diaphragm, a cavity, and an axisymmetric orifice or 2D slot.

The equivalent circuit representation for the actuator illustrated in Figure 2.20 is provided in Figure 2.21. From the equivalent circuit representation in Figure 2.21, the transfer function that governs the output response of the actuator (Q_{out}/V_{ac}) is derived as being equivalent to

$$\frac{Q_{out}}{V_{ac}} = \frac{G/Z_{eCoil}}{\left[G^2/Z_{eCoil} + Z_{eff}\right]} \frac{Z_{aC}}{Z_{aC} + Z_{aO}} \tag{2.39}$$

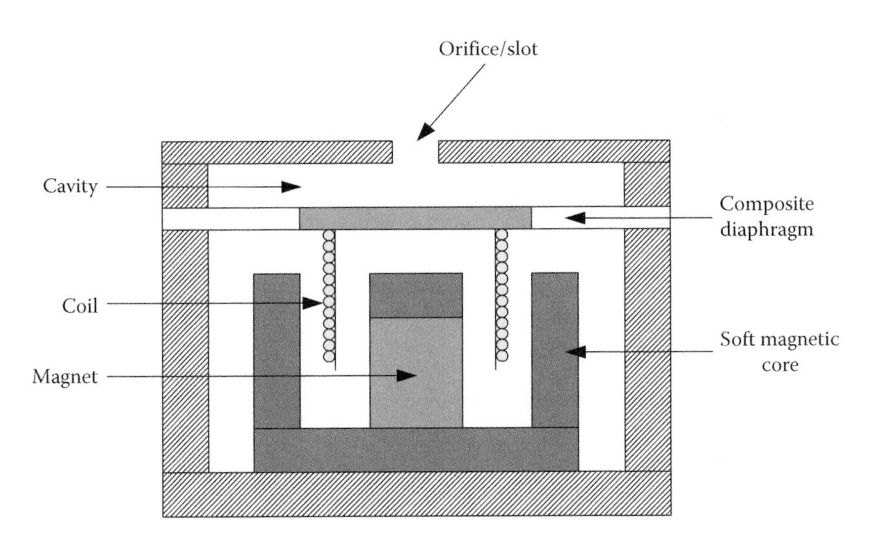

FIGURE 2.20 Schematic of an electrodynamic synthetic jet actuator used by Agashe et al. (From Agashe, J. et al., *AIAA Paper* 2009-1308, 2009.)

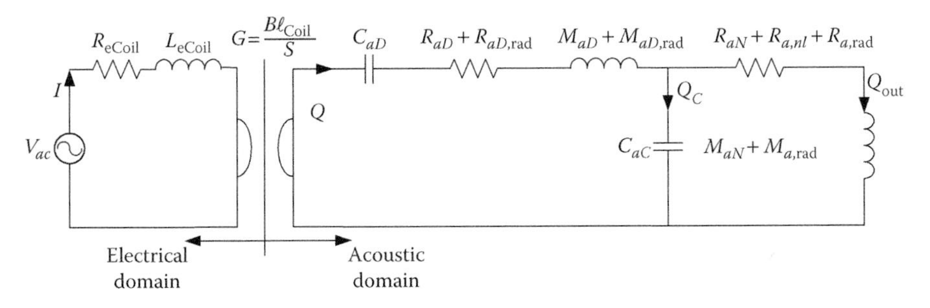

FIGURE 2.21 Equivalent circuit representation for electrodynamic synthetic jet actuator.

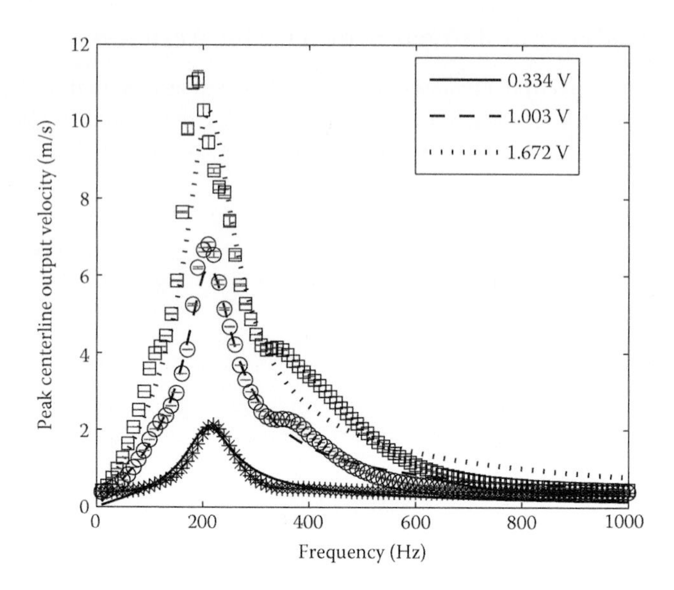

FIGURE 2.22 Comparison of peak centerline output velocity versus frequency between model and experiment for an electrodynamic synthetic jet actuator. Symbols represent experimental values and solid lines represent the model prediction. (From Sawant, S. et al., *AIAA J.*, 50, 1347–1359, 2012.)

where:

$$Z_{eCoil} = R_{eCoil} + j\omega L_{eCoil}$$
$$Z_{eff} = Z_{aD} + Z_{aC} || Z_{aO}$$

Several limiting cases for Equation 2.39 exist and can be found in the work by Agashe et al. (2009) and Sawant et al. (2012).

The model represented by Equation 2.39 is compared to the experimental data of several prototypical actuators by Agashe et al. (2009) and subsequently by Sawant et al. (2012). One such comparison is illustrated in Figure 2.22. The details of the actuator geometry are summarized in Sawant et al. (2012). The results shown in Figure 2.22 are for a 2D slot. The results show a good agreement between the model and experiments for both amplitude and frequency. This particular device exhibits a single resonant frequency that corresponds to the natural frequency of the composite diaphragm.

2.6.3 Dual-Cavity Self-Vented Synthetic Jet Actuator

As a final illustrative case, the lumped element model for the piezoelectric synthetic jet actuator which that was discussed in Section 2.5.1 was extended by Arunajatesan et al. (2009) to an actuator that consists of a single piezoelectric bimorph diaphragm that is bound on either side by a cavity and a slot, the schematic of which is shown in Figure 2.23. The equivalent circuit representation is illustrated in Figure 2.24.

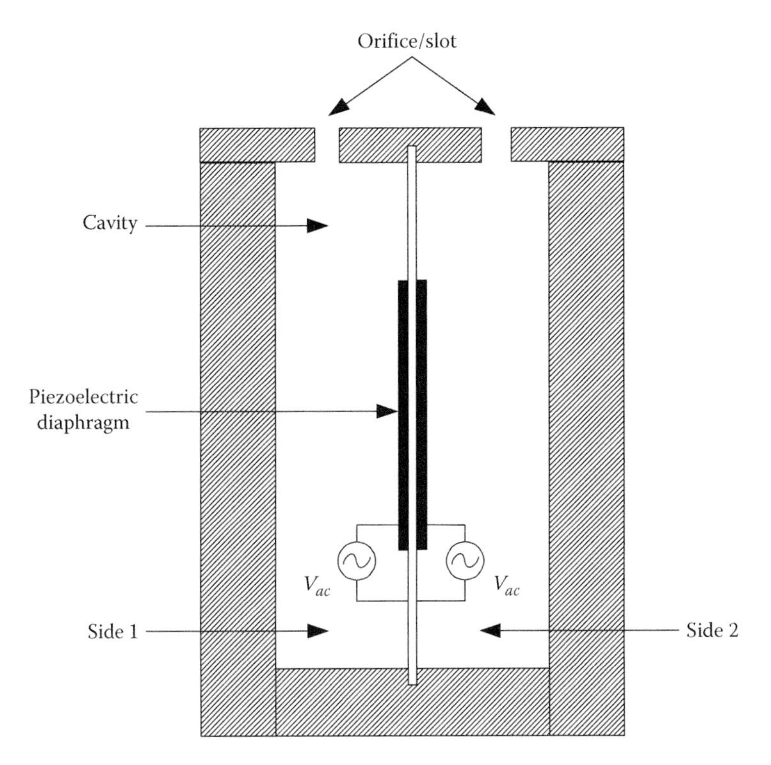

FIGURE 2.23 Schematic of a dual-cavity synthetic jet actuator.

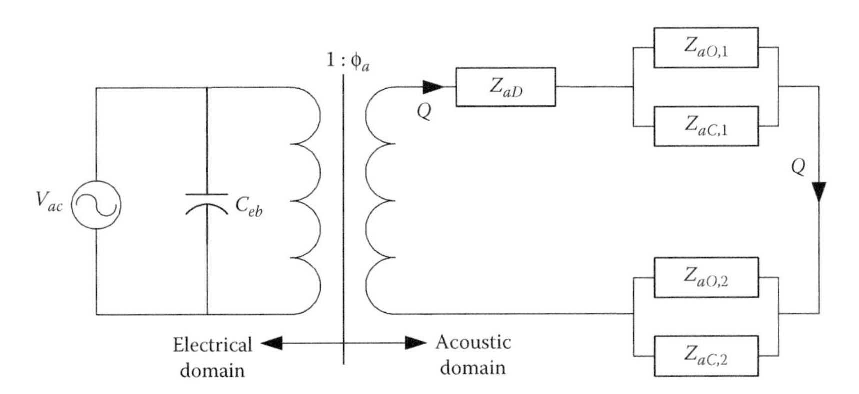

FIGURE 2.24 Equivalent circuit representation for a dual-cavity synthetic jet actuator.

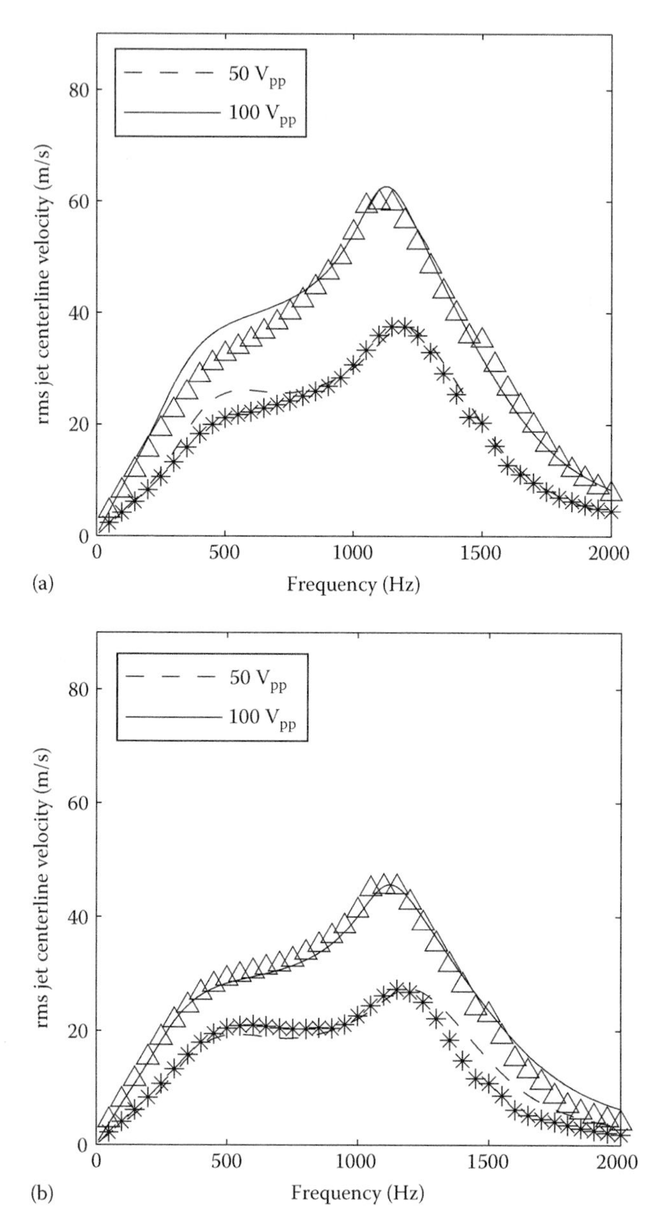

FIGURE 2.25 (a) Side 1 and (b) Side 2 comparisons of rms jet centerline velocity versus frequency between model and experiment. Symbols represent experimental values and solid lines represent the model prediction.

The transfer function of Q_{out}/V_{ac} for sides 1 and 2, respectively, is given by

$$\frac{Q_{\text{out}_{1,2}}}{V_{ac}} = \phi_a \frac{Z_{aC_{1,2}}\left(Z_{aO_2} + Z_{aC_2}\right)}{A + B + C} \tag{2.40}$$

where

$$\begin{aligned} A &= Z_{aD}\left(Z_{aO_1} + Z_{aC_1}\right)\left(Z_{aO_2} + Z_{aC_2}\right) \\ B &= \left(Z_{aO_1} Z_{aC_1}\right)\left(Z_{aO_2} + Z_{aC_2}\right) \\ C &= \left(Z_{aO_2} Z_{aC_2}\right)\left(Z_{aO_1} + Z_{aC_1}\right) \end{aligned} \tag{2.41}$$

For equal cavity and orifice/slot geometries, the transfer function given by Equations 2.40 and 2.41 reduces to

$$\frac{Q_{\text{out}}}{V_{ac}} = \phi_a \frac{Z_{aC}}{Z_{aD}\left(Z_{aO} + Z_{aC}\right) + 2\left(Z_{aO} Z_{aC}\right)} \tag{2.42}$$

This design is particularly useful because of its self-venting nature, which can be beneficial to prevent static deflection of the diaphragm in environments where the pressure is significantly different than ambient conditions. The model was subsequently validated for a prototypical actuator. The details of the actuator geometry and experiments can be found in the work by Arunajatesan et al. (2009). The comparison of the model with experimental results is illustrated in Figure 2.25 for the case in which the cavity volumes V_1 and V_2 are different ($V_1 < V_2$), with identical slot geometries.

2.7 Lumped Element Modeling Limitations

Even though LEM has been shown to be a powerful predictive tool for the performance of synthetic jet actuators, in the hierarchy of modeling fidelity, LEM is the most simplistic approach, with computational fluid dynamics (CFD) being the most high fidelity approach. LEM is sufficient for prediction and design purposes, but lacks the ability to provide appropriate time-dependent boundary conditions for CFD. High fidelity models short of CFD have been developed by several researchers. For example, Yamaleev and Carpenter (2006) solve the time-dependent compressible quasi-1D Euler equations, while the diaphragm is modeled as a moving boundary. The inviscid approach satisfies conservation of mass, momentum, and energy, but it is unclear whether this approach captures the physics of the unsteady vena contracta in the orifice (Raju et al. 2007). Rumsey et al. (2006) summarize the results of the workshop on CFD Validation of Synthetic Jets and Turbulent Separation Control in their paper. The consensus was that despite significant simplifications in the CFD model of the actuator geometry and diaphragm motion, a good agreement between the experiments and the simulations was obtained. An important current research topic is to develop reduced-order models of a ZNMF actuator(s) suitable for use in a controlled flow field, and recent efforts in this regard are described in Raju et al. (2009).

Kim et al. (2005) extended the lumped element technique to account for delays and spatial variations associated with long lengths of tubing between the driver and the orifice/slot. The model was able to predict resonance to within 10% and the velocity at the actuator exit to within 40%. The agreement between the model and experiments improved as the transmission tube length increased. Tang and Zhong (2009) developed a lumped element model in which the diaphragm motion was decoupled from the cavity. The model for the "minor losses" in the orifice exit was linked to the jet exit velocity profile, rather than being constant such as in previous lumped element models (McCormick 2000; Gallas et al. 2003a; Agashe et al. 2009). The results of the model were validated with the existing experimental data from the literature and CFD simulations. The model was shown to have a good agreement in terms of temporal variation for both magnitude and phase when the actuator is operated far away from the Helmholtz frequency. This is not an unexpected result since the driver was decoupled from the cavity. The model was then compared to two reduced-order models: (1) static-incompressible and (2) dynamic-incompressible models. The lumped element model was shown to be the most accurate and suitable approach for design purposes.

As was illustrated by the previous sample designs, there are some inherent limitations associated with the lumped element model. The first limitation is that the lumped assumption is only valid for sufficiently compact devices, meaning that $kL \ll 1$. Kim et al. (2005) incorporated a distributed model into the lumped element model in an effort to account for delays and spatial variations associated with long lengths of tubing.

The lumped element model of the piezoelectric diaphragm is limited up to and just beyond the first resonant frequency of the diaphragm and well before the second resonant mode. This means that the model has an upper frequency limit, beyond which it is no longer valid. Furthermore, the diaphragm model, being a linear model, is limited to small transverse displacements $[w(r)]$. This limitation translates into the fact that the diaphragm center deflection (w_0) must be much less than the total thickness of the diaphragm (t). The purpose of the diaphragm in the synthetic jet actuator is to produce large volume displacement, hence large w_0. Therefore, this limitation can be potentially problematic. A nonlinear model is required for large displacements.

The orifice/slot model assumes that the flow inside of the orifice/slot is fully developed. This condition is only achieved when the dimensionless stroke length ($\bar{L}/h$) is much less than unity, meaning that the particle displacement is much less than the orifice/slot height. If this is the case, the fluid particle sees the orifice/slot as infinitely long, and the flow will be fully developed for most of the orifice/slot height. This limits the model. Furthermore, the model is limited in terms of the ratio of $h/2a$ for which there is a good agreement between the model and experimental measurements. It has been shown previously by Gallas et al. (2003a) that a good agreement between the model and experiments is obtained for values of $h/2a_0$ that exceed approximately unity.

The model is also limited by the accuracy of the nonlinear orifice/slot resistance ($R_{a,nl}$). The accurate computation of $R_{a,nl}$ relies heavily on the value of K_d. As was mentioned previously, a value of $K_d = 1$ was used in previous lumped element models. The nonlinear dimensionless orifice dump loss coefficient represents the pressure drop necessary to accelerate a uniform flow into its fully developed velocity profile. Even though the value of K_d is treated as a constant, in reality it is a function of the Reynolds number (Re), Stokes number (S), and Strouhal number (St); therefore, a more realistic value of K_d, that is, a function of geometry, frequency, and so on, is necessary to increase the fidelity of the model.*

Other models for K_d that depend on the Stokes number (S) exist in the literature. For example, Raju et al. (2007) proposed a phenomenological model based on the exact solution for oscillatory flow in a 2D channel (Panton 1996). In their model,

$$K_d = (2/\pi)^2 \Delta C_{p,\text{mom}} \tag{2.43}$$

where

$$\Delta C_{p,\text{mom}} = 2\int_{-1/2}^{1/2} \left(\frac{u_x\,|u_x| - U\,|U|}{\bar{U}_j^2} \right) d\left(\frac{y}{d}\right)$$

The result is a model for the rms value of K_d, that is, a function of Stokes number, illustrated in Figure 2.26. The authors noted that the frequency-dependent model of K_d significantly reduced the rms error when compared to the traditional model employed by lumped element models.

Tang and Zhong (2009) also developed a model based on the jet exit velocity profile. A fully developed oscillatory flow is assumed at the orifice exit. The method for calculating K_d

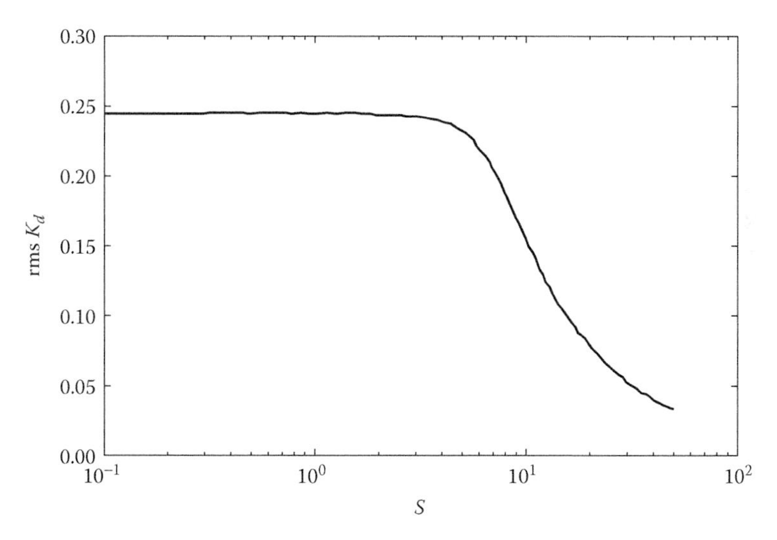

FIGURE 2.26 rms value of loss coefficient K_d versus S. (Adapted from Raju, R. et al., *Phys. Fluids*, 19, 78107, 2007. With permission.)

* In reality, out of the three dimensionless quantities, Re, St, and S, only 2 of the 3 are independent with $1/St = Re/S^2$ (Holman 2006); therefore, K_d is only a function of 2 of the 3.

follows the analysis of Wakeland and Keolian (2002). Based on the jet exit velocity profile, K_d is computed using

$$K_d = \frac{\int \left| \int u^3(r,t)\, dA \right| \mathrm{d}t}{\int \left| \int u(r,t)\, dA \right|^3 \mathrm{d}t} \tag{2.44}$$

where:

$u(r, t)$ is the spatially varying velocity profile at the orifice exit
A is the cross-sectional area of the orifice

The model for K_d is a function of Stokes number and was shown to asymptote to a value of 2 at low frequencies and to a value of 1 as the Stokes number approaches infinity.

Oyarzun and Cattafesta (2010) demonstrated the influence that different models for K_d have on the agreement between the lumped element model and experimental results. The results are illustrated in Figure 2.27. The model used for K_d is that of Raju et al. (2007), illustrated in Figure 2.26, which is properly rescaled from its rms values. The actuator that was tested consisted of a single piezoelectric bimorph and a 2D slot measuring $30 \times 0.50 \times 0.50$ mm (aspect ratio of 60). The details of the actuator design can be found the work by in Oyarzun and Cattafesta (2010).

From Figure 2.27, it is clear from the experimental values that the actuator has two resonant frequencies, yet the lumped element model with a constant value of $K_d = 1$ fails to

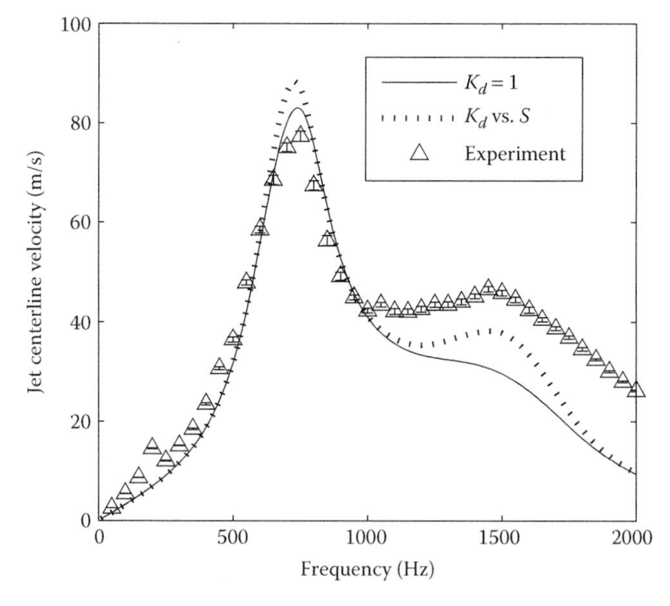

FIGURE 2.27 Influence of model for K_d on agreement between model and experiments. (From Oyarzun, M. A. and L. Cattafesta, Design and optimization of piezoceramic zero-net mass-flux actuators. *The 5th Flow Control Conference*, Chicago, IL, 2010.)

capture this feature. The constant model for K_d is able to predict the actuator response up to and just beyond the first resonant frequency, but breaks down thereafter. The frequency-dependent model for K_d overpredicts the velocity at the first resonant frequency, but significantly improves the prediction in the higher frequency range, with a clearly visible second resonant frequency. The orifice nonlinearity, and therefore K_d, plays a significant role in the lumped element model and a better understanding is necessary to improve the fidelity of the lumped element model.

Finally, the model is restricted to straight orifice/slot geometries. The present model cannot predict the performance of a device with a nozzle or diffuser-shaped orifice/slot, or one in which the edges are rounded or beveled. In the case of rounded or beveled geometries, this should primarily affect the minor losses, thus the nonlinear resistance term.

2.8 Design Trade-Offs and Optimization

A key conclusion from the lumped element analysis of the equivalent circuits illustrated in Figures 2.18 and 2.21, supported by experiments, is that a ZNMF actuator is a coupled, nonlinear, two degree-of-freedom oscillator with two characteristic frequencies. One is the natural frequency of the driver mechanism ($f_D = 1/2\pi \sqrt{1/(M_{aD} + M_{aD,\mathrm{rad}})C_{aD}}$) and the other is the Helmholtz frequency of the cavity/orifice/slot ($f_H = 1/2\pi\sqrt{1/(M_{aN} + M_{a,\mathrm{rad}})C_{aC}}$). In this particular case, the device is capacitively coupled by the acoustic compliance of the cavity (C_{aC}). As such, the system has two resonant frequencies, f_1 and f_2, where $f_1 < f_2$. The two sets of frequencies are related by $f_1 f_2 = f_D f_H$. A common misconception is that the two resonant frequencies of the coupled oscillator are identical to f_D and f_H. However, this may or may not be the case and, in fact, depends on the coupling between the diaphragm and the cavity, as well as the orifice/slot damping. This phenomenon is discussed for linear, undamped, coupled oscillators in Fischer (1955) and can be used to advantage when designing ZNMF actuators to obtain devices with single or dual resonance frequencies.

Some additional physics can be extracted by taking a simpler look at the equivalent circuit for a generic synthetic jet actuator, illustrated in Figure 2.28. Using Figure 2.28, the

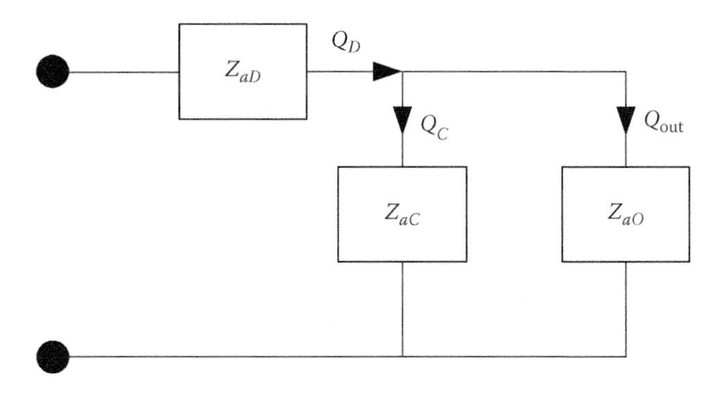

FIGURE 2.28 Simplified circuit representation of ZNMF actuator.

driver dynamics can be decoupled and a relationship between the jet volume flow rate (Q_{out}) and the driver volume flow rate (Q_D) can be derived. From Figure 2.28, it is clear that $Q_D = Q_C + Q_{out}$, where Q_C is the cavity volume flow rate. The relationship between Q_{out} and Q_D is

$$\frac{Q_{out}(\omega)}{Q_D(\omega)} = \frac{Z_{aC}}{Z_{aC} + Z_{aO}} = \frac{1/j\omega C_{aC}}{1/j\omega C_{aC} + R_{aN} + j\omega M_{aN}} \tag{2.45}$$

The above equation assumes only linear resistance and mass terms, R_{aN} and M_{aN}. In the incompressible case, $Q_D = Q_{out}$, which occurs when $Z_{aC} \gg Z_{aO}$. The incompressible case therefore occurs at low frequencies and for small cavity volumes. The ratio of the cavity impedance to the orifice impedance can be easily obtained and has been shown to reduce to the following (Gallas 2005)

$$\frac{Z_{aC}}{Z_{aO}} \propto \frac{1}{\left(\frac{\omega}{\omega_H}\right)^2 \left(j\frac{1}{S^2} - 1\right)} \tag{2.46}$$

where:

ω is the driver frequency
ω_H is the Helmholtz frequency
S is the Stokes number

From Equation 2.46, it is clear that the incompressibility limit is a function of the driver to Helmholtz frequency and is valid when $\omega/\omega_H \ll 1$. Similarly, Equation 2.45 reduces to the following, assuming only linear resistance and mass terms

$$\frac{Q_{out}}{Q_D} = \frac{Q_{out}(\omega)}{j\omega\Delta\text{Vol}} = \frac{Z_{aC}}{Z_{aC} + Z_{aO}} \approx \frac{1}{\left[1 - \left(\frac{\omega}{\omega_H}\right)^2\right] + j\left[\frac{1}{S}\left(\frac{\omega}{\omega_H}\right)\right]^2} \tag{2.47}$$

The above equation provides some interesting results. First, in the incompressible limit (i.e., $\omega/\omega_H \ll 1$), the actuator output (Q_{out}/Q_D) tends to be 1, meaning that the jet flow rate is directly proportional to the driver performance. The more interesting case, however, is the compressible case, in which C_{aC} is finite, and hence, ω_H is finite as well. When the actuator is driven at cavity resonance ($\omega = \omega_H$), the complex impedance in the loop formed with the cavity and orifice/slot branches is identically zero and the impedance is purely resistive. This allows Q_{out} to be greater than Q_D via an acoustic lever arm and increases with the increasing Stokes number (S). At frequencies much greater than cavity resonance, or for very large cavities, the acoustic impedance of the cavity goes to zero allowing the cavity volume velocity (Q_C) to be nonnegligible, thus yielding a small Q_{out} compared to Q_D. It is important to note that these trends were derived from a linear transfer function and are only valid at low forcing levels. The nonlinear resistance ($R_{a,nl}$), which was neglected in the analysis, tends to decrease the output as the input amplitude increases. Nonetheless, these results indicate that there is a trade-off between the cavity design and the driver amplitude, depending on the desired output to be achieved and on the desired bandwidth of operation, that is, gain versus bandwidth.

This gain versus bandwidth design trade-off is also evident in the design of the piezoelectric composite diaphragm. In general, gain and bandwidth are usually at odds with one another, meaning that an actuator that produces higher gain typically has a lower bandwidth. This fact was demonstrated in the optimization work of Papila et al. (2008). In their work, the design/optimization of clamped circular piezoelectric unimorph and bimorph diaphragms was investigated, particularly the conflicting requirements of maximum volume displacement for a prescribed bandwidth. The results indicate that the optimal volume displacement, that is, gain, is related to the bandwidth by a universal power law. This power law states that the product of the square of the natural frequency of the diaphragm and the displaced volume ($\Delta\text{Vol} \times f_D^2$), or gain–bandwidth product, is a constant. This power law clearly demonstrates the trade-off between the diaphragm gain and the resulting bandwidth.

Another trade-off is the generalized displacement versus generalized force. An actuator will provide its maximum output displacement when the load seen by the actuator is minimized and the actuator is free. The output of an actuator often becomes nonlinear as its maximum output is approached. On the other hand, the actuator will produce its maximum force when the load is maximized and the actuator is blocked. This trade-off is best illustrated by Equation 2.8, which is the linear relation between the volume displaced by the piezoelectric diaphragm ($Q = j\omega\Delta\text{Vol}$) loaded by a differential pressure (P) across the diaphragm and excited by an applied voltage (V)

$$Q = j\omega C_{aD}P + j\omega d_a V \tag{2.48}$$

The above equation demonstrates the trade-off between maximum or free displacement (e.g., $P = 0$) and force or pressure (e.g., $\Delta\text{Vol} = 0$). The maximum or free displacement is $\Delta\text{Vol}_{\text{free}} = d_a V$, where $V = E_c t_p$ is limited by the product of the coercive electric field of the piezoelectric ceramic (E_c) and the piezoelectric thickness (t_p). The blocked pressure is $P_{\text{blocked}} = -d_a V / C_{aD}$. The operating point of the actuator lies along the line given by Equation 2.48, somewhere between these two extremes. The design goal, then, is to increase $\Delta\text{Vol}_{\text{free}}$ and P_{blocked} subject to design constraints (Cattafesta and Sheplak 2011).

The same types of design trade-offs apply for the electrodynamic case. The magnetic assembly is what determines the maximum blocked force produced by the actuator. Recall, the blocked force was previously defined as $F_b = B\ell_{\text{coil}}I$, where B is the magnetic flux density in the coil, ℓ_{coil} is the coil length, and I is the current through the coil. A closed-form equation for the magnetic flux density was derived by Agashe et al. (2009), and when substituted into the equation for F_b, it was found that maximizing F_b involves a coupled optimization of the coil design and the magnetic assembly design. It was determined that if a very small gap between the magnetic assembly and the coil was chosen, higher magnetic flux density could be achieved, but the resulting coil length will be smaller, and vice versa.

Therefore, there is a trade-off between magnetic flux density and coil length, the product of which ($B\ell_{\text{coil}}$) is the important quantity.

For the electrodynamic case, the magnetic and coil assemblies are responsible for determining the blocked force that the actuator can achieve. However, the composite diaphragm has a profound impact on the dynamics that the actuator exhibits. Recall, the overall output response of an electrodynamic ZNMF actuator is given by Equation 2.39

Equation 2.39 shows that Q_{out}/V_{ac} is directly proportional to G/Z_{eCoil}, where $G = B\ell_{\text{coil}}/S_{\text{eff}}$ is the electroacoustic transduction coefficient and S_{eff} is the piston/diaphragm effective area. The diaphragm effective area is related to the solidity ratio (b/a), which is illustrated in Figure 2.10 as the ratio of the radius of the rigid central boss region to the radius of the compliant annular region. To achieve a large transduction coefficient (G), smaller solidity ratios are preferred, meaning a larger annular compliant region (Agashe et al. 2009). A smaller solidity ratio also leads to larger volume velocity, but this comes at a cost. As the solidity ratio decreases, so does the resonant frequency of the diaphragm (i.e., decreased bandwidth). Therefore, as was the case with the piezoelectric composite diaphragm, there is a trade-off between large G and/or large ΔVol, and high bandwidth.

The design of the orifice/slot also exhibits design trade-offs as well. Recall Equations 2.24 through 2.27 and Equations 2.32 through 2.36 that define the acoustic resistance and mass terms of the orifice/slot. Increasing the orifice radius (a) decreases the total acoustic mass. This has the effect of increasing the Helmholtz frequency. Conversely, increasing the orifice height (h) increases the acoustic mass of the orifice and actually decreases the Helmholtz frequency of the actuator. The orifice resistance terms (R_{aN} and $R_{a,nl}$) do not have an effect on the dynamics of the actuator response. However, they have a large impact on the magnitude of the jet velocity that exits the orifice. In general, a smaller orifice radius leads to larger jet velocities because a large amount of flow is exiting through a smaller aperture. This is valid up to a certain point. The linear resistance term is proportional to $1/a^4$. Therefore, decreasing the orifice radius by a small amount causes a significant increase in the orifice resistance, that is, damping. A less significant increase in orifice damping occurs by increasing the orifice height, since the orifice damping is directly proportional to the orifice height. Furthermore, the orifice nonlinearity is proportional to Q_{out}, meaning that as the output volume flow rate increases so does the nonlinear damping. This nonlinear damping term can, in fact, saturate the actuator response, and therefore, there is a trade-off between Q_{out} and $R_{a,nl}$.

Taking all of these design trade-offs into account, it is evident that there is no clear procedure for how to optimize a particular synthetic jet actuator, especially for the objective functions described in Section 2.3.2. Given the number of design variables that can be varied, a trial-and-error-based design procedure is not a viable option. Several optimization studies have been performed in the literature. In the work of Gallas et al. (2003b), a lumped element model for a piezoelectric ZNMF actuator was incorporated into an optimization routine in MATLAB® using the built-in *fmincon* function. The optimization was split into two parts, namely, (1) optimization of the cavity/orifice and (2) optimization of the driver, and different objective functions were used to investigate design trade-offs. The experimental optimization of synthetic jet actuators by parametrically varying some of the geometric variables was performed by Gomes et al. (2006). In this study, the driver geometry was fixed, and the cavity volume and orifice height were varied using different shims and orifice plates for a given fixed orifice diameter (1.2 mm). The work of Papila et al. (2008), as mentioned earlier, focused on the optimization of clamped circular piezoceramic actuators in both unimorph and bimorph configurations and specifically addressed the conflicting requirement between maximum volume displacement and bandwidth.

More recently, Oyarzun and Cattafesta (2010) developed a validated design tool that was used to optimize several actuator designs subject to multiple objective functions. Like in the work by Gallas et al. (2003b), the lumped element model was incorporated into a MATLAB optimization routine, using *fmincon*, and was subjected to several design and model constraints as well as several objective functions. The resulting optimized designs were then fabricated and characterized and the results were compared with the lumped element model prediction. For illustrative purposes, two of the optimized designs are presented in the subsequent paragraphs.

The first optimization case attempted to optimize a baseline actuator design called "Case 1" and reported on in Gallas et al. (2003a). For this particular optimization, the driver geometry was fixed, and the cavity and orifice geometries were allowed to vary in an attempt to improve the performance of the baseline device. The results of the optimization are illustrated in Figure 2.29.

As can be seen from Figure 2.29, there is a significant increase in the jet centerline velocity, approximately twofold, in the optimized design compared to the baseline design. Furthermore, the baseline design has two distinct resonant frequencies, while the optimized design has a single dominant peak in the actuator response. This can be attributed

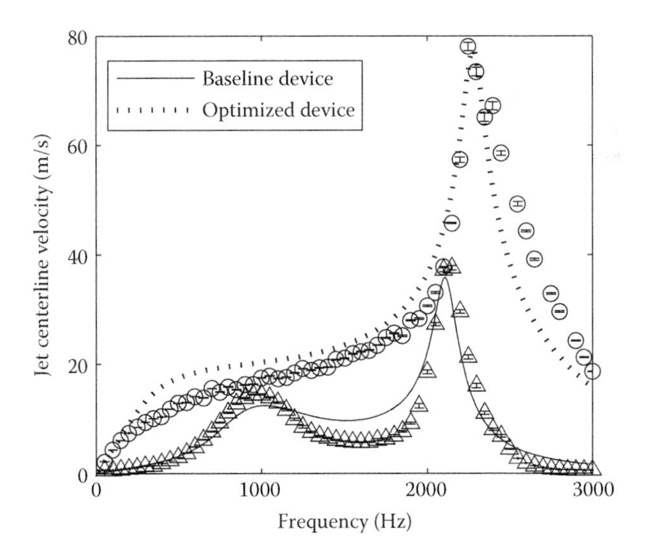

FIGURE 2.29 Comparison of performance between baseline actuator design. Symbols represent experimental values and solid lines represent the model prediction. [Case 1 (From Gallas, Q. et al. 2003a, *AIAA J.*, 41, 240–247.) and optimized actuator design (From Oyarzun, M. A. and L. Cattafesta, Design and optimization of piezoceramic zero-net mass-flux actuators. *The 5th Flow Control Conference*, Chicago, IL, 2010.) for V_{ac} = 25 V, ζ = 0.03, and K_d = 1.]

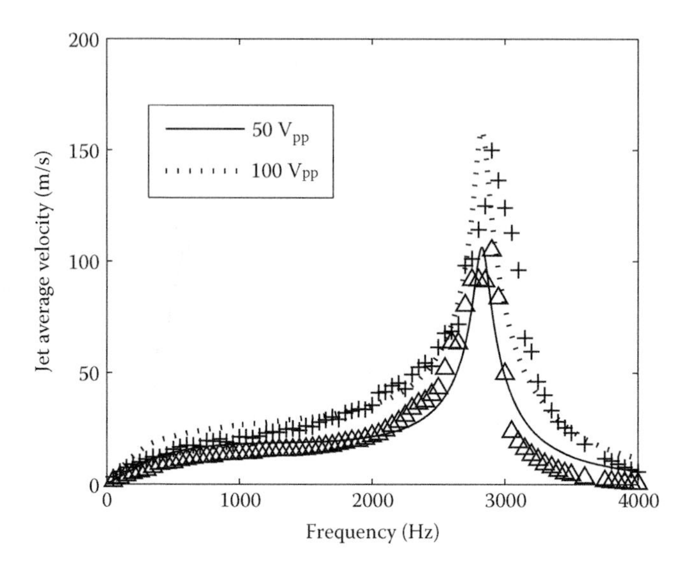

FIGURE 2.30 Comparison between model and experiments for a fully optimized actuator design for $\zeta = 0.03$ and $K_d = 1$. Symbols represent experimental values and solid lines represent the model prediction. (From Oyarzun, M. A. and L. Cattafesta, Design and optimization of piezoceramic zero-net mass-flux actuators. *The 5th Flow Control Conference*, Chicago, IL, 2010.)

to an increase in the damping in the orifice due to a smaller orifice radius as well as an increase in the nonlinear damping term due to the larger output velocities, which together dampen the first resonant peak.

The second optimization case considered was the full optimization of the entire actuator topology (i.e., piezoelectric diaphragm, cavity, and orifice). In this case, all of the actuator geometric variables were allowed to vary, while the material properties of the piezoelectric ceramic were fixed (PZT-5A). The optimization was subject to several constraints, both geometric and model, and several different objective functions, which are summarized in Oyarzun and Cattafesta (2010). The custom piezoelectric diaphragm was then fabricated by American Piezo Ceramics (APC) International Ltd (Mackeyville, PA). The results of the actuator characterization are illustrated in Figure 2.30.

The comparison between the model and the experiments shows a good agreement in terms of both amplitude and frequency, which is a promising result as this indicates that the lumped element model is well suited and capable of producing optimized actuators that are validated by experimental results.

A final design consideration that should not be overlooked is to ensure that a synthetic jet is actually formed. A synthetic jet results from the interaction of a train of vortices that are periodically shed from the orifice/slot with the surrounding fluid. The train of vortices entrain the surrounding fluid and create a jet in a time-averaged sense (Glezer and Amitay 2002). It was previously shown by Holman et al. (2005) that for a synthetic jet to form, the following criteria must be met:

$$\frac{1}{St} = \frac{Re_{\bar{U}}}{S^2} = \frac{\bar{L}/d}{\pi} > C \quad (2.49)$$

where:

St is the Strouhal number ($St = \omega d/\bar{U}$)
Re is the Reynolds number ($Re = \rho\bar{U}/\nu$)
S is the Stokes number ($S = \sqrt{\omega a^2/\nu}$)
d is the orifice/slot diameter
C is a constant (0.16 for an axisymmetric orifice and 1.0 for a 2D slot)
$\bar{U}$ is the time- and spatial-averaged exit velocity

The time- and spatial-averaged exit velocity is given by

$$\bar{U} = \frac{2}{T}\frac{1}{A}\int_A \int_0^{T/2} u\left(t, y\right) \mathrm{d}t\mathrm{d}A \quad (2.50)$$

where:

T is the period
A is the orifice/slot area
$u(t, y)$ is the time- and spatially varying velocity in the orifice/slot

Therefore, the design of a synthetic jet actuator should be such that the aforementioned formation criteria are met.

2.9 Future Trends

Synthetic jet actuators have shown great promise in a number of active research areas and have proven to be beneficial. The current state of the art in synthetic jet actuators is limited to actuators that generate jets on the order of ~100 m/s at frequencies up to a few kilohertz. Shaw et al. (2006) reported the development of synthetic jets with peak values in excess of 250 m/s at approximately 850 Hz. However, due to the proprietary nature of the research, design details are not available. In order for synthetic jets to transition from the benchtop to demonstrate benefits at realistic vehicle flight Mach numbers and Reynolds numbers, new advancements are required.

Advancements in driver technology show promise in developing high control authority and high bandwidth actuators. Devices such as kinematic amplifiers (e.g., cymbals), snap-through designs, and multilayer diaphragms have been developed that show great promise. Dogan et al. (2001) provide a review of different solid-state ceramic actuator designs.

One potential advancement in driver technology is the "cymbal." A schematic of a cymbal actuator is illustrated in Figure 2.31. Cymbal actuators consist of a piezoelectric disc that is sandwiched between two metal conical endcaps (Dogan et al. 1997; Fernandez et al. 1998; Lam et al. 2006). The basic operating principle is that when an electric field is applied across the piezoelectric, small radial displacements lead to large transverse displacements due to the shape of the conical endcaps, acting like a kinematic amplifier. The motion of the endcap results in large displacements over a larger area than typical

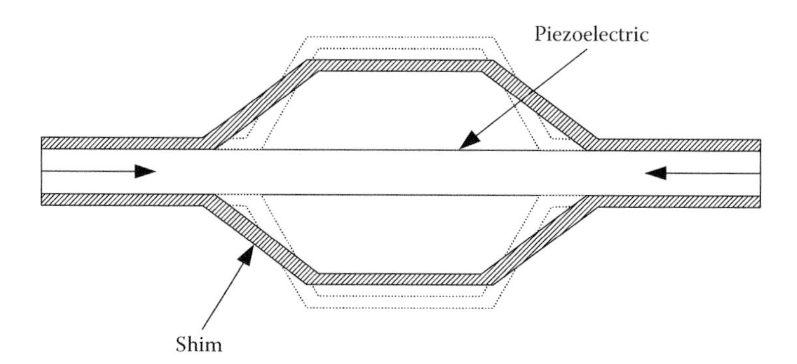

FIGURE 2.31 Schematic of a cymbal actuator.

piezoelectric diaphragms. Cymbal type actuators are particularly attractive because they are relatively inexpensive to fabricate, have a fast time response, and typically generate larger displacements at dc than the current state of the art. Furthermore, these devices can be tailored for different applications.

Another type of driver, developed by the National Aeronautics and Space Administration (NASA), that has garnered interest is the so-called radial field diaphragm (RFD) (Bryant et al. 2004). RFDs consist of several geometrically defined inter-circulating electrodes (ICE) and interdigitated ring electrodes (IRE). Unlike traditional unimorph or bimorph type actuators, when a voltage potential is applied to the electrodes, a radially distributed electric field mechanically strains the piezoelectric along the z-axis. The resulting transverse displacement profile is conical in shape. RFDs typically exhibit high displacements, several times larger than equivalent unimorphs, with displacements as large as 3 mm at dc having been reported (Bryant et al. 2004).

Multilayer piezoelectric diaphragms, such as RAINBOW and THUNDER, have also been developed in an effort to generate larger displacements (Dausch and Wise 1998; Mossi et al. 1999; Dogan et al. 2001; Mossi et al. 2005). The THUNDER design is an improved version of the RAINBOW design that was developed at the NASA Langley Research Center. The THUNDER actuator consists of three layers that are sandwiched together with epoxy. The first and third layers are prestressed metal layers that are epoxied to a piezoelectric layer. The prestressed layers cause the overall structure to be dome shaped. The advantage of such devices is that they are highly durable due to the way they are manufactured and can withstand mechanical impacts and high voltage levels. Voltages in excess of 800 V can be applied without causing damage, which results in significant displacement levels.

Other driver designs include macro-fiber composite (MFC) actuators (Sodano et al. 2004) and snap-through actuator designs (Schultz and Hyer 2003, 2004). An MFC actuator consists of thin PZT fibers that are embedded in Kapton film and covered with an interdigitated electrode pattern. The benefit of these devices is that the overall strength is greatly enhanced and at the same time is flexible. Furthermore, the electrode pattern takes advantage of the higher d_{33} coefficient as opposed to the d_{31} coefficient that disc benders make use of. As a result, the MFC has larger electromechanical coupling and produces larger force and free displacement. Snap-through designs use a piezoceramic actuator bonded to one side of a two-layer unsymmetric cross ply to snap the laminate from one stable shape to another, which has the potential to produce higher forces and displacements.

Several of these designs have the potential to be implemented into actuator designs to achieve improvements in existing actuator performance and help synthetic jets transition from the laboratory to the real-world, full-scale applications.

References

Agashe, J. 2009. Design and optimization of electrodynamic zero-net mass-flux (ZNMF) actuators. PhD thesis, University of Florida, Gainesville, FL.

Agashe, J., D. Arnold, and L. Cattafesta. 2009. Development of compact electrodynamic zero-net mass-flux actuators. *AIAA Paper* 1308, Orlando, FL, January 5–8.

Amitay, M. and F. Cannelle. 2006. Evolution of finite span synthetic jets. *Physics of Fluids*, 18(5): 054101.

Amitay, M. and A. Glezer. 2006. Aerodynamic flow control using synthetic jet actuators. In *Control of Fluid Flow*. eds. P. D. Koumoutsakos and I. Mezic, Berlin, Germany: Springer, pp. 45–73.

Amitay, M., D. Smith, V. Kibens, D. Parekh, and Glezer, A. 2001. Aerodynamic flow control over an unconventional airfoil using synthetic jet actuators. *AIAA Journal*, 39(3): 361–370.

Arunajatesan, S., M. Oyarzun, M. Palaviccini, and L. Cattafesta. 2009. Modeling of zero-net mass-flux actuators for feedback flow control. *AIAA Paper* 743, Orlando, FL, January 5–8.

Bendat, J. and A. Piersol. 2011. *Random Data: Analysis and Measurement Procedures*. Hoboken, NJ: John Wiley & Sons.

Bera, J., M. Michard, N. Grosjean, and G. Comte-Bellot. 2001. Flow analysis of two-dimensional pulsed jets by particle image velocimetry. *Experiments in Fluids*, 31(5): 519–532.

Blackstock, D. 2000. *Fundamentals of Physical Acoustics*. Hoboken, NJ: John Wiley & Sons.

Bryant, R., R. Effinger, I. Aranda, B. Copeland, E. Covington, and J. Hogge. 2004. Radial field piezoelectric diaphragms. *Journal of Intelligent Material Systems and Structures*, 15(7): 527–538.

Cater, J. and J. Soria. 2002. The evolution of round zero-net-mass-flux jets. *Journal of Fluid Mechanics*, 472: 167–200.

Cattafesta, L. N. and M. Sheplak. 2008. Actuators and sensors. In *Fundamentals and Applications of Modern Flow Control*. eds. R. D. Joslin and D. N. Miller. Reston, VA: American Institute of Aeronautics and Astronautics, pp. 149–176.

Cattafesta, L. N. and M. Sheplak. 2011. Actuators for active flow control. *Annual Review of Fluid Mechanics*, 43: 247–272.

Cattafesta, L. N., Q. Song, D. Williams, C. Rowley, and F. Alvi. 2008. Active control of flow-induced cavity oscillations. *Progress in Aerospace Sciences*, 44(7): 479–502.

Chen, F., C. Yao, G. Beeler, R. Bryant, and R. Fox. 2000. Development of synthetic jet actuators for active flow control at NASA Langley Research Center. *AIAA Paper* 2405, Denver, CO, June 19–22.

Chen, Y., S. Liang, K. Aung, A. Glezer, and J. Jagoda. 1999. Enhanced mixing in a simulated combustor using synthetic jet actuators. *AIAA Paper* 449, Reno, NV, January 11–14.

Chiekh, M., J. Béra, and M. Sunyach. 2003. Synthetic jet control for flows in a diffuser: Vectoring, spreading and mixing enhancement. *Journal of Turbulence*, 4: 26.

Dausch, D. and S. Wise. 1998. *Compositional Effects on Electromechanical Degradation of RAINBOW Actuators*. Hampton, VA: National Aeronautics and Space Administration, Langley Research Center.

Didden, N. 1979. On the formation of vortex rings: Rolling-up and production of circulation. *Journal of Applied Mathematics and Physics*, 30: 101–116.

Dogan, A., J. Tressler, and R. Newnham. 2001. Solid-state ceramic actuator designs. *AIAA Journal*, 39(7): 1354–1362.

Dogan, A., K. Uchino, and R. Newnham. 1997. Composite piezoelectric transducer with truncated conical endcaps. *IEEE Transactions on Ultrasonics Ferroelectrics and Frequency Control*, 44(3): 597–605.

Fernandez, J., A. Dogan, J. Fielding, K. Uchino, and R. Newnham. 1998. Tailoring the performance of ceramic-metal piezocomposite actuators, "cymbals." *Sensors and Actuators A: Physical*, 65(2): 228–237.

Fischer, F. 1955. *Fundamentals of Electroacoustics*. New York: Interscience Publishers.

Fox, R., A. McDonald, and P. Pritchard. 1998. *Introduction to Fluid Mechanics*. Hoboken, NJ: John Wiley & Sons.

Gallas, Q. 2005. On the modeling and design of zero-net mass flux actuators. PhD thesis, University of Florida, Gainesville, FL.

Gallas, Q., R. Holman, T. Nishida, B. Carroll, M. Sheplak, and L. Cattafesta. 2003a. Lumped element modeling of piezoelectric-driven synthetic jet actuators. *AIAA Journal*, 41(2): 240–247.

Gallas, Q., G. Wang, M. Papila, M. Sheplak, and L. Cattafesta. 2003b. Optimization of synthetic jet actuators. *AIAA Paper* 635, Reno, NV, January 6–9.

Glezer, A. and M. Amitay. 2002. Synthetic jets. *Annual Review of Fluid Mechanics*, 34(1): 503–529.

Gomes, L., W. Crowther, and N. Wood. 2006. Towards a practical piezoceramic diaphragm based synthetic jet actuator for high subsonic applications-effect of chamber and orifice depth on actuator peak velocity. *The 3rd AIAA Flow Control Conference*, San Francisco, CA, June 5–8, 2006. pp. 267–283.

Henderson, I. 2004. *Piezoelectric Ceramics: Principles and Applications*. Mackeyville, PA: APC International Ltd.

Holman, R. 2006. An experimental investigation of flows from zero-net mass-flux actuators. PhD thesis, University of Florida, Gainesville, FL.

Holman, R., Q. Gallas, B. Carroll, and L. Cattafesta. 2003. Interaction of adjacent synthetic jets in an airfoil separation control application. *AIAA Paper* 3709, Orlando, FL, June 23–26.

Holman, R., Y. Utturkar, R. Mittal, B. L. Smith, and L. Cattafesta. 2005. Formation criterion for synthetic jets. *AIAA Journal*, 43(10): 2110–2116.

Joslin, R. D. and D. N. Miller. 2009. *Fundamentals and Applications of Modern Flow Control.* Reston, VA: American Institute of Aeronautics and Astronautics.

Kim, B.-H., D. R. Williams, S. Emo, and M. Acharya. 2005. Modeling pulsed-blowing systems for flow control. *AIAA Journal*, 43(2): 314–325.

Krishnan, G. and K. Mohseni. 2009. Axisymmetric synthetic jets: An experimental and theoretical examination. *AIAA Journal*, 47(10): 2273–2283.

Lam, K., X. Wang, and H. Chan. 2006. Lead-free piezoceramic cymbal actuator. *Sensors and Actuators A: Physical*, 125(2): 393–397.

Mahalingam, R., and A. Glezer. 2005. Design and thermal characteristics of a synthetic jet ejector heat sink. *Journal of Electronic Packaging*, 127(2): 172.

McCormick, D. 2000. Boundary layer separation control with directed synthetic jets. *AIAA Paper* 519, Reno, NV, January 10–13.

Meirovitch, L. 2001. *Fundamentals of Vibrations.* Boston, MA: McGraw-Hill.

Meissner, M. 1987. Self-sustained deep cavity oscillations induced by grazing flow. *Acustica*, 62: 220–228.

Merhaut, J. and Gerber, R. 1981. *Theory of Electroacoustics.* New York: McGraw-Hill.

Mittal, R. and P. Rampunggoon. 2002. On the virtual aeroshaping effect of synthetic jets. *Physics of Fluids*, 14(4): 1533–1536.

Mossi, K., P. Mane, and R. Bryant. 2005. Velocity profiles for synthetic jets using piezoelectric circular actuators. *AIAA Paper* 2341, Austin, TX, April 18–21.

Mossi, K. M., R. P. Bishop, R. C. Smith, and H. T. Banks. 1999. Evaluation criteria for thunder actuators. *Proceedings of the SPIE, Smart Structures and Materials: Mathematics and Control in Smart Structures*, Newport Beach, CA, pp. 738–743.

Oyarzun, M. 2013. On the design and optimization of zero-net mass-flux actuators for active flow control applications. PhD thesis, University of Florida, Gainesville, FL.

Oyarzun, M. A. and Cattafesta, L. 2010. Design and optimization of piezoceramic zero-net mass-flux actuators. *The 5th Flow Control Conference*, Chicago, IL, June 28–July 1.

Panton, R. L. 1996. *Incompressible Flow*. 2nd Edition. Hoboken, NJ: John Wiley & Sons.

Papila, M., M. Sheplak, and L. N. Cattafesta III. 2008. Optimization of clamped circular piezoelectric composite actuators. *Sensors and Actuators A: Physical*, 147(1): 310–323.

Pavlova, A. and M. Amitay. 2006. Electronic cooling using synthetic jet impingement. *Journal of Heat Transfer*, 128(9): 897–907.

Prasad, S. A., Q. Gallas, S. B. Horowitz, B. D. Homeijer, B. V. Sankar, L. N. Cattafesta, and M. Sheplak. 2006. Analytical electroacoustic model of a piezoelectric composite circular plate. *AIAA Journal*, 44(10): 2311–2318.

Raju, R., E. Aram, R. Mittal, and L. Cattafesta. 2009. Simple models of zero-net mass-flux jets for flow control simulations. *International Journal of Flow Control*, 1(3): 179–197.

Raju, R., Q. Gallas, R. Mittal, and L. Cattafesta. 2007. Scaling of pressure drop for oscillatory flow through a slot. *Physics of Fluids*, 19(7): 78107.

Rao, S. S. 2007. *Vibration of Continuous Systems.* Hoboken, NJ: John Wiley & Sons.

Reddy, J. N. 1997. *Mechanics of Laminated Composite Plates: Theory and Analysis.* Boca Raton, FL: CRC press.

Rossi, M. and P. R. W. Roe. 1988. *Acoustics and Electroacoustics.* Norwood, MA: Artech house.

Rowley, C. W. and Batten, B. A. 2008. Dynamic and closed-loop control. In *Fundamentals and Applications of Modern Flow Control.* eds. R. D. Joslin and D. N. Miller. Reston, VA: American Institute of Aeronautics and Astronautics, pp. 115–148.

Rumsey, C. L., T. Gatski, W. Sellers III, V. Vasta, and S. Viken. 2006. Summary of the 2004 computational fluid dynamics validation workshop on synthetic jets. *AIAA Journal*, 44(2): 194–207.

Sawant, S. G., M. Oyarzun, M. Sheplak, L. N. Cattafesta, and D. Arnold. 2012. Modeling of electrodynamic zero-net mass-flux actuators. *AIAA Journal*, 50(6): 1347–1359.

Schultz, M. R. and M. W. Hyer. 2003. Snap-through of unsymmetric cross-ply laminates using piezoceramic actuators. *Journal of Intelligent Material Systems and Structures*, 14(12): 795–814.

Schultz, M. R. and M. W. Hyer. 2004. A morphing concept based on unsymmetric composite laminates and piezoceramic mfc actuators. *AIAA Paper* 1806, Palm Springs, CA, April 19–22.

Seifert, A. 2007. Closed-loop active flow control systems: Actuators. In *Active Flow Control.* ed. R. King. Berlin, Germany: Springer, pp. 85–102.

Shaw, L. L., B. R. Smith, and S. Saddoughi. 2006. Full-scale flight demonstration of active control of a pod wake. *AIAA Paper* 3185, San Francisco, CA, June 5–8.

Shuster, J. M. and D. R. Smith. 2004. A study of the formation and scaling of a synthetic jet. *AIAA Paper* 90, Reno, NV, January 5–8.

Shuster, J. M. and D. R. Smith. 2007. Experimental study of the formation and scaling of a round synthetic jet. *Physics of Fluids*, 19: 045109.

Smith, B. L. and A. Glezer. 1998. The formation and evolution of synthetic jets. *Physics of Fluids*, 10: 2281–2297.

Smith, B. L. and A. Glezer. 2002. Jet vectoring using synthetic jets. *Journal of Fluid Mechanics*, 458(1): 1–34.

Smith, B. L. and A. Glezer. 2005. Vectoring of adjacent synthetic jets. *AIAA Journal*, 43(10): 2117–2124.

Sodano, H. A., G. Park, and D. J. Inman. 2004. An investigation into the performance of macro-fiber composites for sensing and structural vibration applications. *Mechanical Systems and Signal Processing*, 18(3): 683–697.

Solecki, R. and R. J. Conant. 2003. *Advanced Mechanics of Materials.* New York: Oxford University Press.

Tang, H. and S. Zhong. 2009. Lumped element modelling of synthetic jet actuators. *Aerospace Science and Technology*, 13(6): 331–339.

Tian, Y., L. Cattafesta, and R. Mittal. 2006. Adaptive control of separated flow. *AIAA Paper* 1401, Reno, NV, January 9–12.

Vukasinovic, B., D. Brzozowski, and A. Glezer. 2009. Fluidic control of separation over a hemispherical turret. *AIAA Journal*, 47(9): 2212–2222.

Vukasinovic, B., A. Glezer, S. Gordeyev, E. Jumper, and V. Kibens. 2010. Fluidic control of a turret wake: Aerodynamic and aero-optical effects. *AIAA Journal*, 48(8): 1686–1699.

Wakeland, R. S. and R. M. Keolian. 2002. Influence of velocity profile nonuniformity on minor losses for flow exiting thermoacoustic heat exchangers. *The Journal of the Acoustical Society of America*, 112: 1249–1252.

White, F. M. 2006. *Viscous Fluid Flow*. New York: McGraw-Hill.

Yamaleev, N. K. and M. H. Carpenter. 2006. Quasi-ne-dimsensional model for realistic three-dimensional synthetic jet actuators. *AIAA Journal*, 44(2): 208–216.

Yao, C., F. J. Chen, D. Neuhart, and J. Harris. 2004. Synthetic jet flow field database for CFD validation. *AIAA Paper* 2218, Portland, OR, June 28–July 1.

Zhang, T.-Y. and C. Gao. 2004. Fracture behaviors of piezoelectric materials. *Theoretical and Applied Fracture Mechanics*, 41(1): 339–379.

II

Techniques

3

Measurement Techniques for Synthetic Jets

Barton L. Smith
Utah State University

Douglas R. Smith
Air Force Office of Scientific Research

3.1 Introduction

Synthetic jet measurements present numerous challenges. The flow is highly unsteady, requires a large dynamic range, and contains regions of high vorticity and high shear. This chapter will discuss various approaches to making measurements in these challenging flow fields. Since any experimental study of synthetic jets must start with quantification of its flow parameters, this chapter will start with an introduction to the definitions of important flow parameters and scaling of a synthetic jet flow. The wide range of devices used to generate synthetic jets in liquids and gasses will be reviewed. Some flow visualization techniques that have been employed to learn about synthetic jets will then be discussed. Subsequently, we will detail ways to address the measurement challenges mentioned in Chapter 1 using traditional methods, such as hotwire anemometry (HWA), as well as unobtrusive techniques involving seed particles, such as laser Doppler anemometry (LDA), also called laser Doppler velocimetry (LDV), and particle image velocimetry (PIV).

3.2 Flow Parameters

Synthetic jets can be characterized in several ways, depending on the application, and sometimes even depending on author preference. Generally, a dimensionless number describing the frequency and some sort of Reynolds number are required to characterize a synthetic jet. The frequency is often scaled by the displacement amplitude normalized by the exit diameter (L_0/D). If the synthetic jet is formed from a cylindrical cavity open on one end, so the exit diameter is the same as the cylinder, then L_0 is the distance the piston travels during the expulsion phase of each cycle. The displacement amplitude is also often called the stroke length (a term borrowed from vortex ring studies). More generally,

$$L_0 = \int_0^{T/2} u(t)\mathrm{d}t \tag{3.1}$$

where:
- $u(t)$ is the time-varying exit-averaged velocity
- $T = 1/f$

This parameter represents the length of the "slug" of fluid that is ejected from the orifice during every cycle. If the cross-stream average velocity is sinusoidal,

$$L_0/D = u_{\max}/\pi f D \tag{3.2}$$

where:
- D is the diameter of the orifice
- $u_{\max}$ is the cross-stream average streamwise velocity
- f is the driving frequency

This is clearly similar to the inverse of the Strouhal number, and many researchers prefer to think in terms of a dimensionless frequency rather than a dimensionless amplitude. The preferred quantity may depend on how the synthetic jet actuator is being employed, that is, is the synthetic jet being used to force a flow at an unstable frequency (Amitay et al. 2001) (for which Strouhal number is more applicable) or is it being used to impart momentum or to divert a flow (Smith and Swift 2003) (for which displacement amplitude is more descriptive). We also note that nonsinusoidal flow, which is common for synthetic jets, is easily accounted for in the definition of L_0 (by using Equation 3.1) but not for Strouhal number.

The Reynolds number can take several forms including ones based on the maximum velocity and either the viscous penetration depth (Re_δ) or the exit dimension D (Re_D). These Reynolds numbers are defined as

$$Re_\delta = \frac{u_{\max}\delta}{\nu};\ Re_D = \frac{D u_{\max}}{\nu} \tag{3.3}$$

where:

ν is the kinematic viscosity

the Stokes-layer thickness $\delta = \sqrt{\nu/\pi f}$

The Reynolds number based on Stokes-layer thickness best describes the laminar/turbulent state of oscillating flow (Hino et al. 1976, 1983; Ohmi et al. 1982; Lodahl et al. 1998) and is thus most useful for cases with long exits where the flow can transition from laminar to turbulent, resulting in major changes in the exit velocity profile. The two Reynolds numbers are related to one another and the displacement amplitude by

$$\frac{Re_\delta^2}{2} = \left(\frac{L_0}{D}\right)\frac{Re_D}{2} \tag{3.4}$$

The power required to produce the synthetic jet is an important consideration in flow control applications. While discussions of the efficiencies of drivers are beyond the scope of this chapter, it is straightforward to determine the acoustic power imparted to any synthetic jet flow

$$\dot{E} = 1/T \int_0^T pQdt \tag{3.5}$$

where:

p is the time-varying pressure inside the cavity

Q is the volume flow rate at the exit

The above-mentioned parameters are sufficient to characterize a synthetic jet in a quiescent environment. If the synthetic jet is used as an actuator in a cross-stream flow (Seifert et al. 1996; Amitay et al. 2001; Bridges and Smith 2003), then one more parameter is required to compare the momentum of the synthetic jet to the momentum of the crossflow. In looking back over the last two decades of active flow control with synthetic jet devices, the typical parameter used to capture this ratio is the momentum coefficient (C_μ), defined as

$$C_\mu \equiv \frac{\rho u_{max}^2}{\rho_\infty U_\infty^2}\frac{L_{act}}{L} \tag{3.6}$$

where:

u_{max} is the maximum velocity measured at the orifice

L_{act} and L are the characteristic length scales associated with the actuator and the flow field, respectively

Typical C_μ values in flow control applications fall in a range from $O(10^{-2})$ to $O(10^{-4})$, a magnitude that frequently reflects the ratio of relevant length scales in the problem more so than the actual momentum flux ratio. If the working fluid is incompressible ($\rho/\rho_\infty = 1$) and the synthetic jet issues into a crossflow without a meaningful geometric length scale, then the momentum coefficient reduces to a velocity ratio parameter, not unlike the parameter used to characterize conventional jets in crossflow.

Although the momentum coefficient has seen wide use, it is not clear that it is a meaningful parameter for scaling the effect of the control input on the larger, controlled flow. Recent work by Stalnov and Seifert (2010) evaluated the utility of five measures of the synthetic jet strength at scaling the effect of the jet on boundary layer separation control for Reynolds numbers less than 10^6. These five measures included the two described above, momentum coefficient and velocity ratio, as well as a Reynolds number-corrected momentum coefficient, a frequency-corrected velocity ratio, and a ratio of actuator vorticity to baseline flow circulation. The comparison explored the efficacy of the control on wing sections at both low and high reduced frequencies [F^+ values of $O(1)$ and $O(10)$] and over a range of angles of attack. Curiously, the uncorrected momentum coefficient (i.e., the one in widespread use) failed to collapse the data sets in the cases considered. Only the Reynolds number-corrected momentum coefficient succeed in scaling the data sets. Here, the Reynolds number-corrected coefficient uses the incoming momentum thickness as the freestream length scale. Since the momentum thickness depends on the state of the boundary layer, it naturally introduces a Reynolds number dependence into the scaling.

Finally, when the synthetic jet orifice has a large spanwise extent, as is often the case in flow control applications, spanwise nonuniformity of the momentum coefficient can arise, and it is important to carefully verify the synthetic jet performance at more than one location along the span. This issue relates to some of the challenges discussed in Section 3.3 on synthetic jet generator devices.

An additional issue for consideration when estimating the momentum coefficient of a synthetic jet actuator is how a crossflow might change the operational characteristics of the actuator and the flow at the orifice. To minimize this concern, the researcher can choose an actuator design where the length scale for the actuator orifice in the direction of the local crossflow is small relative to other length scales in the flow.

3.3 Devices

Synthetic jets have been formed using several mechanisms. Some typical devices exploit resonant behavior in the design and are driven by either piezoelectric disks or electromagnetic devices such as speakers. Other devices use simple mechanical actuation to generate the necessary periodic motion. Examples include crank-driven pistons and programmable linear slide-driven pistons. In Section 3.3.1, we will cover resonant cavity devices, piston-cylinder devices, and the impact of the generator shape on the ensuing flow field.

3.3.1 Resonant Cavity Devices

Most early synthetic jet work was performed using a resonator device of some type. For example, Ingard and Labate's (1950) jet was formed at the end of a standing wave resonator, and Smith and Glezer's (1998, 2002) jet was formed by a Helmholtz resonator driven by a piezoceramic disk. Many other examples of these devices can be found including the work by Krishnan and Mohseni (2009b). The relative compactness of the Helmholtz resonator-based devices for high frequency operation made them very popular. Their main disadvantage, however, was narrow bandwidth. Their operation in a resonant condition tends to make them prone to failure (Cattafesta and Sheplak 2011). The voltage requirements for piezoelectric devices are also relatively high (100 V peak-to-peak is not uncommon), so specialized amplifiers are often required.

The resonant frequency of a Helmholtz resonator is a function of the orifice length, area, and the cavity volume. It is not a function of cavity shape, leaving a synthetic jet designer free to shape the cavity anyway that is convenient without altering its frequency behavior. The actuator used by Smith and Glezer (2002) is shown in Figure 3.1 with the bottom cover removed. The design of this actuator evolved from others that were placed inside of airfoils or cylinders (Amitay et al. 2001) and thus had a very shallow cavity, which is shaded. The total cavity volume roughly matches that of a Helmholtz resonator with this actuator's orifice dimensions and a resonant frequency matching that of the drivers. In the photograph, one piezoelectric driver is installed with the remaining three removed. The driver had its own orifice (made of black plastic in this case), but this had no effect since the displacement amplitude in this orifice was too small to form a synthetic jet. Four of these drivers together had sufficient displacement to form a synthetic jet at the slot exit of the cavity.

The slot orifice was formed between the orifice plate and the bottom cover. A piezoresistive pressure sensor was installed to monitor the cavity pressure fluctuations, which were calibrated to the orifice velocity. It was later demonstrated (Persoons and O'Donovan 2007) that there is a predictable relationship between the cavity pressure and exit velocity in the synthetic jet.

Several studies have shown that when the cavity is short (meaning that the distance from the orifice to the back of the device is small), there can be an interaction between the vorticity generated from the inflow and the outgoing vortex ring or pair (Rizzetta

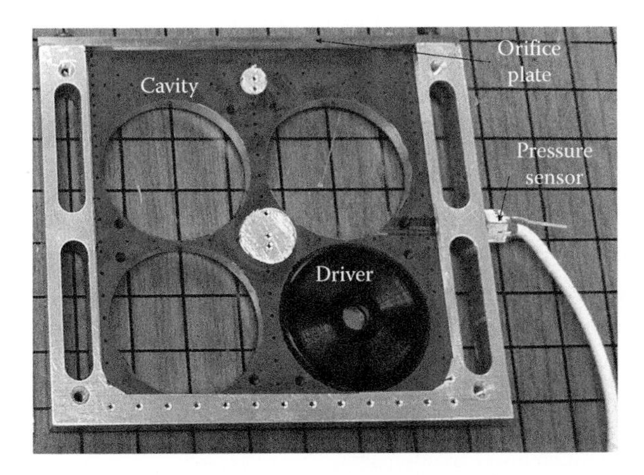

FIGURE 3.1 The synthetic jet actuator used in Smith and Glezer. (From Smith, B. L. and A. Glezer, *J. Fluid. Mech.*, 458, 1–34, 2002.) The many small holes accommodated pressure taps that extended through the cover plate into the conduit, for which this actuator formed one wall. (Photo courtesy of B. L. Smith.)

FIGURE 3.2 The synthetic jet actuator used in Bridges and Smith. (From Bridges, A. and D. R. Smith, *AIAA J.*, 41, 2394–2402, 2003.) The brass disk is 50 mm in diameter. (Photo courtesy of B. L. Smith.)

et al. 1999). As synthetic jets became popular flow control devices in the late 1990s, many studies attempted to numerically or experimentally determine the effect of various cavity parameters (Mane et al. 2007, 2008; Tang et al. 2007). Some of the numerical studies (Rizzetta et al. 1999) did so without accounting for the impact of these parameters on the cavity resonance.

A second example of a resonator device is shown in Figure 3.2 (Bridges and Smith 2003). This actuator, which features opposed piezoelectric disks, was used in a study of the interaction between a single synthetic jet and a turbulent boundary layer. As discussed earlier, Helmholtz devices are insensitive to the cavity shape, and here a small circular cylinder cavity was employed. An attempt was made to closely match the cavity volume to the resonant frequency of the piezoelectric disk. Two driver disks were used in this design to increase the output momentum of the device. As is apparent in Figure 3.2, the driver disk was formed by bonding the piezoelectric material to a thin circular brass disk. Brass has appeal here because it is a relatively pliable material that responds well to the deformation imposed by the piezoelectric material. The author has investigated similarly shaped actuators employing smaller diameter stainless steel disks and measured lower momentum outputs at the orifice. This observation suggests that the device output is significantly affected by the mechanical displacement of fluid by the disk.

In some past work, electromagnetic drivers have been used in resonant devices. These drivers have the advantage of operating at lower frequencies and being capable of larger displacements than the piezoelectric disks (Cattafesta and Sheplak 2011). They are, on the contrary, much less efficient (and therefore generate heat), larger, and heavier (due to their large magnets) than the piezoelectric disks. In general, actuators based on piezoelectric devices operate at very high frequency (200–4000 Hz), while electromagnetically driven devices operate at much lower frequencies.

Additional drawbacks of resonator devices become apparent when attempting to employ them in arrays for a flow control application. It is not unusual for the operating performance of the actuator to be narrowly focused around the resonant condition with steep performance declines for even small frequency departures. Machining variations during actuator fabrication can lead to subtle changes in the resonant characteristics of the device. As a result, one can see significant output variations across an array leading to an inhomogeneous control input to the base flow. No clear method for quantifying this nonuniformity has been found for flow control applications with synthetic jets, and the remedy is simply fabricating more actuators than needed and choosing those actuators with the closest frequency and output momentum characteristics.

Helmholtz resonators and the piezodriver can be modeled as a lumped-element system consisting of analogous electrical components. Starting with Breuer's group (Rathnasingham and Breuer 1997) and followed by Cattafesta and coworkers (Gallas et al. 2003; Raju et al. 2009), a series of papers were published outlining the methods for designing such devices to obtain the desired output at a given frequency. This topic is discussed extensively in Chapter 2.

3.3.2 Piston-Cylinder Devices

Piston-cylinder devices offer several advantages to the experimentalist. For example, it becomes simple to choose driving parameters that result in a desired displacement amplitude and Reynolds number, which is a difficult iterative process for resonant devices. Piston-cylinders also offer large cavity pressures (and thus exit velocities). The devices are readily assembled from simple materials and are easily driven with a stepper motor–linear slide combination or some other linear actuator. The actuators can be scaled easily while preserving symmetry that precludes complications when interpreting the measurements. If a slug model is used, a simple conservation of mass analysis can be used to estimate the operating point of the actuator. Examples of use of such a system include Bera et al. (2001) and Cater and Soria (2002). The device used in Shuster and Smith (2007) is shown in Figure 3.3.

While most were purely laboratory devices, at least one attempt was made to develop a crank-driven piston-cylinder device for full-scale flow control applications (Gilarranz et al. 2005a, 2005b). Crittenden and Glezer (2006) used a modified, commercially produced

FIGURE 3.3 The piston-cylinder synthetic jet actuator used in Shuster and Smith (2007). (From Shuster, J. M. and D. R. Smith, *Phys. Fluids*, 19, 045109, 2007. With permission.)

model airplane engine to generate a synthetic jet. While most synthetic jet flows are dominated near the exit by vortex rings, due to the very large displacement amplitude, there was very little influence of the vortex rings for this flow.

3.3.3 Synthetic Jet Generator Shape

Generally speaking, studies of synthetic jets, whether in flow control studies or on jets in isolation, have focussed on synthetic jets that were either rectangular, circular, or annular (Travnicek and Tesar 2003; Tesar and Travnicek 2008) at their origin. It is important not to overlook the role of the initial generator shape on the subsequent jet behavior. For circular generators, the evolution is unremarkable, starting with a vortex ring that rapidly develops circumferential instabilities before breaking down to form a statistically steady turbulent jet. For rectangular generators, the synthetic jet development is more complicated. Most studies have used a rectangular orifice with a high aspect ratio, that is, an orifice that resembles a narrow, long slot. This design was initially chosen to span a continuous rectangular jet (Smith and Glezer 2002). It also worked well for other early flow control experiments where a synthetic jet actuator was placed at the leading edge of an airfoil and used to control the flow separation at post-stall angles of attack. However, measurements have shown that the jet rapidly narrows (Smith and Glezer 1998) and can even undergo an axis-switching phenomenon characteristic of elliptic jets (Amitay and Cannelle 2006). With such synthetic jets, simple centerline measurements with any of the techniques mentioned will almost certainly lead to incorrect inferences regarding the behavior of the jet. Other orifice geometries, for example, square and triangular (Oren et al. 2009), have received cursory attention, but the role of orifice shape on synthetic jet formation and evolution is an area open to further investigation.

3.4 Flow Visualization

Flow visualization is a crucial first step toward understanding the behavior of an unknown flow, and this is especially true for synthetic jets. For a comprehensive review of synthetic jet visualizations, see Alvi and Cattafesta (2010). These authors provide examples of compressible and incompressible synthetic jets in quiescent surroundings, as well as visualizations of synthetic jets used as actuators. The techniques reviewed include use of smoke (both time-averaged and phase-locked), Schlieren, PIV, and laser-induced fluorescence (LIF). Section 3.4.1 will discuss some of the advantages and limitations of some of these techniques.

3.4.1 Schlieren Flow Visualization

Although Ingard and Labate in the 1950s made smoke photographs of what was later termed a synthetic jet, the first information about the time and space development of synthetic jets was obtained using phase-locked schlieren images as shown in Figure 3.4.

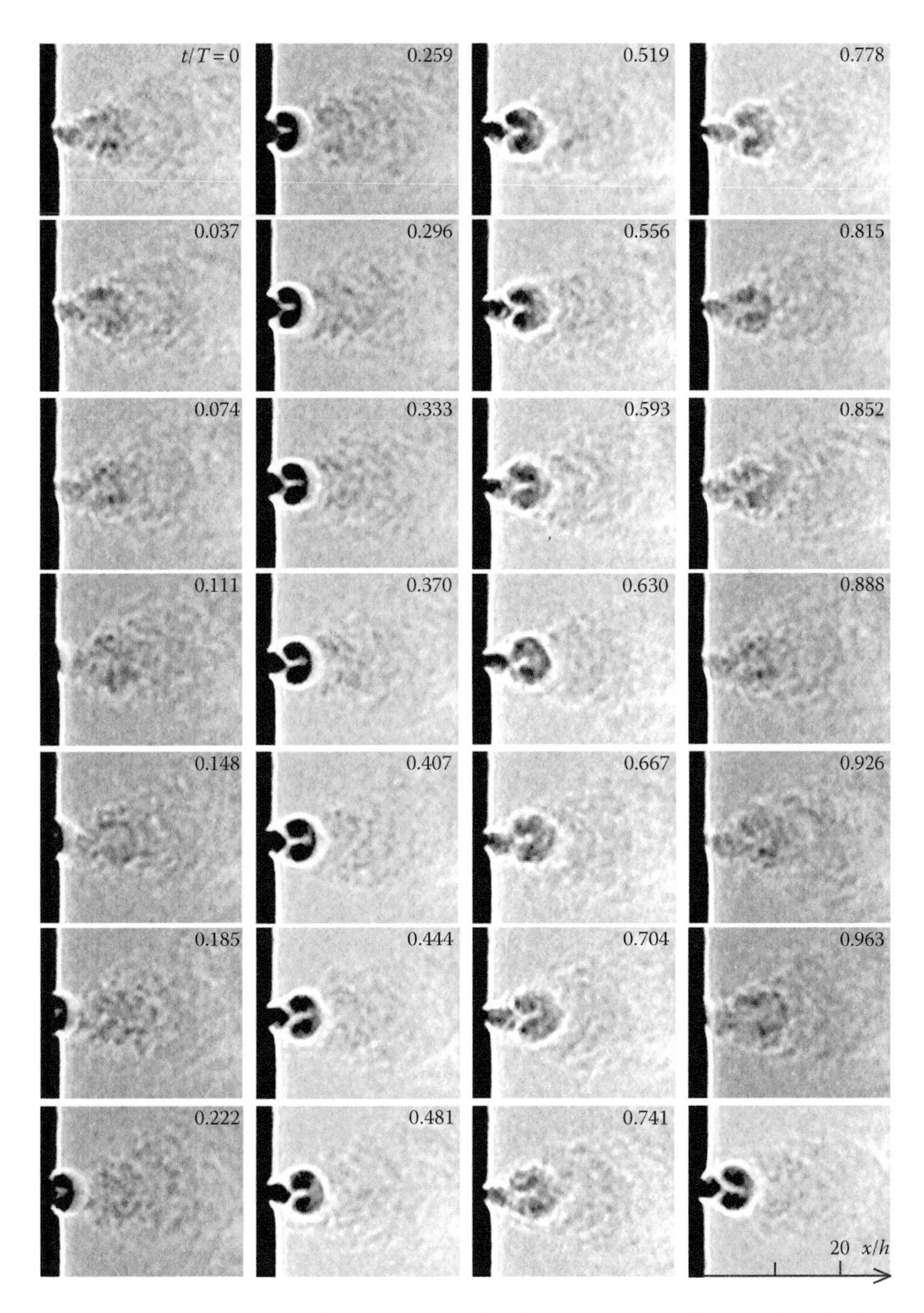

FIGURE 3.4 Phase-locked schlieren images of the planar synthetic jet from Smith and Glezer (1998) acquired at 27 equal instances during the cycle. The forward and backward exit velocities begin at $t/T = 0$ and $t/T = 0.5$, respectively. (Used with permission from Smith, B. L. and A. Glezer, *Phys. Fluids*, 10, 2281–2297, 1998.)

These images were acquired using a double-pass schlieren system, which is twice as sensitive as a typical system, since the light passes through the flow twice. The density gradients necessary for the images in Smith and Glezer (1998) were obtained simply by heating the actuator cavity to about 30°C above the ambient temperature. The system does require a large space, since the focal length of the mirror is long to minimize the angle of the light passing through the flow. The lens in the system reduces this angle while shortening the light path.

To further emphasize the importance of flow visualization early in an experimental program, these images were instrumental in the design of the hotwire measurements that followed since they demonstrated that for the displacement amplitudes in that work

1. Planar synthetic jets are formed from vortex pairs.
2. Synthetic jets are initially much wider than a conventional jet.
3. Vortex pairs are initially laminar, but become turbulent and break down.

These observations were also important for interpreting the response of the single-sensor hotwire to the synthetic jet flow, which appeared like a rectified sine wave near the exit. As the probe was traversed downstream, one peak grew while the other shrunk, until the peak corresponding to reverse flow disappeared at about one displacement amplitude downstream (see Section 3.5.1).

The schlieren system is essentially the same as the one used at Georgia Tech (Smith and Glezer 1998). A point source of light (generated by a fiber-optic cable at the bottom center) points toward a spherical mirror at the top center. The oscillating flow facility sits between the point source and the mirror. The light returning from the mirror is diverted toward the camera by a cube beam splitter. The light focussing between the splitter and the camera grazes a two-dimensional (2D) aperture formed by two razor blades.

In the schlieren technique, a point source of light was passed through the heated synthetic jet flow and was reflected back through the flow by a spherical mirror. A "knife" edge was placed at the location where the light returning from the mirror focuses and served to cut out light that had been deflected downward by the density gradients in the flow. A second knife at the same location created a 2D aperture that caused density gradients on either side of the jet to appear dark. Use of a system such as this is only effective for rectangular geometries since the effect is integrated along the span of the jet. A similar setup was later used by Smith and Swift (2003) to compare the development of longer displacement amplitude synthetic jets to continuous jets formed in the same facility and is shown in Figure 3.5. A similar schlieren system is shown in an excellent book by Settles (2001). The major component of this system, the spherical mirror, is a popular item for telescope hobbyists and is thus available relatively cheaply.

3.4.2 LIF Visualization

Synthetic jets formed in liquids have often been visualized using LIF. The earliest example of this usage was a study of an axisymmetric synthetic jet by Cater and Soria (2002). One of their images is shown in Figure 3.6. They used a continuous laser at 532 nm and a dye

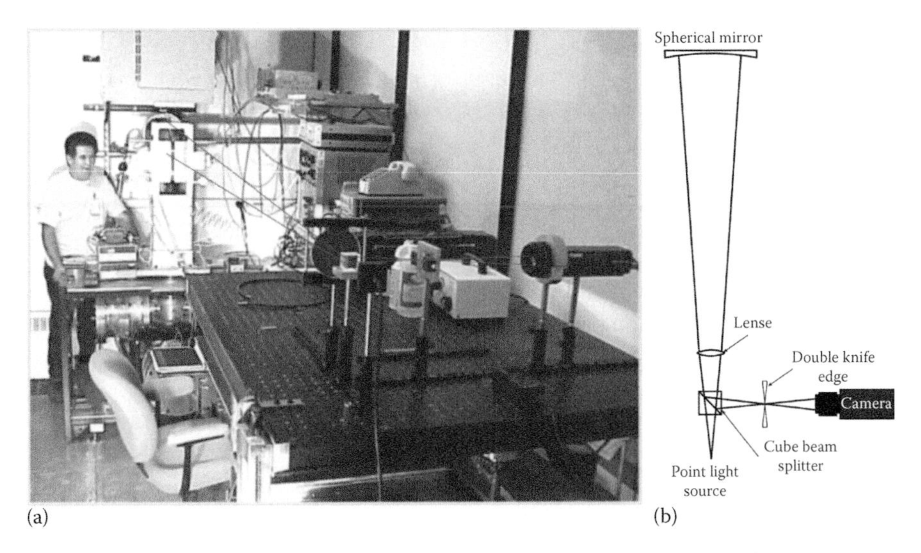

FIGURE 3.5 (a) Photograph of the double-pass schlieren system used to study synthetic jets and oscillating exit flow at Los Alamos. (Photo courtesy of B. L. Smith.) (From Smith, B. L. and G. W. Swift, *Exp. Fluids*, 34, 467–472, 2003a; Smith, B. L. and G. W. Swift, *J. Acoust. Soc. Am.*, 113, 2455–2463, 2003b.) (b) Schematic of the same system. (From Smith, B. L. and A. Glezer, *Phys. Fluids*, 10, 2281–2297, 1998. With permission.)

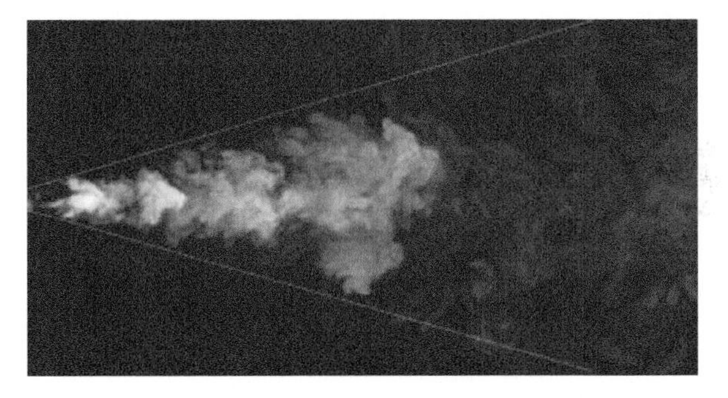

FIGURE 3.6 A synthetic jet from Cater and Soria (2002) imaged using LIF. $Re_0 = 10^4$; $St_0 = 0.0015$. This image was used to demonstrate that synthetic jets grow faster than conventional jets. (Reprinted with permission from Cater, J. E. and J. Soria, *J. Fluid Mech.*, 472, 167–200, 2002.)

that shifted the wavelength to 620 nm. Use of a fluorescent dye allows one to filter out all other light wavelengths, including room light and the light of the laser, thereby eliminating reflections and background light. The same group (Gordon et al. 2004) later used the same dye and laser along with a digital camera to estimate the mixing of a synthetic jet with a crossflow. Since the concentration of the dye at any location can be related to the intensity of light, this technique allowed these researchers to determine the effect of the synthetic jet on mixing. A similar setup with a high-speed camera has been used to view the evolution of coaxial synthetic jets (Ahmed and Bangash 2009).

3.5 Velocity Measurements

Most velocity measurements require significant time to acquire. Many methods of generating synthetic jets depend on a driver whose response may vary over time. Piezoelectric devices tend to loose capacity over time. The performance of electromagnetic devices varies with temperature, and since they tend to self-heat when driven hard, their output varies over time. For these reasons, synthetic jet experiments that do not use positive-displacement drivers should incorporate a means of determining the driver motion. Failure to do so makes performing a repeatable measurement nearly impossible.

Several schemes have been used to monitor synthetic jet output. Smith and Glezer (2005) calibrated the exit velocity (measured by a single-sensor hotwire) against the amplitude of pressure measured in each cavity. Yao and his coworkers (2006) used both a pressure sensor and a displacement gage. Krishnan and Mohseni (2009b) used a laser-based sensor to measure the driver displacement.

Synthetic jet velocity measurements can be very challenging, as the flow contains high shear rates, rotation, and flow reversal. Unlike conventional jets and due to their strong entrainment, the static pressure in and near a synthetic jet orifice is below ambient, making pitot probe measurements impossible. Additionally, the drivers that generate synthetic jets often generate heat, causing the outflow to be at a temperature different to the inflow (Yao et al. 2006). In spite of these issues, many good measurements have been accomplished.

One particularly interesting synthetic jet measurement study used all of the measurement techniques discussed here on the same experiment: HWA, LDV, and PIV. The Synthetic Jet Flowfield Database developed by NASA Langley Research Center (Rumsey 2004; Schaeffer and Jenkins 2006; Yao et al. 2006) was acquired for the purpose of computational fluid dynamics (CFD) validation, but provided a view into the relative advantages of all of the measurement techniques described here specific to synthetic jets. This study included three cases, the first of which is a synthetic jet in a quiescent surroundings. Only Case 1 will be discussed here and is described in some detail in Yao et al. (2006). Since none of the measurements were capable of measuring the exit velocity, exit parameters (such as L_0/h and *Re*) were not determined. It can be inferred from the behavior of the centerline velocity, which peaks at $x/h = 10$, that $L_0/h \approx 10$ for this case (Fugal et al. 2005). This study is unique in that the motion of the driver was recorded. This information, coupled with details of ambient conditions and geometry, can be used to generate a CFD model of the driver, which was the purpose of these data. While pairs of the instruments agreed in different parts of the flowfield, the authors do not draw a conclusion about what instruments are most reliable at which locations.

Some examples and issues with respect to synthetic jet measurements using several techniques will now be discussed in more detail. First, we will discuss HWA, followed by LDA. Each of these produce velocity as a function of time at a single point. LDA is an unobtrusive, laser-based technique, while HWA requires a probe. The section wraps up with a discussion of phase-locked PIV measurements (which produce velocity as a function of space at an instant in time), followed by time-resolved PIV, which provides time and space information on velocity.

3.5.1 Hotwire Anemometry

HWA, while limited in its ability to measure multiple components of velocity and reversing flow, nevertheless, presents some interesting advantages to synthetic jet measurements. First among these is frequency response. Hotwires can have very high frequency response (10^4–10^5 Hz). Since most synthetic jet actuators operate between 20 and 4000 Hz, hotwires should be able to easily capture the fluid motion as well as its higher harmonics.

An often-overlooked advantage is the spatial resolution of a single hotwire. For measurements on the centerline of a rectangular synthetic jet, it may not be possible to achieve a more accurate measurement than with a single hotwire. While common, commercially available probes are typically on the order of 1 mm wide (wider than many laboratory synthetic jet actuator dimensions), the wire diameter is usually a few microns. If the sensor is aligned with the long dimension of planar synthetic jet exit, one achieves excellent resolution in the important cross-stream direction.

While successful measurements have been made in planar synthetic jets using hotwires, the technique is less well suited for axisymmetric synthetic jets unless the jet diameter is much larger than the largest probe dimension, since significant averaging will occur over the probe.

The hotwire signal generated in the near field of a synthetic jet (before vortex breakdown) is rectified, since the hotwire cannot sense flow direction (Smith and Glezer 1998; Yao et al. 2006; Chaudhari et al. 2009). The local time that the flow reverses tends to vary from one cycle of the actuator to the next. An example of a raw hotwire signal at several downstream locations is shown in Figure 3.7. It is therefore necessary to "de-rectify" the hotwire signal before phase averaging is performed. Even so, the velocity zero crossings are very poorly measured since the frequency response of a hotwire anemometer drops off with the flow velocity. This can be seen in the results of derectifying the signal and then phase averaging for the same data, as shown in Figure 3.8. In this case, reverse flow occurred up to $x/h = 3$, which corresponds to $x/L_0 = 0.18$.

In the works by Yao et al. (2006), issues with derectification (which was performed after phase averaging) were blamed for the lack of agreement between the hotwire measurements and other techniques. However, due to spatial averaging suffered by the PIV and LDV measurements, only the hotwire measurements showed that the average velocity goes to zero at the synthetic jet exit.

An example of a hotwire setup, which was used in Krishnan and Mohseni (2009a) for the measurement of a synthetic jet, is shown in Figure 3.9. In addition to measuring the velocity downstream of the actuator, this setup (which is identical to the one used in Krishnan and Mohseni 2009b) was capable of measuring the driver displacement. The purpose of these measurements was to show that synthetic jets can be modeled similarly to steady jets if the turbulent viscosity is experimentally determined through measurements of the jet spreading and centerline velocity decay. These authors claimed a time-averaged velocity uncertainty of 2%, which is a typical figure for hotwire measurements.

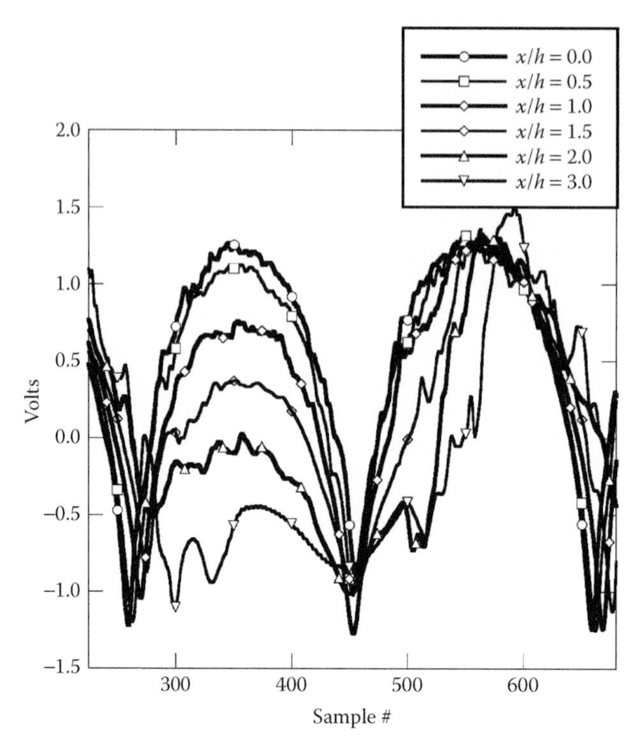

FIGURE 3.7 Raw, unpublished hotwire data acquired for the study reported in Smith and Swift (2003a) for a planar synthetic jet with $Re = u_{max}, h/\nu = 7000$, and $L_0/h = 17$. Data were acquired at 40,000 Hz at the locations $x/h = 0, 0.5, 1, 1.5, 2$, and 3. (Data taken from Smith, B. L. and G. W. Swift, *Exp. Fluids*, 34, 467–472, 2003a.)

Smith and Glezer (1998) used two-component HWA to measure the downstream characteristics of a rectangular synthetic jet. These measurements revealed the jet to be very similar to steady jets. While the measurements were able to demonstrate interesting features of the jet, due to the relatively poor resolution of the X-wire probe, there is little doubt that these measurements could have been made more easily and probably more accurately using PIV or other techniques.

For flow control applications that rely on the unsteady forcing generated by synthetic jets, a single hotwire on the centerline of a synthetic jet can be used to determine many quantities that may be important (Smith and Glezer 1998; Yao et al. 2006; Krishnan and Mohseni 2009) including the following:

- The distance over which the vortex pair breaks down
- The speed of the vortex pair
- The spectra of the jet

In addition, many have used single hotwires to determine the performance of their actuator before or during use in flow control applications (Smith and Glezer 2002; Amitay and Cannelle 2006). This is discussed further in Section 3.5.5.

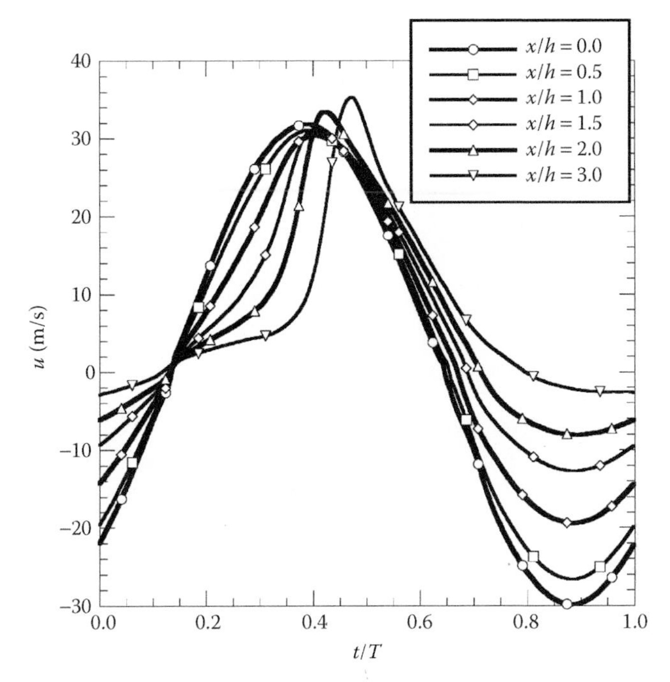

FIGURE 3.8 Phase-averaged hotwire data acquired for the study reported in Smith and Swift (2003a) for a planar synthetic jet with $Re = u_{\max}, h/\nu = 7000$, and $L_0/h = 17$. (Data taken from Smith, B. L. and G. W. Swift, *Exp. Fluids*, 34, 467–472, 2003a.)

3.5.2 Laser Doppler Velocimetry

LDV (also called LDA) is a technique where the motion of seed particles is detected by interrogating the reflected light from the particles as they pass through a region of light "fringes" that result from crossing the two laser beams. As with HWA, the result is the velocity as a function of time at a single location. The technique is often preferred to HWA in that it is optically based and thus unobtrusive, can detect two or three components of velocity, requires no calibration, and is considered more accurate than any other velocity measuring techniques. The expense of an LDV system is similar to a PIV system, and perhaps twice that of a HWA system.

One interesting issue with using seeded flows for synthetic jet measurements is that seed particles tend to pool near the exit. This can be an issue, especially for planar synthetic jets with a small slot size. It should be noted that this same issue exists for PIV.

LDV has been used less often in synthetic jet measurements than other techniques. With only one exception (Gallas et al. 2003) known to the authors, LDV measurements were used primarily to compare with hotwire measurements (Yao et al. 2006) or to "validate" PIV measurements (Holman et al. 2005; Schaeffer and Jenkins 2006; Goepfert et al. 2010). In the work discussed in Holman et al. (2005), PIV and LDV were used to measure the exit velocity profiles to determine the outflow waveform. The authors were able to achieve

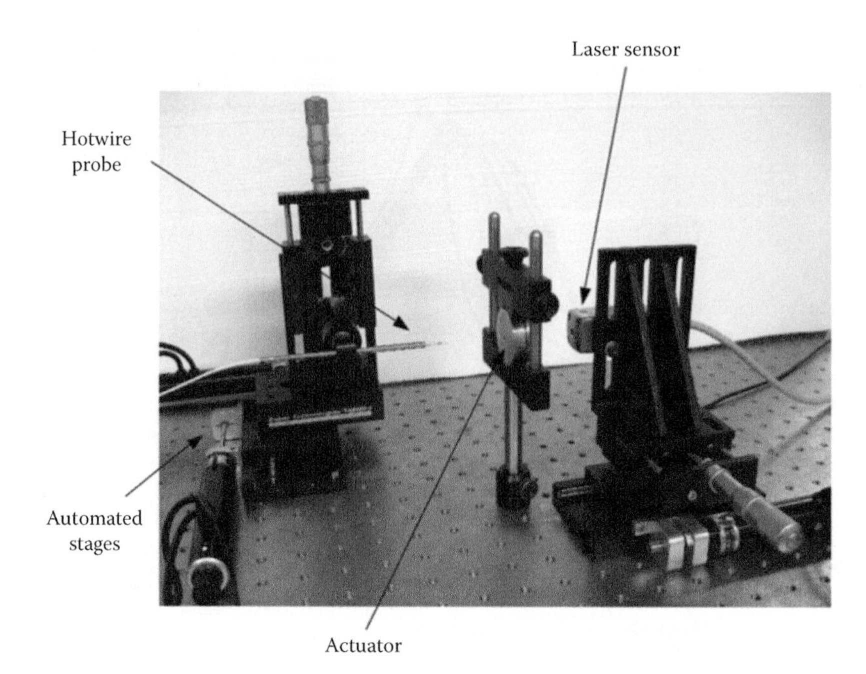

FIGURE 3.9 The measurement system used in Krishnan and Mohseni (2009a) that incorporates a single-sensor hotwire probe, a traverse stage for the probe, and a laser-based measurement of the driver displacement. (From Krishnan, G. and K. Mohseni, *AIAA J.*, 47, 2273–2283, 2009a. Reprinted by permission of the American Institute of Aeronatuics and Astronautics, Inc.)

better resolution using their LDV system than they could with the PIV setup that they used. While the PIV interrogation region can be made arbitrarily small by using higher-magnification optics, the laser sheet thickness is usually at least 1 mm. Therefore, in a narrow set of circumstances, LDV measurement may be able to achieve higher resolution than PIV. These authors concluded that, in their case, both techniques resulted in similar exit volume flow rate waveforms.

Gallas et al. (2003) provide a detailed explanation about some of the limitations of LDV for a synthetic jet measurement. In particular, they show that the required angle between the two LDV laser beams limits how close the measurement volume can be placed to the exit plane. In their case, they managed to measure within the diameter of the exit. As is shown in Section 3.5.5, while a measurement downstream of the exit may have a minor error relative to the exit plane on the outflow (in which these authors were interested), the error in the inflow can be very large.

3.5.3 Phase-Locked PIV Measurements

Many of the features of synthetic jets read like a list of the weaknesses of PIV. Synthetic jets contain regions of strong rotation and very high shear rates. Measurement of the full cycle requires very large dynamic range, since, even at an instant, the flow profile ranges from the peak velocity down to zero velocity at the edges of the jet, and even more so since, in time, the flow velocity varies from a maximum down to zero and then to a negative maximum. Since, for a fixed time interval dt between the two PIV images, the dynamic range of PIV (defined as the largest measurable velocity divided by the velocity resolution) is 100–200 (Adrian 2005), the dt between images must be chosen to match the maximum velocity for each phase of the cycle to achieve sufficient dynamic range. In spite of this, and most likely due to the power, convenience, and the wide range of possible spatial resolutions, many PIV measurements have been made.

In the very near field of a typically sized synthetic jet, PIV measurements are problematic. The main issue is the very large shear rates that are generated and the high spatial resolution required. Early measurements (Smith and Glezer 2005) did not account for these issues and may have had large errors as a result. This jet was formed by a 0.5-mm slot. While it is, in principle, possible to image regions this small, or even smaller, doing so requires a very short standoff distance for the camera (on the order of a few centimeters). Moreover, the Nd-YAG laser sheet that is commonly used for PIV measurements is normally about 1–3 mm thick. This precludes the use of PIV for axisymmetric smaller than 5 mm, since the velocity would be averaged in the direction perpendicular to the camera.

The periodic nature of a synthetic jet makes it particularly amenable to phase-locked PIV measurements. This approach has been used successfully by many researchers (Bera et al. 2001; Amitay and Cannelle 2006), among them Shuster and Smith (2007) worked in water with a relatively large piston-cylinder apparatus (Figure 3.3). By choosing to work with a large actuator, and hence large synthetic jet, they hoped to mitigate some of the spatial resolution issues associated with common actuators.

A similar-scale synthetic jet experiment in air has recently been undertaken by Nani and Smith (2012). The purpose of this experiment was to demonstrate the effect of the shape of the inner orifice lip shape on synthetic jet efficiency. An image of the experimental setup is shown in Figure 3.10. The actuator consisted of two large loudspeakers arranged in series (to improve the maximum cavity pressure that could be produced) attached to a round exit plane with a removable center. The lip shape was altered by using various center pieces. Visible in the photograph are the actuator cavity and exit plane, the PIV camera and laser, and an enclosure used to contain the seeding particles necessary for PIV measurements.

Unfortunately, since digital camera resolution is limited, there is a trade-off between spatial resolution and field of view. This is particularly challenging with synthetic jets because the vortices present at the actuator orifice require good resolution to accurately compute the velocity gradients, but resolving these features necessitates limiting the field of view to within a few exit dimensions. This has led investigators to contrive elaborate experimental apparatuses that translate the laser sheet and cameras through the field of interest in an effort to gain sufficient spatial data to make reasonable statements about the behavior of the jet. Figure 3.11 shows a representative system used by Tamburello and Amitay (2006) to collect velocity field data in a steady jet flow subjected to synthetic jet control. In these experiments, the investigators were obliged to have a small field of view to resolve the interaction of the synthetic jet with the larger steady jet. To access the larger jet flow field, both the imaging system and the laser sheet optics were mounted on computer-controlled traversing mechanisms. By assembling the individual smaller fields of view, a larger picture of the flow field was obtained.

FIGURE 3.10 The actuator and PIV measurement system used in Nani and Smith. (From Nani, D. J. and B. L. Smith, *Phys. Fluids*, 24, 115110, 2012. Reprinted with permission.) The laser pair is located at the bottom left, while the camera, fitted with a 105 mm lens, is at the center right. The white exit plan of the synthetic jet is visible in the center of the image with a plug (used to visualize the sides of the jet exit) inserted into the orifice.

One can also use multiple fields of view to collect all of the necessary information in a synthetic jet study. Nani and Smith (2012) used a field of view in which the synthetic jet exit filled nearly the entire frame to measure the volume flow rate waveform (necessary to compute acoustic power, see Figure 3.13 in Section 3.5.5). A second field of view spanning $20D$ was used to determine the jet momentum flux.

As a consequence of the low temporal resolution in PIV measurements, the method for obtaining phase-locked PIV data differs from the corresponding measurement with a hotwire. In the latter case, one could simply sample both the actuator driver signal and the hotwire signal simultaneously and then extract the phase-averaged information in post-processing. With PIV, the synthetic jet actuator driver voltage (or a sync signal corresponding to it) must be used to trigger the PIV acquisition. Generally, one would choose a range of phase angles at which to make the measurements then systematically work through those angles acquiring the needed data at a single phase before moving on to the next phase angle. Ideally, the dt between the two PIV images would be chosen for each phase based on the phase-averaged velocity.

In the water tank measurements by Shuster and Smith (2007), a position encoder was used to provide feedback on the actuator phase. The triggering was based on precise encoder count values; however, these values were not consistently reached at the top and bottom dead center positions in the piston-cylinder device making phase-locking at these phase angles difficult. Reliable triggering could only be achieved at $\pm 10^\circ$ of these extreme positions.

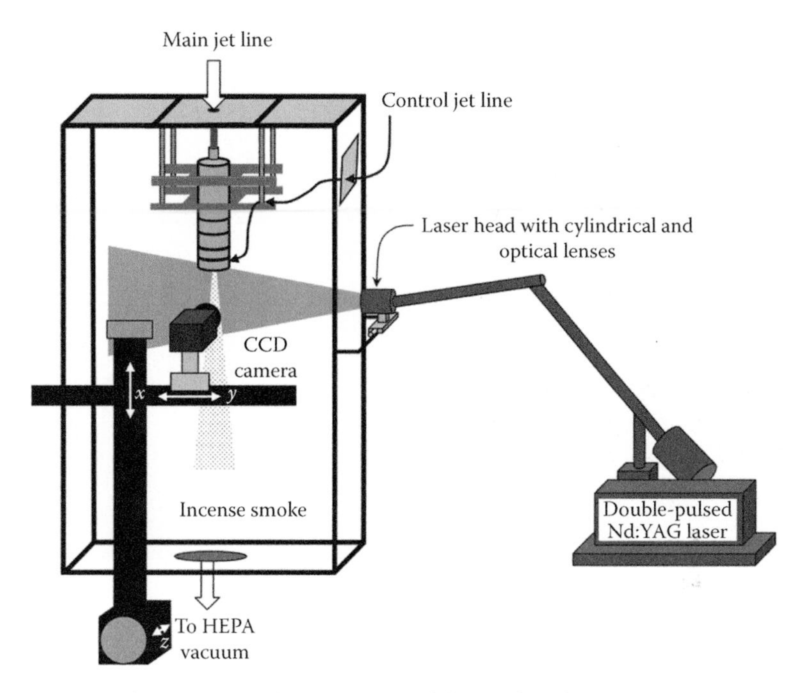

FIGURE 3.11 The experimental PIV setup used by Tamburello and Amitay (2006). (From Tamburello, D. A. and M. Amitay, Manipulation of an axisymmetric jet using a single perpendicular control jet. *The AIAA 3rd Flow Control Conference*, San Francisco, CA, 2006. Reprinted with permission.) The laser sheet optics and the imaging system can be traversed throughout the field of interest, permitting both high spatial resolution and a wide ensemble field of view.

The uncertainty of PIV is often quoted as a fixed particle displacement value, such as in Gordon et al. (2004) who claim 0.1 pixels. It should be noted that this represents a best-case scenario and is probably invalid in regions of high shear (Timmins et al. 2012; Wilson and Smith 2013). The uncertainty of any PIV measurement is an active research topic (Sciacchitano et al. 2013). Many of the issues known to cause error in PIV measurements are present in synthetic jets, most notably shear and regions of very slow moving flow. For large tracer particles, a synthetic jet vortex ring or pair may have sufficient acceleration to cause particle slip. An example is shown in Figure 3.12 where the vortex core is completely void of particles. The particles are olive oil droplets formed in a Laskin nozzle and are commonly reported to be about 0.5 μm in diameter. This issue is likely present in all air-based synthetic jet measurements, and the impact of this loss of particles has never been assessed.

3.5.4 Time-Resolved PIV Measurements

Until very recently, for air-based synthetic jet measurements, PIV offered no real temporal resolution. In water, the situation was not as dire, since water allowed the actuator to be operated at much lower frequencies, and one could capture several vector fields during a single actuator cycle. While the authors could find no examples of a time-resolved

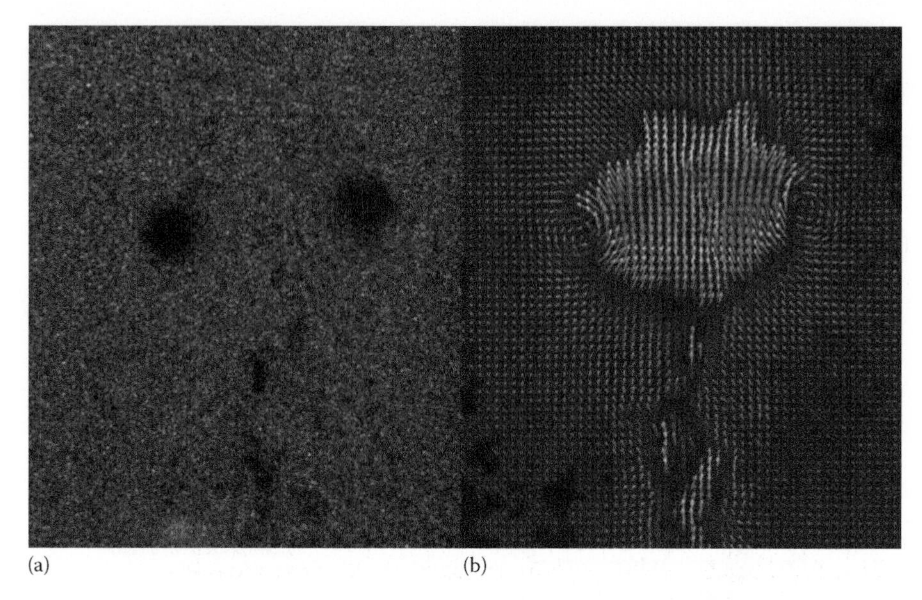

FIGURE 3.12 Example from Nani and Smith (2012) of (a) a raw PIV image and (b) the resultant vector field of an axisymmetric synthetic jet with $L_0/D = 10$ and $Re_\delta = 500$. (From Nani, D. J. and B. L. Smith, *Phys. Fluids*, 24, 115110, 2012. Reprinted with permission.)

PIV measurement having been performed for a synthetic jet, doing so would have many advantages over phase-locked measurements. First among these is that the actual evolution of the vortex rings (pairs) could be examined and Fourier transforms could be performed in space and/or time. More importantly, high-speed PIV presents the opportunity to solve the dynamic range problem associated with PIV measurement, which is especially problematic for jets (which have regions of very high speed in the core and low speed in the entrainment regions), and even more so with synthetic jets, which vary from zero velocity to large values.

The fundamentals of dynamic range improvement through the use of multiple frames are given by Hain and Kahler (2007). A similar scheme was implemented by Persoons et al. (2010) in a study to determine how impinging synthetic jets impact heat transfer. They showed that high dynamic range PIV is required to accurately resolve turbulence intensity profiles near the heat transfer surface (where velocities are small), while capturing the entire flow field and to be able to correlate this with local heat transfer coefficient data.

3.5.5 Measurements of Exit Flow Parameters

As discussed in Section 3.2, synthetic jet flows can be characterized in several different parameters. Most of these (such as L_0/D, Re, C_μ) rely on a measurement of the outflow waveform or maximum outflow velocity, while others (such as $\dot{E}$) require an accurate measurement of the flow rate over the full period. As will be shown, the outflow is much more easily measured than the inflow.

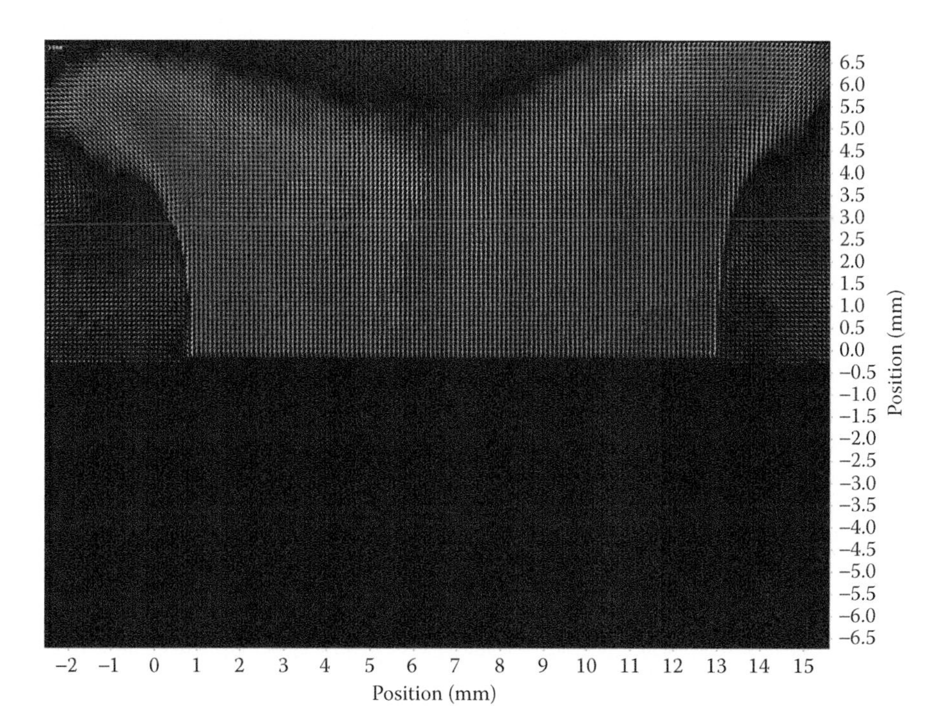

FIGURE 3.13 **(See color insert.)** Example from Nani and Smith (2012) of a phase-averaged vector field used to determine the volume flow rate waveform in an axisymmetric synthetic jet with $L_0/D = 7$ and $Re_\delta = 500$. (From Nani, D. J. and B. L. Smith, *Phys. Fluids*, 24, 115110, 2012. Reprinted with permission.)

The axisymmetric synthetic jet reported in Nani and Smith (2012) will be used as an example. The camera field of view was set such that the exit plane was at the center and filled most of the frame as shown in Figure 3.13. This arrangement made determination of the edges of the exit less ambiguous and more accurate. Since the region near the edges represents a larger area than the center in an axisymmetric geometry, accurately determining the orifice location is much more important than that of planar geometries. A tight-fitting plug was placed into the orifice (visible in Figure 3.10) and photographed with the PIV camera so that the location of the edges could be found.

The streamwise velocity profiles for peak suction and peak blowing are shown in Figure 3.14 and have some surprising features. Note that all of these profiles are acquired at a very short distance from the jet exit. The outflow profiles are invariant over the streamwise distances shown, while the inflow shows significantly less inflow at even a very short distance downstream. Much of the flow entering a synthetic jet orifice originates very close to the exit plane and is therefore missed by an integration downstream of the exit. This has important ramifications for researchers interested in determining their exit flow parameters. If the full waveform (in and outflow) is required, measurements must be made at very high resolution. If only the outflow is of interest (as for a measurement of L_0), much

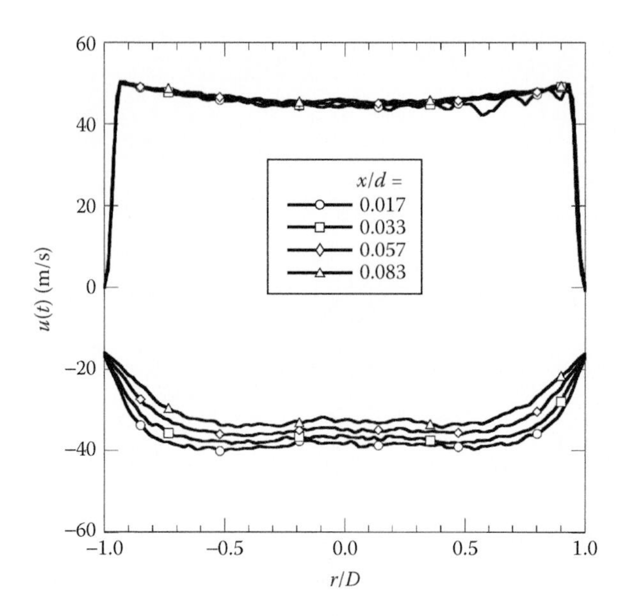

FIGURE 3.14 Phase-averaged (N = 50) streamwise velocity profiles for the maximum outflow and maximum inflow from Nani and Smith (2012) for a round inner orifice lip with $L_0/D = 7$ and $Re_\delta = 500$. (From Nani, D. J. and B. L. Smith, *Phys. Fluids*, 24, 115110, 2012. Reprinted with permission.)

less resolution is required. This is consistent with Gallas et al. (2003), who concluded that an LDV measurement at $x/D = 0.18$ was sufficient for determining the peak outflow velocity.

It is common to assess the flow parameters of synthetic jets using a centerline measurement. The appropriateness of this method depends heavily on the outflow velocity profile. Two extreme profiles are shown in Figure 3.15a for two synthetic jets with identical displacement amplitudes and Reynolds numbers, but different orifice shapes. One had sharp, square edges on the inlet and outlet, while the second had a radius equal to the orifice diameter and exit plane thickness on the inside lip. The square edge led to flow separation on the inner orifice lip. The flow profile had much smaller velocities near the edges and higher velocities in the center when compared to the profile from the rounded inner orifice lip.

The error incurred by using a centerline velocity to estimate the average is demonstrated in Figure 3.15b, where the centerline velocity waveform is compared to the average velocity waveform. For the rounded lip, the difference between the centerline and average velocity is small (about 5% at the peak velocity), while for the sharp edge, the error is very large (about 25%). We note that for quantities involving u^2, such as C_μ, this error is doubled. Furthermore, it appears that the average and centerline velocity waveforms have different phases.

The volume flow rate waveform was measured by using phase-locked PIV measurements on a plane that crossed the center of the jet exit. A "row" of these data is integrated to find volume flow rate.

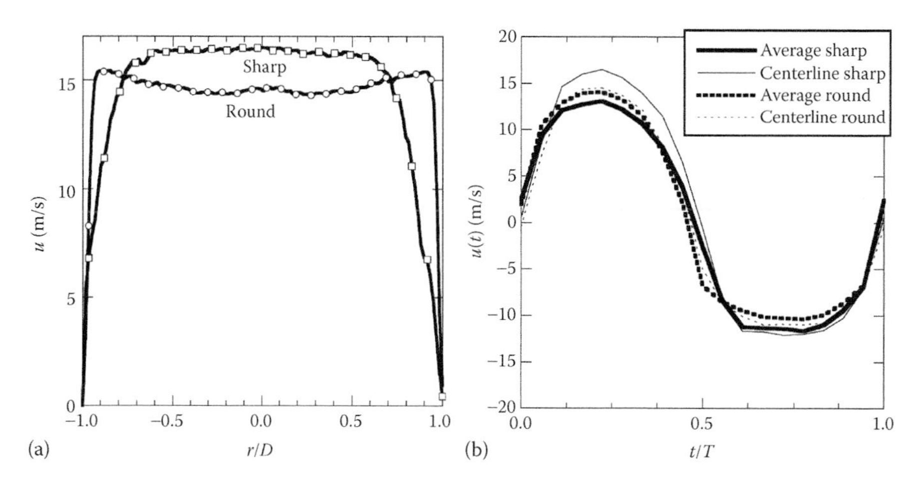

FIGURE 3.15 Comparison of an axisymmetric synthetic jet with $L_0/D = 20$ and $Re_\delta = 500$ for a sharp-edged inner orifice lip and a rounded inner orifice lip: (a) phase-averaged ($N = 50$) streamwise velocity profiles for the maximum outflow and maximum inflow and (b) the average exit velocity and the centerline exit velocity. (From Nani, D. J. and B. L. Smith, *Phys. Fluids*, 24, 115110, 2012. Reprinted with permission.)

$$Q_1(x_{\mathrm{int}},t) = \int_0^{D/2} 2\pi u r dr \approx \frac{2\pi\left[\sum_{j=1}^{N_y} u(i_{\mathrm{int}},j,t)y(j)\Delta y + \sum_{j=-N_y}^{0} u(i_{\mathrm{int}},j,t)y(j)\Delta y\right]}{2} \tag{3.7}$$

where:

$j = 0$ is the index of the vector at the radial center of the exit
$u(i_{\mathrm{int}}, j, t)$ is the streamwise velocity on a line above the exit at $x_{\mathrm{int}} = x(i_{\mathrm{int}})$
N_y is the number of vectors from the center of the exit to the edge
Δy is the vector spacing

The terms inside the square brackets represent the average of the result from integrating the center of the exit to the right edge and from the center to the left edge along a line similar to surface 1 in Figure 3.16. If this operation could be performed at $x = 0$, the flow was axisymmetric, and the center and wall locations were precisely known; this would be an accurate representation of the flow rate. In Nani and Smith (2012), the first row of data was located near $x/D = 0.008$, and the first row that was not affected by reflections off the exit plane was $x/D = 0.016$. Somewhat surprisingly, it was evident that even at $x/D = 0.016$, a substantial amount of inflow moves "under" this integration line resulting in an underestimate of the volume flow rate and an artificial "DC" velocity (see Figure 3.14). As a result, it is necessary to also integrate along vertical lines that extended from the exit plane to the measurement line, labeled as 2 in Figure 3.16.

$$Q_2(x_{\mathrm{int}},t) = \pi D\int_0^{x_{\mathrm{int}}} v dx \approx \frac{\pi D\left[\sum_{i=1}^{i_{\mathrm{int}}} v(i,j_{-D/2},t)\Delta x + \sum_{i=1}^{i_{\mathrm{int}}} v(i,j_{D/2},t)\Delta x\right]}{2} \tag{3.8}$$

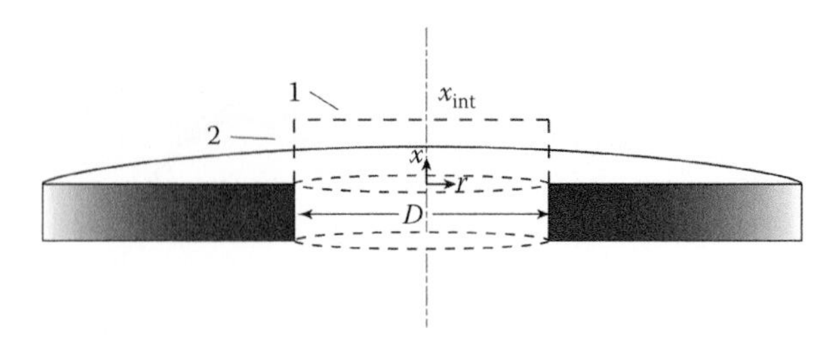

FIGURE 3.16 Lines over which PIV data are averaged to obtain the volume flow rate as a function of time.

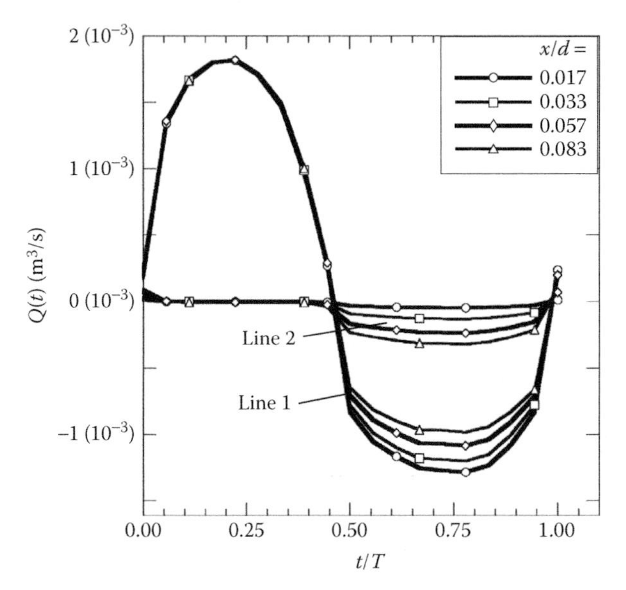

FIGURE 3.17 Volume flow rate as a function of time through surfaces 1 and 2 (Figure 3.16) with surface 1 at various downstream distances for an axisymmetric synthetic jet with $L_0/D = 20$ and $Re_\delta = 500$. (From Nani, D. J. and B. L. Smith, *Phys. Fluids*, 24, 115110, 2012. Reprinted with permission.)

where:

i_{int} is the index of the streamwise position where the integration is performed

$j_{-D/2}$ and $j_{D/2}$ are the indices corresponding to the left and right side of the orifice, respectively.

For any choice of x_{int}, the total flow rate is

$$Q = Q_1 + Q_2$$

The manner in which Q_1 and Q_2 vary with the integration location is shown in Figure 3.17. As shown in Figure 3.16, the flow rate captured on line 1 is invariant with downstream distance near the exit for the outward flow. It follows from this that no flow crosses line 2 during this period, and this is confirmed in Figure 3.17. However, Q_1 decreases rapidly with downstream distance during the inflow portion of the cycle, and this decrease is matched by an increase in Q_2. The sum of Q_1 and Q_2 remains constant with downstream distance.

References

Adrian, R. J. 2005. Twenty years of particle image velocimetry. *Experiments in Fluids*, 39(2): 159–169.

Ahmed, A. and Z. A. Bangash. 2009. Experimental investigation of axisymmetric coaxial synthetic jets. *Experimental Thermal and Fluid Science*, 22: 1142–1148.

Alvi, F. S. and L. N. Cattafesta III. 2010. The art and science of flow control-case studies using flow visualization methods. *European Physical Journal—Special Topics*, 182(1): 97–112.

Amitay, M. and F. Cannelle. 2006. Evolution of finite span synthetic jets. *Physics of Fluids*, 18(5): 054101.

Amitay, M., D. R. Smith, V. Kibens, D. E. Parekh, and A. Glezer. 2001. Aerodynamic flow control over an unconventional airfoil using synthetic jet actuators. *AIAA Journal*, 39(3): 361–370.

Bera, J. C., M. Michard, N. Grosjean, and G. Comte-Bellot. 2001. Flow analysis of two-dimensional pulsed jets by particle image velocimetry. *Experiments in Fluids*, 31(5): 519–532.

Bridges, A. and D. R. Smith. 2003. Influence of orifice orientation on a synthetic jet-boundary-layer interaction. *AIAA Journal*, 41(12): 2394–2402.

Cater, J. E. and J. Soria. 2002. The evolution of round zero-net-mass-flux jets. *Journal of Fluid Mechanics*, 472: 167–200.

Cattafesta III, L.N. and M. Sheplak. 2011. Actuators for active flow control. *Annual Review of Fluid Mechanics*, 43: 247–272.

Chaudhari, M., G. Verma, B. Puranik, and A. Agrawal. 2009. Frequency response of a synthetic jet cavity. *Experimental Thermal and Fluid Science*, 33(3): 439–448.

Crittenden, T. M. and A. Glezer. 2006. A high-speed, compressible synthetic jet. *Physics of Fluids*, 18(1): 017107.

Fugal, S. R., B. L. Smith, and R. E. Spall. 2005. Displacement amplitude scaling of a 2-d synthetic jet. *Physics of Fluids*, 17: 045103.

Gallas, Q., R. Holman, T. Nishida, B. Carroll, M. Sheplak, and L. Cattafesta. 2003. Lumped element modeling of piezoelectric-driven synthetic jet actuators. *AIAA Journal*, 41(2): 240–247.

Gilarranz, J. L., L. W. Traub, and O. K. Rediniotis. 2005a. A new class of synthetic jet actuators—Part I: Design, fabrication and bench top characterization. *Journal of Fluids Engineering—Transactions of the ASME*, 127(2): 367–376.

Gilarranz, J. L., L. W. Traub, and O. K. Rediniotis. 2005b. A new class of synthetic jet actuators—Part II: Application to flow separation control. *Journal of Fluids Engineering—Transactions of the ASME*, 127(2): 377–387.

Goepfert, C., J.-L. Marie, D. Chareyron, and M. Lance. 2010. Characterization of a system generating a homogeneous isotropic turbulence field by free synthetic jets. *Experiments in Fluids*, 48: 809–822.

Gordon, M., J. E. Cater, and J. Soria. 2004. Investigation of the mean passive scalar field in zero-net-mass-flux jets in cross-flow using planar-laser-induced fluorescence. *Physics of Fluids*, 16(3): 794–808.

Hain, R. and C. J. Kahler. 2007. Fundamentals of multiframe particle image velocimetry (PIV). *Experiments in Fluids*, 42: 575–587.

Hino, M., M. Kashiwayanagi, A. Nakayama, and T. Hara. 1983. Experiments on the turbulence statistics and the structure of reciprocating oscillatory flow. *Journal of Fluid Mechanics*, 131: 363–400.

Hino, S. M., M. Sawamoto, and S. Takasu. 1976. Experiments on transition to turbulence in an oscillatory pipe flow. *Journal of Fluid Mechanics*, 75: 193–207.

Holman, R., Y. Utturkar, R. Mittal, B. L. Smith, and L. Cattafesta. 2005. Formation criterion for synthetic jets. *AIAA Journal*, 43(10): 2110–2116.

Ingard, U. and S. Labate. 1950. Acoustic circulation effects and the nonlinear impedance of orifices. *Journal of the Acoustical Society of America*, 22: 211–218.

Krishnan, G. and K. Mohseni. 2009a. Axisymmetric synthetic jets: An experimental and theoretical examination. *AIAA Journal*, 47(10): 2273–2283.

Krishnan, G. and K. Mohseni. 2009b. An experimental and analytical investigation of rectangular synthetic jets. *Journal of Fluids Engineering—Transactions of the ASME*, 131: 121101.

Lodahl, C. R., B. M. Sumer, and J. Fredsoe. 1998. Turbulent combined oscillatory flow and current in a pipe. *Journal of Fluid Mechanics*, 373: 313–348.

Mane, P., K. Mossi, and R. Bryant. 2008. Experimental design and analysis for piezoelectric circular actuators in flow control applications. *Smart Materials & Structures*, 17(1): 015013.

Mane, P., K. Mossi, A. Rostami, R. Bryant, and N. Castro. 2007. Piezoelectric actuators as synthetic jets: Cavity dimension effects. *Journal of Intelligent Material Systems and Structures*, 18(11): 1175–1190.

Nani, D. J. and B. L. Smith. 2012. Effect of orifice inner lip radius on synthetic jet efficiency. *Physics of Fluids*, 24(11): 115110.

Ohmi, M., M. Iguchi, K. Kakehashi, and M. Tetsuya. 1982. Transition to turbulence and velocity distribution in an oscillating pipe flow. *Bulletin of the JSME*, 25: 365–371.

Oren, L., E. Gutmark, S. Muragappan, and S. Khosla. 2009. Flow characteristics of non circular synthetic jets. *The 47th AIAA Aerospace Sciences Meeting*, Orlando, FL, January 5–8.

Persoons, T., R. Farrelly, A. McGuinn, and D. B. Murray. 2010. High dynamic range whole-field turbulence measurements in impinging synthetic jets for heat transfer applications. *The 15th International Symposium on Applications of Laser Techniques to Fluid Mechanics*, Libson, Portugal, July 5–8.

Persoons, T. and O'Donovan, T. S. 2007. A pressure-based estimate of synthetic jet velocity. *Physics of Fluids*, 19(12): 128104.

Raju, R., E. Aram, R. Mittal, and L. Cattafesta. 2009. Numerical investigation of synthetic jet flowfields. *International Journal of Flow Control*, 1: 179–197.

Rathnasingham, R. R. and K. S. Breuer. 1997. Coupled fluid-structural characteristics of actuators for flow control. *AIAA Journal*, 35(5): 832–837.

Rizzetta, D. P., M. R. Visbal, and M. J. Stanek. 1999. Numerical investigation of synthetic jet flowfields. *AIAA Journal*, 37(8): 919–927.

Rumsey, C. 2004. CFD validation of synthetic jets and turbulent separation control. http://cfdval2004.larc.nasa.gov.

Schaeffer, N. W. and L. N. Jenkins. 2006. Isolated synthetic jet in crossflow: Experimental protocols for a validation dataset. *AIAA Journal*, 44(12): 2846–2856.

Sciacchitano, A., B. Wieneke, and F. Scarano. 2013. Piv uncertainty quantification by image matching. *Measurement Science & Technology*, 24(4): 045302.

Seifert, A., A. Darabi, and I. Wygnanski. 1996. Delay of airfoil stall by periodic excitation. *Journal of Aircraft*, 33(4): 691–698.

Settles, G. S. 2001. *Schlieren and Shadowgraph Techniques*. Berlin, Germany: Springer.

Shuster, J. M. and D. R. Smith. 2007. Experimental study of the formation and scaling of a round synthetic jet. *Physics of Fluids*, 19(4): 045109.

Smith, B. L. and A. Glezer. 1998. The formation and evolution of synthetic jets. *Physics of Fluids*, 10(9): 2281–2297.

Smith, B. L. and A. Glezer. 2002. Jet vectoring using synthetic jets. *Journal of Fluid Mechanics*, 458: 1–34.

Smith, B. L. and A. Glezer. 2005. Vectoring of adjacent synthetic jets. *AIAA Journal*, 43: 2117–2124.

Smith, B. L. and G. W. Swift. 2003a. A comparison between synthetic jets and continuous jets. *Experiments in Fluids*, 34(4): 467–472.

Smith, B. L. and G. W. Swift. 2003b. Power dissipation and time-averaged pressure in oscillating flow through a sudden area change. *Journal of the Acoustical Society of America*, 113(5): 2455–2463.

Stalnov, O. and A. Seifert. 2010. On amplitude scaling of active separation control. In: *Active Flow Control II*, ed. R. King. Berlin, Germany: Springer, pp. 63–80.

Tamburello, D. A. and M. Amitay. 2006. Manipulation of an axisymmetric jet using a single perpendicular control jet. *The AIAA 3rd Flow Control Conference*, San Francisco, CA, June 5–8.

Tang, T. S., Zhong, M. Jabbal, L. Garcillan, F. Guo, N. Wood, and C. Warsop. 2007. Towards the design of synthetic-jet actuators for full-scale flight conditions—Part 2: Low-dimensional performance prediction models and actuator design method. *Flow, Turbulence and Combustion*, 78(3/4): 309–329.

Tesar, V. and Z. Travnicek. 2008. Excitational metamorphosis of surface flowfield under an impinging annular jet. *Chemical Engineering Journal*, 144(2): 312–316.

Timmins, B. H., B. M. Wilson, B. L. Smith, and P. P. Vlachos. 2012. A method for automatic estimation of instantaneous local uncertainty in particle image velocimetry measurements. *Experiments in Fluids*, 53: 1133–1147.

Travnicek, Z. and V. Tesar. 2003. Annular synthetic jet used for impinging flow mass-transfer. *International Journal of Heat and Mass Transfer*, 46(17): 3291–3297.

Wilson, B. M. and B. L. Smith. 2013. Uncertainty on piv mean and fluctuating velocity due to bias and random errors. *Measurement Science & Technology*, 24(3): 035302.

Yao, C.-S., Chen, F. J., and D. Neuhart. 2006. Synthetic jet flowfield database for computational fluid dynamics validation. *AIAA Journal*, 44(12): 3153–3157.

4

Computational Modeling of Synthetic Jets

Rajat Mittal
The Johns Hopkins University

Shawn Aram
The Johns Hopkins University

Reni Raju
The George Washington University

4.1 Introduction

Computational modeling of synthetic jets (SJs) has played a significant role in improving our understanding of the function and performance of these devices. There are a number of features of these devices that make computational modeling a particularly effective approach for analysis; first, these devices are small, thereby making it difficult to measure flow and pressure within the device; second, these devices significantly enhance the range of the scales (both spatial and temporal) into the flow, which increase the challenge for experimental measurements. Analytical models of these devices have to contend with complex flow phenomena including unsteadiness, compressibility, flow separation, boundary layer formation, and complex vortex dynamics. These very same features also create challenges for computational modeling, and in this chapter we summarize the key contributions as well as the challenges and limitations associated with computational flow modeling of these devices.

4.1.1 Characterization of SJs

Figure 4.1 shows a schematic of a prototypical SJ in a flow control application. The SJ is comprised of a cavity, a slot, and a driver (diaphragm, piston, etc.) and is flush mounted to the control surface on which there might be a grazing flow boundary layer. The slot as well as the cavity can have a variety of cross-sectional shapes, and a few common shapes are also shown in the schematic. Similarly, the driver, which has an amplitude a and frequency f, might be installed in many different configurations. The simplest way to characterize the output of the jet is in terms of the mean jet expulsion velocity, which is defined as

$$\overline{V_J} = \frac{1}{A}\frac{2}{T}\int_0^{T/2}\int^{A} v(x, z, t)\mathrm{d}A\mathrm{d}t \tag{4.1}$$

where:
$T = (1/f)$ is the time-period of the driver
A is the area of the slot
$v(x, z, t)$ is the jet exit velocity

If the total volume change in the cavity due to the driver during the expulsion phase is $\Delta\forall$, then

$$\overline{V_J} = \frac{2\Delta\forall}{TA} \tag{4.2}$$

if the flow inside the actuator can be considered incompressible.

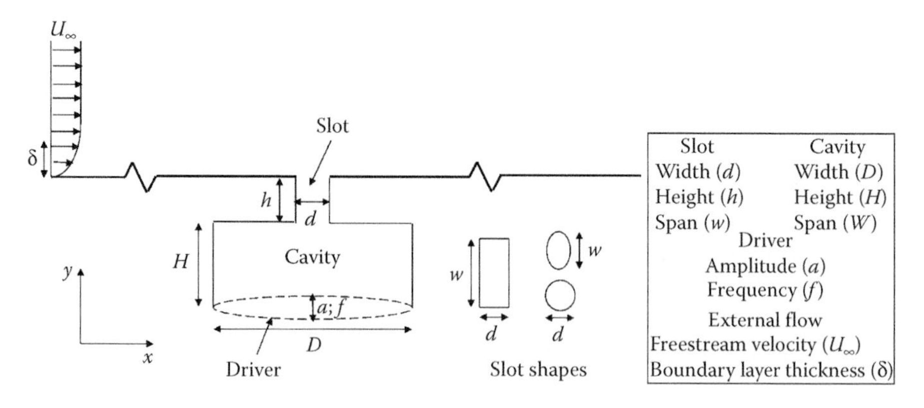

FIGURE 4.1 A schematic of a prototypical SJ with key parameters.

Compressibility effects can modulate the jet velocity in two different ways; significant compression/rarefaction of the flow inside the cavity can occur if the driver frequency is comparable to the Helmholtz frequency (F_H) of the cavity (Gallas et al. 2003a), which scales with ($c/\forall^{1/3}$), where $\forall$ is the volume of the cavity and c is the speed of sound for the flow inside the cavity. In addition, sufficiently high flow velocities inside the slot could in principle lead to choked flow. While the former mechanism can be predicted with some level of accuracy by lumped element models (LEMs) (Gallas et al. 2003a), the second mechanism has not been reported in conventional SJs since they are not yet capable of producing high enough jet velocities. Thus in the analysis of flow control using SJs, one may assume that $\overline{V}_J$ is known a priori and it becomes one of the key input parameters in flow control using these actuators. However, while $\overline{V}_J$ is one of the key variables of a SJ, the effect of the SJ on the external flow may depend on other factors such as the momentum, energy, or vorticity imparted to the external flow. These quantities depend not only on $\overline{V}_J$ but also on the spatial and temporal variations of the jet exit velocity profile. For instance, the average momentum flux $\overline{P}_J$ during expulsion is given by

$$\overline{P}_J = \frac{1}{A}\frac{2}{T}\int_0^{T/2}\int^{A} \rho v^2(x,z,t)\mathrm{d}A\mathrm{d}t \tag{4.3}$$

It is equal to $\rho\overline{V}_J$ only for a plug-flow jet profile. Similarly, the kinetic energy $\overline{E}_J$ and vorticity flux $\overline{\Omega}_J$ may not be related in a simple manner to $\overline{V}_J$. In general, the jet profile depends on the actuator (cavity and slot) geometry, jet operational parameters as well as the external flow, and in nondimensional terms, this dependence may be written as follows:

$$\frac{\overline{P}_J}{\rho\overline{V}_J^2};\frac{\overline{E}_J}{\rho\overline{V}_J^3};\frac{\overline{\Omega}_J}{\overline{V}_J^2};\ldots = fn\left(St, Re_J, \frac{h}{d}, \frac{w}{d}, \frac{\overline{V}_J}{U_\infty}, \frac{\delta}{d}, \frac{H}{D}, \frac{W}{D}, \text{Driver Config.}\right) \tag{4.4}$$

where:

$St = 2\pi fD/\overline{V}_J$ and $Re_J = \overline{V}_J D/\nu$ are the jet Strouhal and Reynolds numbers, respectively

Driver Config. refers to the nondimensional parameters that specify the driver configuration

In the above-mentioned expression, parameters one and two on the right-hand side are the jet "operational" parameters, parameters three and four are the jet slot parameters, parameters five and six are jet–boundary layer interaction parameters, and the rest are cavity and driver parameters. Note that this set of nondimensional parameters is not unique and other parameters may be used instead of the ones defined here.

In flow control applications, we are ultimately interested in the effect that the SJ has on the external flow. For instance, in separation control applications, the effect of the SJ may be measured in terms of relative reduction in the size of the separation zone $(\Delta L_{sep}/L_{sep})$. For a wing or an aerodynamic control surface, the effect of interest might be the relative change in lift ($\Delta L/L$) and/or drag ($\Delta D/D$). In heat transfer applications, the effect may be measured in terms of enhancement of heat flux or reduction in temperature. More basic parameters such as jet velocity, spreading rate, and penetration distance might also be of importance in heat transfer and mixing enhancement applications. The effect of the jet on the external flow depends on the jet exit profile as well as the characteristics of the external flow, and this dependence may be expressed as follows:

$$\left(\frac{\Delta L_{sep}}{L_{sep}}\right);\left(\frac{\Delta L}{L}\right);\left(\frac{\Delta D}{D}\right);\ldots = fn\left(St,\ Re_J,\ \frac{\overline{V}_J}{U_\infty},\ \frac{\overline{P}_J}{\rho U_\infty^2},\ \frac{\overline{E}_J}{\rho U_\infty^3},\ \frac{\overline{\Omega}_J}{U_\infty^2},\ \alpha,\ \frac{C}{d},\ \ldots\right) \tag{4.5}$$

where the angle of attack of the incoming flow (α) and chord length (C) are just two of the many parameters that might characterize the external flow.

Within the context of this discussion, it is fair to say that most of the past and ongoing research in SJ attempts to determine the dependence either in Equation 4.4 or in Equation 4.5. Computational modeling has been used extensively to address these very same issues and here we attempt to provide a concise summary of this research. While a jet in quiescent external flow may be viewed as a special case of a jet in grazing flow with zero external flow velocity, both the flow physics and the applications of these two situations are quite distinct and we therefore discuss these separately here.

4.2 SJs in Quiescent Flow

SJs in quiescent media are most relevant to impingement heat transfer applications (see Chapter 9 on application of SJ actuator [SJA] to heat transfer by Utturkar et al.). However, due to the relative simplicity of this case, it also serves as a good case for validation studies as well as for understanding the fundamental flow physics of these actuators.

4.2.1 Modeling and Validation

The earliest studies on modeling and validation of SJA were for jets in quiescent media and were carried out using a 2D incompressible Reynolds-averaged Navier–Stokes (RANS) solver (Kral et al. 1997). In this study, the cavity and jet slot were not included; instead, a normal velocity boundary condition was introduced to model the jet exit velocity. The velocity varied sinusoidally in time as $v(x, y = 0, t) = V_0 f(x)\sin(\omega t)$, with V_0 being the amplitude and $f(x)$ being the prescribed spatial velocity profile. A modified boundary condition for pressure, relating its gradient to the time-harmonic velocity perturbation, was also employed at the jet exit. Several laminar and turbulent simulations were carried out for different spatial jet velocity profiles (including a top-hat function [$f(x) = 1$] and sinusoidal profiles such as $f(x) = \sin(\pi x/d)$; $\sin^2(\pi x/d)$) and Strouhal numbers. Flow simulations with the turbulence model activated and a top-hat function for the velocity profile showed a good match with the experiments of Smith and Glezer (1997). Subsequent attempts were also made to improve the prediction by simulating the cavity and the orifice using unsteady compressible direct numerical simulations (DNSs) (Rizzetta et al. 1999). The oscillatory boundary condition was provided by sinusoidally varying the position of the lower boundary. The velocity profiles computed from the cavity solutions were used as boundary conditions to 3D simulations. Although the 3D calculations showed only a reasonable comparison to the experimental data, they were able to capture the spanwise instabilities leading to the breakup of a coherent vortex, not seen for the 2D-modeled boundary conditions. Another attempt to compare with these experiments using 2D DNS was made by Lee and Goldstein (2002). Although a reasonable match was found for the mean of streamwise velocity profiles, the 2D nature of the jet was unable to capture the fluctuating components due to the turbulent nature of the actual jet.

Subsequently, an attempt to consolidate and compare the performance of computational fluid dynamics (CFD) methodologies in predicting the behavior of the SJs was made through a workshop (Synthetic Jets and Turbulent Separation Control Workshop; CFDVAL 2004) organized by NASA Langley Research Center in 2004. Three different cases were chosen including an isolated SJ in quiescent external flow, a jet in a crossflow, and a SJ-based separation control for flow over a surface-mounted hump. The approaches used included DNS, unsteady Reynolds-averaged Navier–Stokes (URANS), large eddy simulation (LES), detached eddy simulations (DESs), and reduced-order modeling. The summary of all these cases has been presented by Rumsey et al. (2006) and discussed briefly in Sections 4.2.4, 4.3.4, and 4.4.3. Overall, a wide variation in the results was observed and no specific technique was found to be best suited in predicting the experimental data. The possible issues were inconsistencies in some of the experimental data, employment of different boundary conditions, and lack of grid resolution for certain cases. Overall, it was concluded that the existing state of CFD still faced some challenges in terms of accurate modeling of SJs and the corresponding flow predictions.

4.2.2 Criterion for Jet Formation

It had been observed that for certain cases, as seen in Figure 4.2, the vortex dipole formed at the lip of the SJ during expulsion is ingested back into the slot during ingestion, whereas in other cases, the vortex dipole manages to convect away from the jet lip. In the latter cases, the jet is found to consist of a train of vortex dipoles (rings in 3D). The expulsion of vortices and the "formation" of the jet is crucial in a number of applications such as heat transfer as well as separation control, and it is therefore important to determine what parameters control this behavior.

Smith and Swift (2001) presented a criterion for jet formation of 2D jets as a function of stroke length [defined as $L_0 = \int_0^{T/2} v(t)dt$] normalized by the slot width d. According to their criterion, jet formation was observed for values of L_0/d larger than about 6.0. Subsequently, Utturkar et al. (2003) and Holman et al. (2005) derived a simple criterion for the formation of the jet based on scaling analysis and rudimentary vortex dynamics. The premise behind this criterion was that a jet would be formed when the self-induced velocity of the vortex dipole formed at the jet slot during expulsion was larger than the

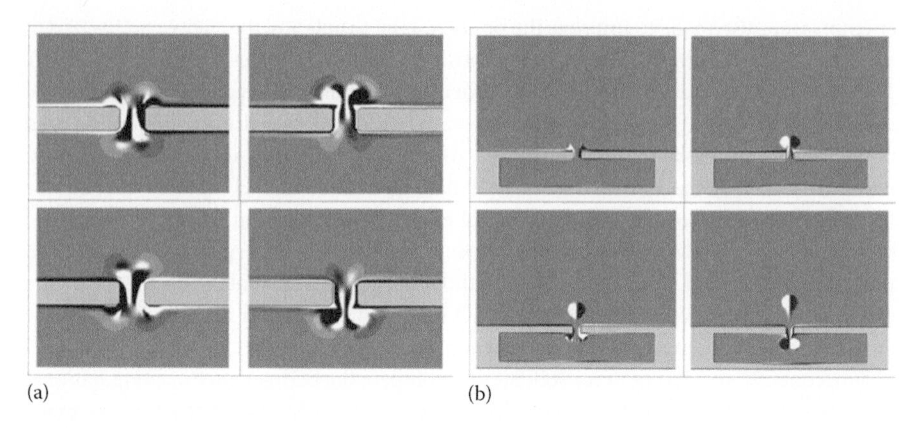

FIGURE 4.2 Vorticity contours of a 2D SJ simulation showing (a) no jet and (b) clear jet formation conditions. (From Utturkar, Y. et al., *AIAA Paper* 2003-0636, 2003.)

ingestion velocity of the jet. The strength of a shed vortex Ω_V relates to the vorticity flux across a 2D planar slice of the slot (Didden 1979) as

$$\Omega_J = \int_0^{T/2}\int_0^{d/2} \xi_z(y,t)v(y,t)\mathrm{d}y\mathrm{d}t$$

where:

$\xi_z(y,t)$ is the spanwise vorticity component in the jet slot at the jet exit
$v(y,t)$ is the jet velocity at the exit plane

The vorticity is primarily contained inside the boundary layer of thickness δ_s that develops inside the slot while the dipole-induced velocity V_I is proportional to Ω_V/d. Thus, an order-of-magnitude analysis of vorticity flux yields the following:

$$\Omega_J \sim \left(\frac{\overline{V}_J}{\delta_s}\right)\overline{V}_J\delta_s\left(\frac{1}{\omega}\right) \sim \frac{\overline{V}_J^2}{\omega} \tag{4.6}$$

where:

$\overline{V}_J$ is the slot-averaged expulsion velocity defined earlier

Now, in order for the jet to form, the self-induced velocity (V_I) should be larger than the suction velocity (V_S), which is of the same order of magnitude as $\overline{V}_J$. This yields the formation criterion as follows:

$$\frac{V_I}{V_S} \sim \frac{\Omega_J/d}{\overline{V}_J} \sim \frac{\overline{V}_J}{\omega d} = \frac{1}{St} = \frac{Re_J}{S^2} > K \tag{4.7}$$

where:

K was expected to be close to unity

Note that the stroke length (L_0) is related to the Strouhal number (Holman et al. 2005) as follows:

$$\frac{1}{St} = \frac{L_0}{d\pi} = \frac{\overline{V}_J}{\omega d}\frac{\nu d}{\nu d} = \frac{Re_J}{S^2} \tag{4.8}$$

where:
$S = \sqrt{\omega d^2/\nu}$ is the Stokes number

Thus, in terms of the criterion of Smith and Swift (2001), the criterion of Holman et al. (2005) would imply $L_0/d > K\pi$, which is slightly lower than the threshold range determined by their experiments.

2D computational studies were conducted for validating the formation criteria as well as for determining the constant K (Utturkar et al. 2003). In this study, the geometrical parameters were held constant while flow parameters varied via the displacement and frequency of the oscillating diaphragm operating conditions (modeled as a moving boundary). Simulations showed (results shown in Figure 4.3a) that $K = 1$ did indeed provide a reasonable demarcation for jet formation. Around this value, the jet remains in transition state where no clear jet formation is seen. For axisymmetric cases, this constant was found to be around 0.16 (Utturkar et al. 2003). Holman et al. (2005) extended this analysis further to consider the effect of orifice radius and proposed a general jet formation criterion as follows:

$$\frac{1}{St} > \frac{32c^2(1+\varepsilon)^p}{k\pi^2} \tag{4.9}$$

The criterion is a function of exit curvature radius ($\varepsilon = 2R/d$); the ratio of spatially averaged and centerline velocities (c); and κ, a constant determined by the ratio of vortex core radius to the distance between two dipole centers. The factor $p < 1$ accounts for flow separation due to curvature. Figure 4.3b shows the experimental data of Smith and Swift (2001) with modified Reynolds number accounting for the exit radius as a function of Stokes number. It can be seen that the general formation criterion shown in Equation 4.7 provides a very good prediction of jet formation.

4.2.3 Design of the SJ Cavity

As outlined in Section 4.1.1, the key geometrical parameters characterizing the SJA include the slot width, length, and, in certain instances, the slot radius and cavity shapes. Jet orifice/slot is typically either rectangular (Yao et al. 2004) or circular (James et al. 1996); on the contrary, the cavity design varies significantly. This is because the cavity design is primarily driven by design constraints including actuator size, type, and placement. For example, in all the three experimental setups of CFDVAL2004, the cavity design varied quite significantly (Rumsey et al. 2006).

The implicit assumption in using what seem to be *ad hoc* cavity designs is that the cavity shape has little influence on the emanating jet. This assumption was examined by Utturkar

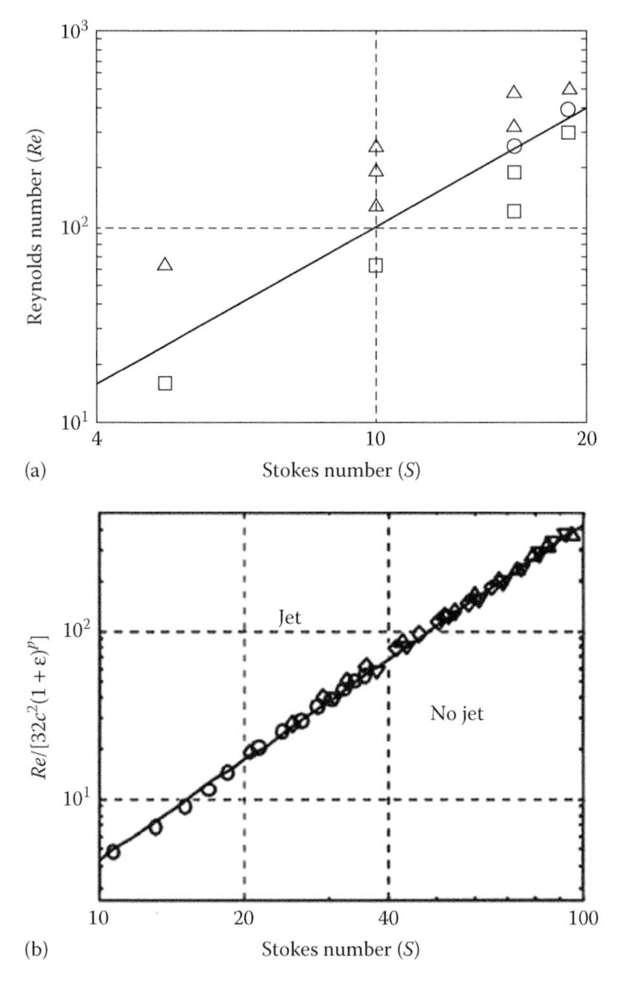

FIGURE 4.3 (a) Jet formation criterion determined by 2D simulations, □, no jet; ○, transition; △, clear jet; —, K = 1. (b) Normalized experimental data of Smith and Swift accounting for the exit radius. Symbols denote experimental measurements and solid line denotes the transition to jet formation. (From Smith, B. L. and G. W. Swift, *AIAA Paper* 2001-3030, 2001; Holman, R. et al., *AIAA J.*, 43, 2110–2116, 2005.)

et al. (2002) using 2D numerical simulations. The variation in the diaphragm placement and cavity aspect ratio D/H were investigated while keeping the rest of the parameters constant at S = 10.0 and h/d = 1.0. Five separate configurations were chosen. Case 1 consisted of a cavity with aspect ratio D/H = 5 and the piezoelectric diaphragm was fixed to the lower wall. For Case 2, the diaphragm was fixed to the right wall with an aspect ratio D/H = 1/5. Two diaphragms on both sidewalls were used for Case 3 that had the same cavity shape as Case 2. To conserve the flux during expulsion, the amplitude of vibration of these diaphragms was kept at one-half that of the previous two cases. For Cases 4 and 5,

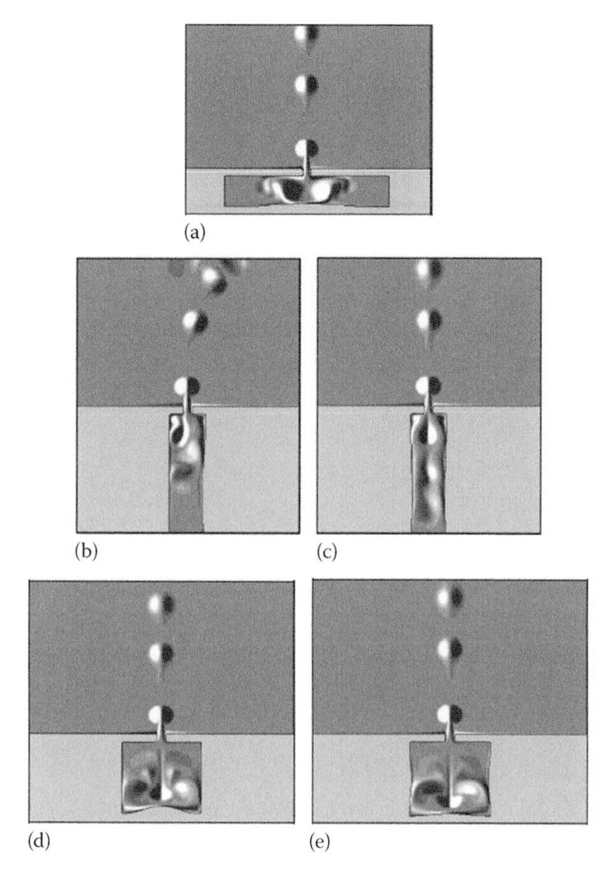

FIGURE 4.4 Plot of vorticity contour for the five cases extracted at the minimum volume phase of the expulsion stroke: (a) Case 1, (b) Case 2, (c) Case 3, (d) Case 4, and (e) Case 5. (From Utturkar, Y. et al., *AIAA Paper* 2002-0124, 2002.)

square cavities were used of which Case 4 has the diaphragm at the bottom wall, while Case 5 had oscillating diaphragms on sidewalls as well as the bottom walls.

Figure 4.4 shows the contours of the spanwise vorticity after the flow has reached a statistical steady state at the maximum expulsion phase for all the five cases. Despite the significant differences in the cavity design, all cases except for Case 2 produced virtually the same external flow. Examination of velocity profiles at jet exit shows similar profiles indicating that large differences in the internal cavity flow do not translate into similar differences in the external flow. This remains true even in the presence of external flows, simulations show that the integral measure of jet momentum coefficients shows less than a 7% deviation for all the cases.

The diaphragm placement for Case 2 is asymmetrical about the vertical centerline of the cavity. This arrangement produces a highly asymmetric flow within the cavity and in turn produces a vortex dipole outside the slot, which is also slightly asymmetric.

This asymmetry produces a self-induced velocity on the dipole that has a small horizontal component thereby resulting in a noticeable horizontal drift in the dipoles as they convect upward. As has been noted before (Rampunggoon 2001), the extent to which asymmetries in the cavity flow can affect the external flow depends primarily on the slot aspect ratio h/d. It has been found that at least in the range of parameters investigated, asymmetries produced in the jet cavity tend to dampen as the flow passes through the slot. Therefore, higher values of h/d typically produce more symmetric external jet flows. The horizontal drift in the vortex dipole might have some important implications in impingement heat transfer type applications of SJs (Campbell et al. 1998; Guarino and Manno 2001) where precise "targeting" of high temperature sites by the dipoles would be desirable.

4.2.4 Transitional SJ in Quiescent Flow: CFDVAL Case 1

The purpose behind the CFDVAL2004 workshop was to determine whether the existing CFD tools and approaches could successfully model the complex flow physics associated with these actuators. A detailed summary for all the cases tested at the workshop was presented by Rumsey et al. (2006) and subsequently in a follow-up paper (Rumsey 2009). There were about 75 participants in the workshop from seven different countries who attempted to solve three separate cases for which experimental data were provided. Case 1 consisted of a SJ in a quiescent medium, Case 2 was a SJ in a crossflow, and Case 3 was based on the flow over a wall-mounted hump.

The experiments for Case 1 were conducted by Yao et al. (2004) using hotwire, laser Doppler velocimetry (LDV), and particle image velocimetry (PIV). During the time of workshop, only the hotwire and PIV results were available. The geometrical configuration consisted of rectangular slot 1.27 mm wide and 35.56 mm long. A circular piezoelectric diaphragm was mounted on one side of the cavity as shown in Figure 4.5a. The cavity configuration along with the slot was asymmetrical with a taper angle of 12° on one sidewall of the slot. Due to the large aspect ratio of the slot, the flow at the center was considered to be nominally two dimensional. The operating frequency was around 445 Hz yielding maximum velocity at the slot exit around 25–30 m/s. This yielded a jet Reynolds number of 1150 and a Strouhal number of 0.04.

For this case, simulation methodologies included DNS, LES, and URANS. A variety of turbulence models ranging from Spalart–Allmaras (SA) to the k–ε, shear stress transport (SST), and Reynolds stress model (RSM) were employed. All simulations used structured grids for the simulations with either a 2D approximation or 3D periodic boundary. Most of the participants used a time-varying velocity to represent the periodic oscillation of the diaphragm corresponding to the location in the experiments; one participant used a quasi-1D model for the motion of the actuator while another used a time-varying boundary condition applied on the bottom wall. Two participants did not model the cavity but instead, used a boundary condition to represent the slot.

The original data using the hotwire and PIV showed some discrepancies that were later reevaluated with both PIV and LDV by Yao et al. (2004). These findings indicated that the

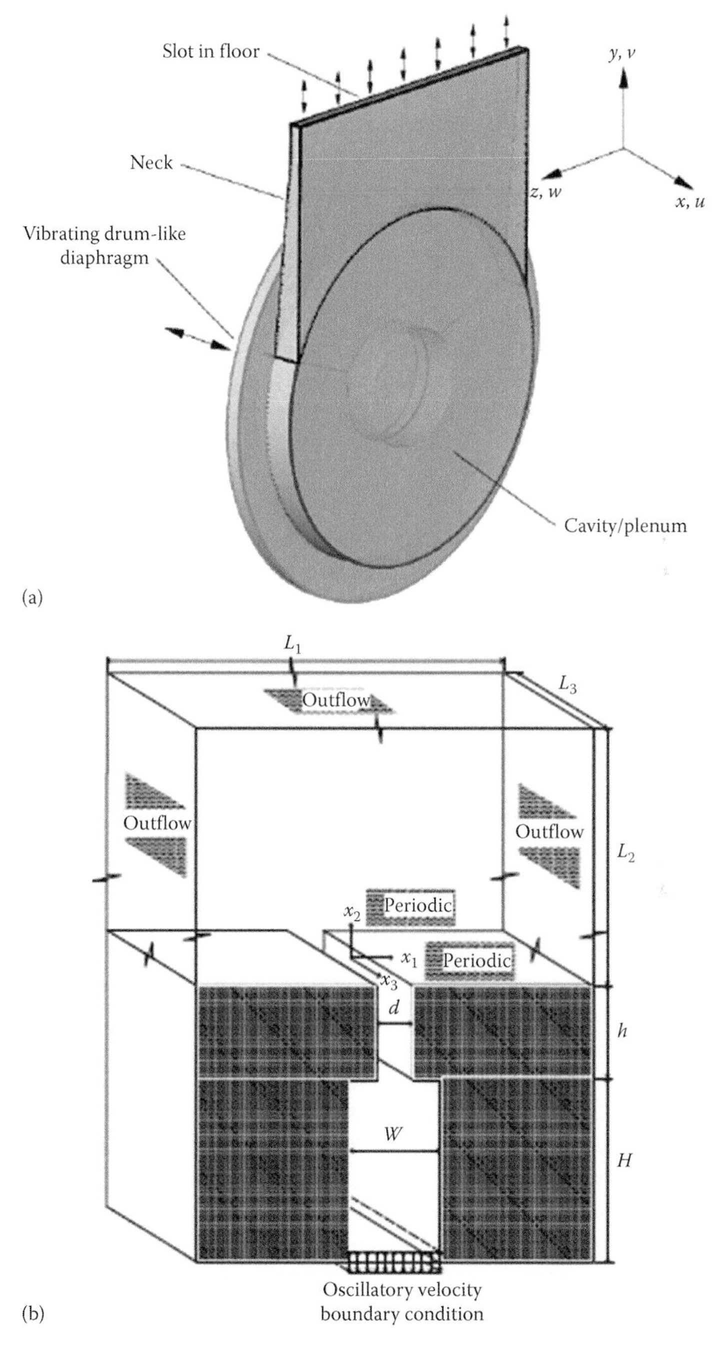

FIGURE 4.5 Case 1 SJA: (a) experimental configuration (Reproduced with permission from Rumsey, C. L. et al., *AIAA J.*, 44, 194–207, 2006.) and (b) a simplified geometry used by Kotapati et al. (From Kotapati, R. B. et al., *J. Fluid. Mech.*, 581, 287–321, 2007.)

hotwire measurements close to the slot exit were not accurate. Between CFD results, there were significant variations even though similar models had been used, in particular for URANS models.

Figure 4.6 shows the comparison of the time history of the velocity (v) at the slot midplane measured just above the slot exit. Notably, the CFD results show a phase difference in the suction and expulsion peaks. This could be attributed to the differences in phase averaging of the transient data. Close to the jet exit, all the calculations, with one exception, underpredicted the time-averaged jet velocity while further away these comparisons improved. However, far away from the jet exit, there was a significant variation in CFD results which could also be attributed to grid resolution away from the exit. Notably, none of the calculations at that time matched the turbulence quantities of the experiments. In hindsight, this was not surprising given the relatively low Reynolds number and transitional nature of the jet that made RANS models somewhat ill-suited for this flow. 3D end effects, which were not included in many of these simulations, were another possible source of this discrepancy.

In a follow-up paper, Rumsey (2009) summarized further attempts to model this case. Continued attempts to enhance the quality of 2D URANS calculations using improved boundary conditions seem to yield better comparison with the experimental data. Modeling attempts also used hybrid models like DES and blended SST-DES with mixed results.

This study also modeled the CFDVAL2004 Case 1. The computational model was built upon a series of assumptions that took into account the observed and expected behavior of the jet and this effort is described briefly here. The jet slot in this case had a large aspect ratio of 28 and the flow could therefore be assumed to be independent of the end effects of slot at least in some vicinity of the slot. The flow was therefore modeled as being homogeneous in the spanwise direction. The driver frequency of 444.7 Hz was close to the natural frequency of the diaphragm determined by a linear composite plate theory (Gallas et al. 2004), whereas the Helmholtz frequency of the cavity was estimated to be about 1900 Hz via LEM (Gallas et al. 2004). When the natural frequency of the diaphragm and the driver frequency are significantly lower than the Helmholtz frequency of the cavity/orifice, compressibility effects do not play a significant role in determining the jet output (Gallas et al. 2003a,b). Furthermore, within the incompressible regime, the cavity shape and size, as well as the diaphragm placement are expected to have little effect on the jet characteristics (Utturkar et al. 2002).

Using these arguments, a simple computational configuration (Figure 4.5b) was chosen for the numerical simulations (Kotapati et al. 2007). The configuration consisted of a finite spanwise length with periodic boundary conditions, and the cavity and slot were assumed to be rectangular in shape. Furthermore, the diaphragm was modeled as an oscillatory boundary condition with $V = V_0 \sin(2\pi ft)$ on the bottom wall of the cavity. The cavity aspect ratio (D/H), slot height-to-width ratio (h/d), and cavity-to-slot width ratio (D/d) were fixed while four different slot spanwise aspect ratios ranging from $3d$ to $9d$ are employed. As mentioned earlier, the jet Reynolds number (Re_J) corresponding to the experiments was determined to be 1150, while the Stokes number was $S = 17$. This implied a jet Strouhal number of 0.04, and a strong jet was expected to form for this case (Holman et al. 2005). It was found that an aspect ratio greater than $4.5d$ was needed to capture the homogeneous core of the flow. The flow is highly 3D as shown in Figure 4.7a.

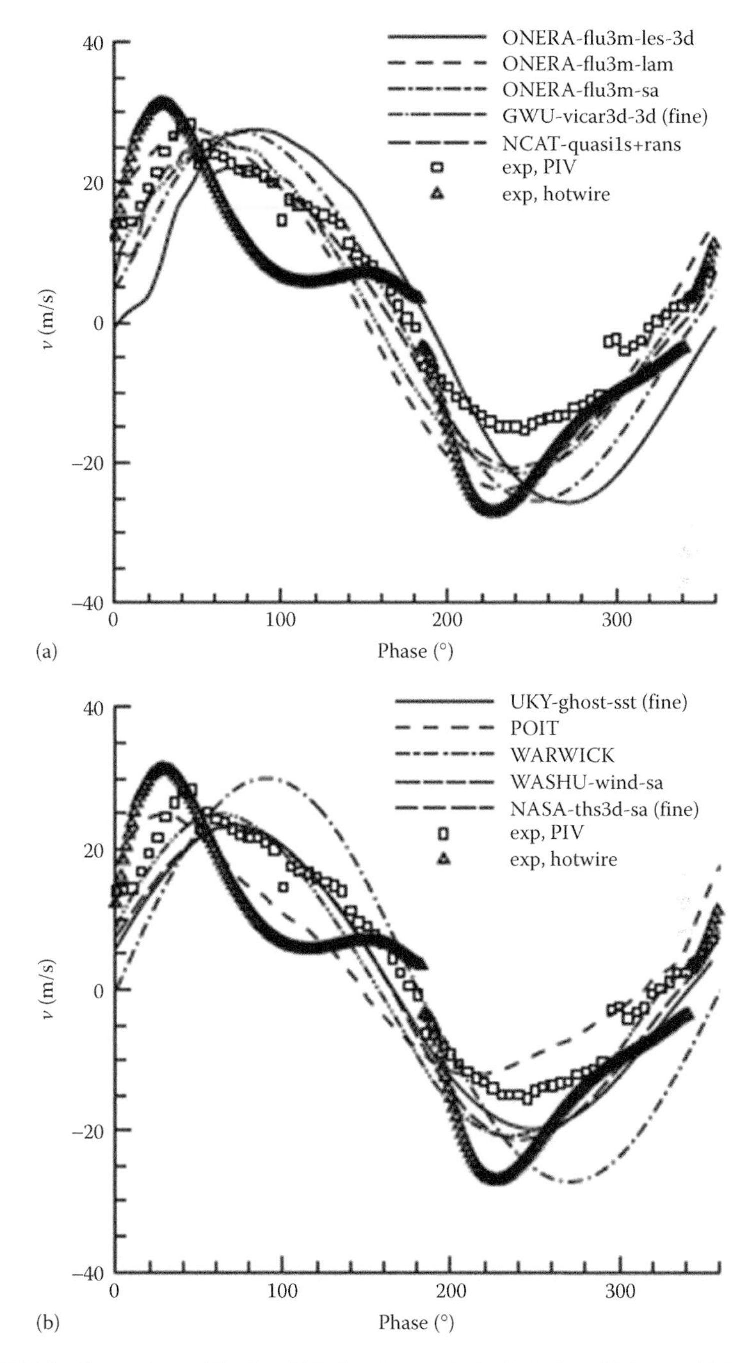

FIGURE 4.6 Comparison of the time histories of normal velocity in the midplane and 0.1 mm away from the slot. Computational results of various groups are split into (a) and (b). (Reproduced with permission from Rumsey, C. L. et al., *AIAA J.*, 44, 194–207, 2006.)

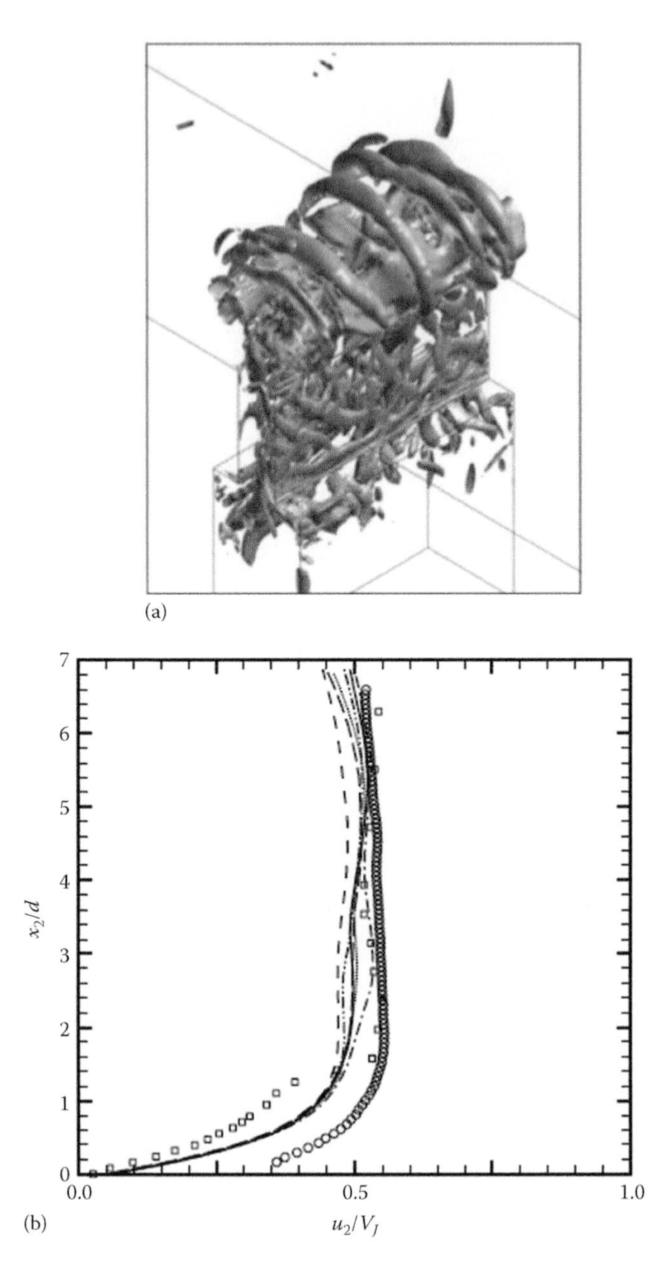

FIGURE 4.7 (a) Vortical structures in the near field at $\varphi \approx 90$, and (b) plot of time-averaged streamwise velocity (u_2/V_J) along the centerline $x_1/d = 0$. □, PIV; ○, hotwire; lines represent numerical data. (From Kotapati, R. B. et al., *J. Fluid. Mech.*, 581, 287–321, 2007.)

The primary vortex pair transitions into a fully developed turbulent jet showing the presence of spanwise-periodic counter rotating rib-like vortical structures in the streamwise direction. As the primary vortex pair advects downstream, a complete mixing of the vortices with the ambient fluid within a short distance from the orifice is seen, which is also consistent with the phase-locked smoke visualizations of a SJ at $Re = 766$ reported by Smith and Glezer (1998).

Measurements of the time-averaged streamwise velocity profiles along the jet centerline above the jet exit plane show a discrepancy up to about $x_2/d = 1.5$ (Figure 4.7b). In particular, the PIV data show significantly higher velocity than the corresponding hotwire measurement. Both the simulations and hotwire measurements show a reasonable agreement near $x_2/d \approx 0$, while the PIV shows an unexpectedly large nonzero velocity. Beyond about $x_2/d = 1.5$, the two experimental measurements are consistent with each other, and in this region, the simulations, although predicting the shape of the profiles reasonably well, slightly underpredict the velocity magnitude. Between the numerical simulations in the region $x_2/d = 2.0$–4.0, there is nearly 5% difference in the local values beyond which spanwise end effects of the slot dominate.

Figure 4.8 compares the measurements of phase-averaged streamwise velocity and cross-stream velocity above the jet exit plane at eight stations with the PIV data. Near the jet exit, there is a good agreement between simulations and experiment wherein the simulations accurately predict the shape and the magnitudes of the jet. Further away around $x_2/d = 1.5$–3.0, the simulations tend to overpredict or underpredict the peak

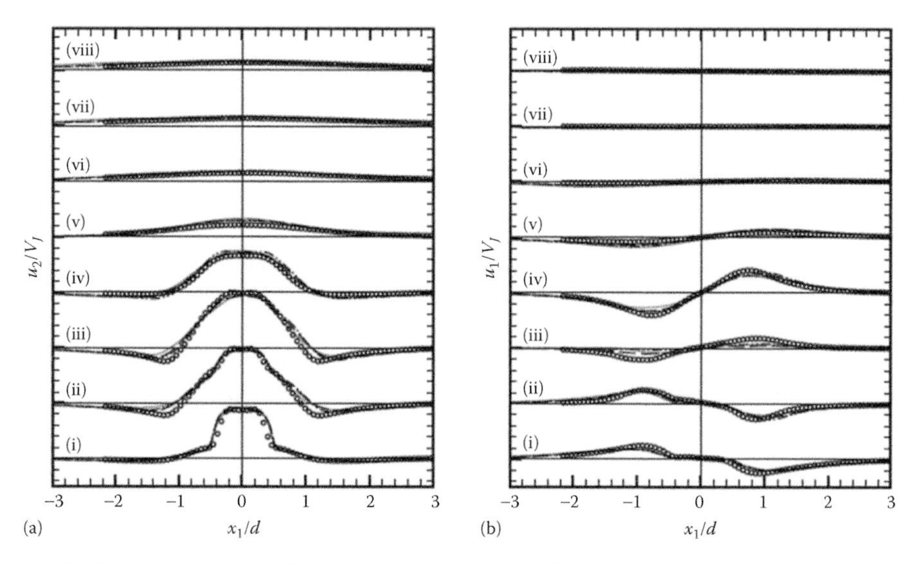

FIGURE 4.8 Cross-stream distributions of phase-averaged (a) streamwise (u_2/V_J) and (b) cross-stream (u_1/V_J) velocities at streamwise stations $x_2/d = 0.5$, 1.0, 1.5, 2.0, 3.0, 4.0, 5.0, and 6.0, respectively, at phase angle $\varphi = 92.6$. ○, PIV; lines represent numerical data. (From Kotapati, R. B. et al., *J. Fluid. Mech.*, 581, 287–321, 2007.)

velocity by nearly 10%. Beyond $x_2/d > 3.0$, there is a significant reduction in the velocity magnitudes for both the streamwise and cross-stream velocities since the induced velocities at these locations are small. Here again a good match is seen between the numerical and experimental data.

The computed results therefore provide strong support to the idea that at least in the incompressible regime, the details of the cavity do not significantly affect the external jet flow and that even a rudimentary representation of the jet slot is sufficient to enable a good prediction of the jet features.

4.3 SJs in Crossflow

4.3.1 Effect of Crossflow

As discussed in Section 4.2.2, one of the main features of a SJ in a quiescent external flow is the formation of vortex dipoles (vortex rings) at the edge of the orifice (slot) which could convect downstream during the expulsion phase by the self-induction. Several scaling analyses of jets in a quiescent flow (Smith and Swift 2001; Utturkar et al. 2002, 2003; Holman et al. 2005) over a wide range of operating conditions have indicated that the jet formation essentially depends on jet Reynolds number and Stokes (or Strouhal) number. In the presence of an external flow, at least two additional parameters have to be considered that include the freestream velocity (U_∞) and boundary layer thickness (δ) (or alternatively, displacement or momentum thickness) of the external flow; the former is usually normalized by a characteristic jet velocity and the latter by a length scale of the jet slot.

One main application of SJAs is the global modification of external flow over control surfaces and particularly separation or turbulence control. While the SJ introduces unsteady mass flow perturbations into the external flow, the effect on the external flow is likely associated with the unsteady jet momentum and vorticity flux imparted to the external flow via the alternating blowing and suction mechanism. The flux of momentum from the jet in particular is considered an important quantity, and Amitay et al. (1998) defined the momentum coefficient as the net jet momentum during one cycle of excitation normalized by momentum flux of the external flow, that is, $C_\mu = (\overline{P}_J/\rho U_\infty^2)$. However, fluid momentum might not be the only important quantity for a jet in crossflow; vorticity and kinetic energy flux could also be important. Mittal et al. (2001) defined the momentum and kinetic energy fluxes of the expulsion and ingestion phases separately and showed the dependence of these quantities on the actuator design and operation parameters.

As in the case of quiescent external flow, a vortex pair also forms in the presence of the crossflow at the edge of the jet exit during the expulsion phase, and this vortex pair is convected downstream. It was shown (Mittal et al. 2001) that depending on the jet-to-freestream velocity ratio and crossflow boundary layer Reynolds number, the jet could penetrate into the freestream during the expulsion phase and entrain higher momentum fluid into the boundary layer. A skewed jet velocity profile at the jet exit during the expulsion phase was observed in the presence of a crossflow that was not the case for the jet in quiescent flow. A parametric study of a 2D SJ in crossflow by Raju (2007) showed a nonmonotonic variation in skewness of velocity profile with nondimensional parameters. It was also noted that depending on the flow conditions, skewness in the velocity profile could develop during the ingestion phase.

4.3.2 Virtual Aeroshaping Effect of SJs

The idea of modifying the apparent aerodynamic shape of control surfaces (referred to as "virtual aeroshaping") with the aim to improve their aerodynamic performance has long been of great interest. SJAs can generate mean recirculation zones on a control surface that are significantly larger than the length scale of the actuator. It has been suggested that these large bubbles effectively modify the shape of the body consequently altering the pressure gradient and the extent of separation (Chatlynne et al. 2001). This capability of SJs is extremely desirable since it would potentially allow for "on-demand" virtual morphing of an aerodynamic surface such as a wing or diffuser.

Honohan et al. (2000) conducted experiments to examine the effect of frequency of SJs on the flow over a 2D circular cylinder. They found formation of a quasi-steady recirculation zone near the control surface at a frequency that was an order of magnitude higher than the shedding frequency of base flow. This phenomenon caused displacement of streamlines close to the surface, increasing the momentum of the flow near the interaction region and consequently delaying flow separation.

Chatlynne et al. (2001) applied an array of SJAs to modify the suction side of a 2D Clark-Y airfoil with the goal of drag reduction. A combination of a miniature surface-mounted passive obstruction placed above the surface of the airfoil with actuators introduced in the recirculation region behind the obstacle modified the streamlines of the flow. Modification of the pressure distribution over the airfoil by the actuator at low angles of attack when the baseline flow was fully attached caused a dramatic reduction in the pressure drag with a slight penalty in lift. It was also observed that by increasing the distance between the jet and obstruction, the streamwise extent of recirculation zone formed near the surface was enhanced while the performance of the SJ degraded in terms of pressure lift-to-drag ratio.

Experimental investigation by Chen and Beeler (2002) also found a localized increase in surface pressure and consequently lift reduction by a SJ due to the virtual shaping effect on a 2D NACA 0015 airfoil at zero angle of attack. It was also realized that the virtual aeroshaping effect of SJ was significantly reduced if it was located inside the separated flow.

Mittal and Rampunggoon (2002) used numerical simulations to examine the effect that key flow parameters have on the formation of mean recirculation zones generated by a 2D SJ interacting with a flat-plate Blasius boundary layer. The Stokes number was kept constant in their study, while the jet-to-crossflow velocity ratio ($\overline{V}_J/U_\infty$), boundary layer thickness to jet width ratio (δ/d), and the jet Reynolds numbers were varied. In the case where the jet velocity was of the order of the crossflow velocity, a small recirculation bubble was formed in the vicinity of the jet slot as is clear from Figure 4.9a, whereas with the higher jet velocity, a bubble that was significantly larger than the jet width was generated (Figure 4.9b). They also showed that the thickness of boundary layer relative to the slot width has a

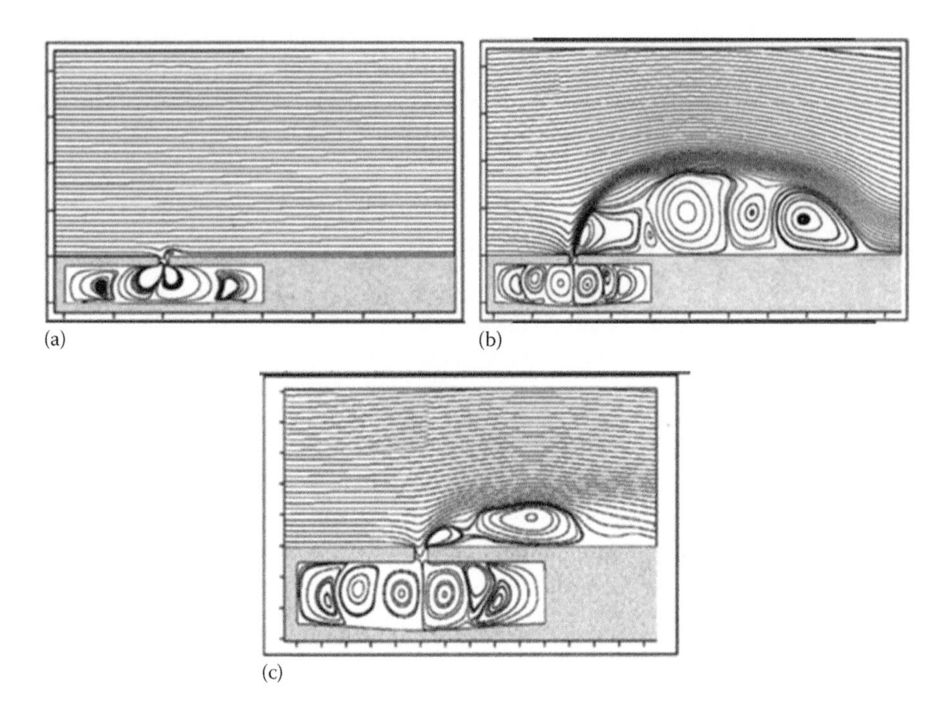

FIGURE 4.9 Streamlines of mean flow for interaction of a SJ with a zero pressure gradient boundary layer at $Re_\delta = 250$: (a) $\delta/d = 5, \overline{V}_J/U_\infty = 1.25$; (b) $\delta/d = 5, \overline{V}_J/U_\infty = 5.0$; and (c) $\delta/d = 2$, $\overline{V}_J/U_\infty = 2.0$. (From Mittal, R. and P. Rampuggoon, *Phys. Fluids*, 14, 1533–1536, 2002.)

significant impact on the bubble size (compare Figure 4.9b,c). A simple scaling law was also extracted that related the streamwise extent of the mean recirculation bubble to the momentum coefficient (C_m). The results indicated that within the range of parameters investigated in this study, the normalized bubble length grew linearly with the momentum coefficient.

4.3.3 Reduced-Order Models of SJs

The dynamics and performance of SJAs depend on several geometrical, structural, and flow parameters (Glezer et al. 2005; Holman et al. 2005). When compared to the global domain, such as an airfoil in which the actuator is imbedded, the scales of the actuator are typically 10^{-2} through 10^{-4} times smaller in size. Due to the range of scales involved, inclusion of a high-fidelity model of a SJA within a macroscale computational flow model is an expensive, if not prohibitive, proposition. The desire to compute the flow physics associated with SJ-based control makes it a practical necessity that simple yet accurate models of these devices be devised and employed in such computations.

In many past simulations, these actuators have been represented via simplified models in one form or another. These include the study of Kral et al. (1997), Rathnasingham and

Breuer (1997), Rizzetta et al. (1999), Lockerby et al. (2002), Rediniotis et al. (2002), Filz et al. (2003), Yamaleev and Carpenter (2003), Gallas et al. (2003a), Kihwan et al. (2005), Ravi (2007), Agashe et al. (2009), Raju et al. (2009), and recently Aram et al. (2010a). Among these models, here we focus primarily on empirical models that attempt to simplify the geometry of actuators.

LEMs are the simplest class of reduced-order models where the actuator components are represented as elements in an equivalent circuit using conjugate power variables. If it is assumed that the characteristic length scales of the governing physical phenomena are much larger than the largest geometric dimension, then the governing partial differential equations of the distributed system are lumped into a set of ordinary differential equations (Gallas et al. 2003a). These models are found to predict the dynamic response of synthetic actuators with reasonable accuracy, and further details regarding these models can be found in Chapter 2.

Neglecting the flow physics associated with the jet slot is one of the primary reasons for the shortcomings of the conventional modified boundary condition models such as those of Kral et al. (1997). It is observed that the flow in the slot tends to separate at the top and bottom lips during the expulsion and ingestion phases, and the formation of secondary vortices near the exit tends to significantly alter the flow field. This motivated Raju et al. (2009) to propose the slot-only (SO) model that considered just the slot with appropriate boundary conditions at the slot entrance. This model is supported by past studies that have shown that the shape of the cavity has little effect on the flow emanating from the jet provided that the incompressible flow assumption inside the actuator cavity is valid. The key features of the model are that it is fully predictive and offers significant simplifications in the inclusion of SJs in large-scale flow simulations.

Three separate flow configurations (or "models") were chosen for comparison, which included the full-cavity (FC), the SO, and the plug-flow configurations seen in Figure 4.7. For FC simulation as seen in Figure 4.10a, only the v component of velocity with sinusoidal time-variant profile and uniform amplitude was applied at the bottom of the cavity for the entire cycle of excitation. Similar boundary conditions were also implemented for the plug-flow configuration in this study over the slot width while using a Neumann boundary condition for u velocity during the ingestion phase (Figure 4.10b). Two variants of the SO models were considered in this study. As seen in Figure 4.10c, for the SO-1 model only the sinusoidal v velocity was prescribed at the bottom of the slot at its junction with the cavity, while for the SO-2 model, presented in Figure 4.10d, boundary conditions for both the u and v velocities during expulsion were applied that were based on a sink-like flow assumption at the slot inlet. Based on this flow behavior, u velocity was prescribed along the slot inlet during the expulsion phase as linearly varying sinusoidal profile with zero value at the slot center and maximum value near the slot walls. A no-slip boundary condition was considered on the slot edges and a Neumann boundary condition for this component of velocity was assumed during the ingestion phase in both models. The v velocity profile in this case was considered to be similar to the plug-flow configurations.

The model assessment was performed over a wide range of dimensionless parameters governing a SJ in grazing flow. A comparison of the integral measures of vorticity, momentum, and kinetic energy fluxes for the aforementioned models showed that the slot-only

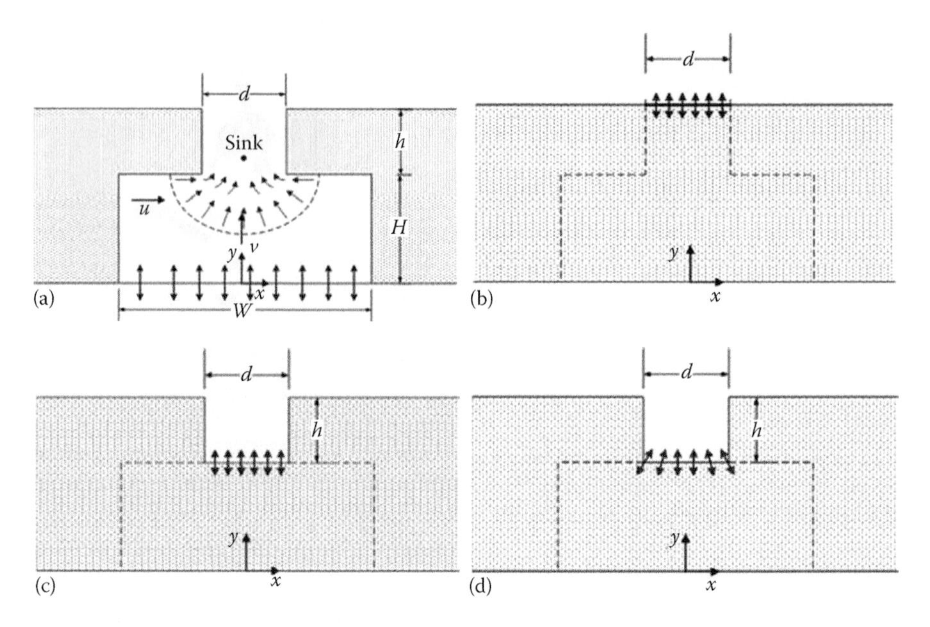

FIGURE 4.10 A schematic showing (a) the typical flow pattern seen in a SJ slot for a FC configuration, (b) the plug-flow model, and (c and d) the variants of proposed SO model. (From Raju, R. et al., *Int. J. Flow Control*, 1, 179–197, 2009.)

SO-2 model provided the best modeling approximation. Figure 4.11 shows the time variation of these quantities obtained downstream of the jet exit for all flow configurations and following flow conditions: $Re_J = 281.25$, $\overline{V}_J/U_\infty = 0.25$, $\delta/d = 2$, and $St = 0.8$. It is observed that the plug-flow and SO-1 models underpredicted all quantities during the expulsion phase. While SO models matched the FC quantities during the ingestion phase, this was not the case for plug-flow configuration.

In addition, the fidelity of the proposed model was also explored for a canonical separated flow at different forcing frequencies and Reynolds numbers. The SO-2 model was able to predict the separation bubble size reduction with greater accuracy than the plug-flow model. Furthermore, while the plug-flow model mostly predicted the correct trend in separation control at various forcing frequencies, the actual effect of the jet on the separation bubble was underpredicted, sometimes quite significantly. Thus, caution needs to be taken in interpreting the results of flow control simulations where very simple representations of the SJs, such as the sinusoidal plug-flow outlet boundary condition, are used.

Although the "slot-only" model represents a significant reduction in the computational complexity required to represent the actuator in flow simulations, it still requires simulation of the unsteady viscous flow in the slot. Furthermore, if the slot location is varied, a new grid may have to be generated. If, however, the entire actuator can be replaced by an appropriate and relatively simple velocity boundary condition at the jet exit, there would be no need to include a slot and would significantly simplify the task of

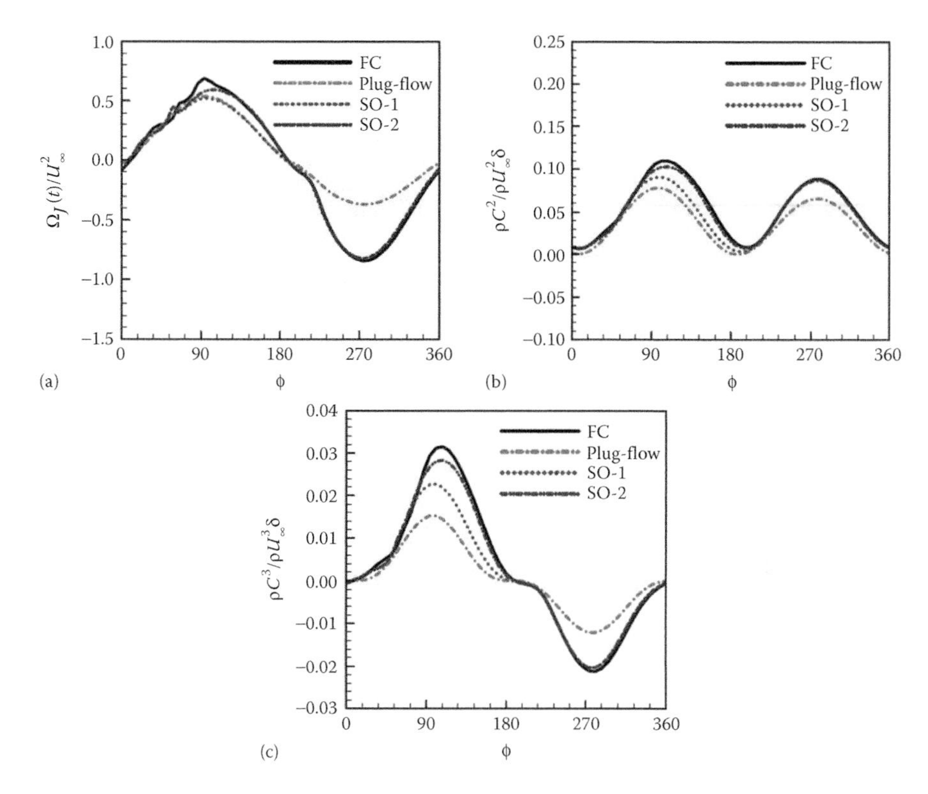

FIGURE 4.11 (a) Comparison of vorticity, (b) momentum, and (c) kinetic energy flux for all models as phase for $Re_J = 281.25$, $\overline{V}_J/U_\infty = 0.25$, $\delta/d = 2$, and $St = 0.8$. (From Raju, R. et al., *Int. J. Flow Control*, 1, 179–197, 2009.)

SJ flow control simulations. Based on this argument, Aram et al. (2010a) proposed a series of modified boundary conditions to reproduce the flow associated with SJs in grazing flows. The proposed models included a nonuniform jet velocity profile with only two spatial degrees of freedom and a uniform slip velocity on the slot-flow boundary. Based on the initial analysis, three of the simplest models were selected for further study including MBC1 (plug-flow model), MBC6 (model with nonuniform vertical velocity and constant slip velocity), and MBC7 (model with nonuniform vertical velocity and zero slip velocity). Figure 4.12a shows a comparison of the spanwise vorticity at the peak expulsion for selected models at the following condition: $Re_J = 250$, $S = 7.07$, $St = 0.2$, $U_\infty/\overline{V}_J = 4$, and $\delta/d = 3$. The strength of the vortex at both edges of the slot exit in MBC1 is lower than the FC model, but MBC6 and MBC7 provide a reasonable qualitative approximation to the FC model. A comparison of key integral quantities associated with v momentum and vorticity fluxes presented in Figure 4.12b showed that the models with nonuniform jet velocity (MBC6 and MBC7) were able to predict these quantities with good accuracy, whereas a simple plug-flow model (MBC1) with zero slip and uniform jet velocity underpredicts these three quantities during the expulsion phase.

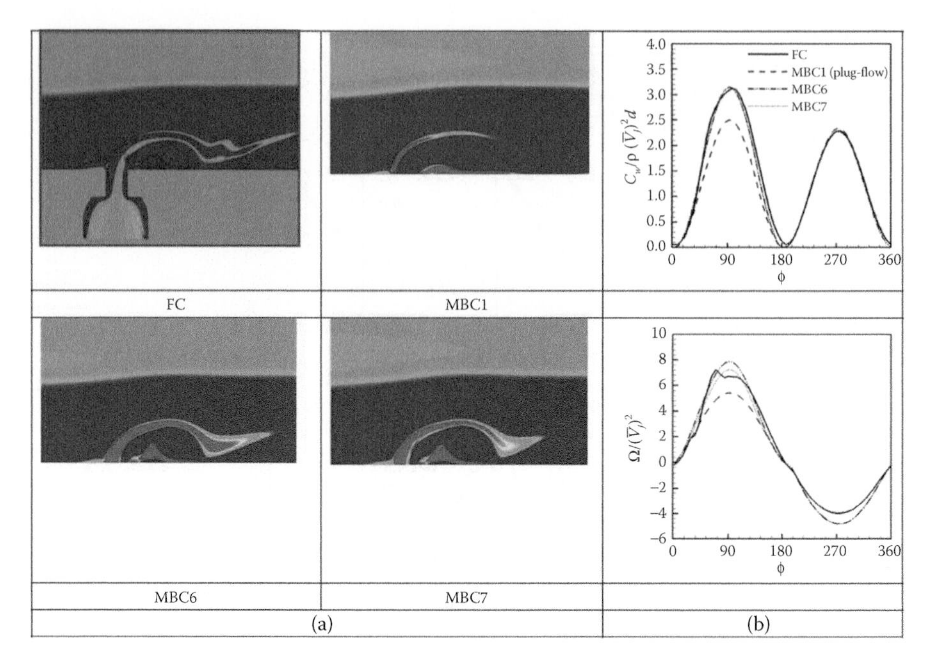

FIGURE 4.12 (a) Instantaneous spanwise vorticity during peak expulsion, (b) v momentum flux (top right) and spanwise vorticity flux (bottom right) for case with $Re_J = 250$, $S = 7.07$, $St = 0.2$, $U_\infty/\overline{V}_J = 4$, and $\delta/d = 3$. (From Aram, E. et al., *Int. J. Flow Control*, 2, 109–125, 2010a.)

A further study was performed to assess the performance of these models for a canonical separated flow at different forcing frequencies. A key finding of this study was that a simple plug-flow type model can predict incorrect trends for separation reduction with jet frequency. It was also found that inclusion of a jet flow asymmetry parameter improved the fidelity of the model significantly and led to better qualitative and quantitative prediction of separation control. Inclusion of a nonzero slip velocity improved the prediction of horizontal momentum flux from the jet but did not significantly improve the fidelity of the model.

Both the selected proposed models contained unknown parameter(s), jet asymmetry and slip, that needed to be determined a priori for a given flow configuration in order to be used in a predictive manner in any simulation. In general, these unknown parameters depend on the external flow as well as jet flow properties, and this dependence is quite difficult to determine a priori. Therefore, a preliminary attempt was made in this study to provide empirical closure to the selected models. Numerical simulations suggested that the two unknown parameters in the models are well represented through a power law dependence on a parameter that quantifies the grazing flow velocity of the boundary layer upstream of the jet relative to the jet velocity. This parameter can be determined concurrently from a flow control simulation and the jet asymmetry model adjusted during the course of the simulation. Alternatively this quantity might be available from a baseline flow simulation (with no flow control) or from a potential flow type model of the configuration under consideration. The dependency of unknown parameters on the jet Strouhal and Reynolds numbers was also examined. It was found that both values were fairly independent of the Strouhal number in the range of study. The jet asymmetry also exhibited a weak power law dependence on the jet Reynolds number.

4.3.4 SJ in a Turbulent Crossflow Boundary Layer: CFDVAL Case 2

As mentioned in Sections 4.2.1 and 4.2.4, in 2004 NASA Langley Research Center held the CFD validation of SJs and turbulent separation control workshop and three cases were investigated. Test case 2, which is the focus of this section, included the interaction of a round SJ, with an orifice diameter 6.35 mm, placed flush with an incompressible turbulent boundary layer of thickness 21 mm. The boundary Reynolds number was about 2.2×10^6 per meter and the jet Reynolds number was about 20,000. A number of groups (National Aeronautics and Space Administration [NASA; two groups], Université des Sciences et de la Technologie d'Oran [USTO], Office National d'Etudes et Recherches Aérospatiales [ONERA]), and The Committee on Space research (COSPAR) International Reference Atmosphere (CIRA) modeled this case using LES (ONERA) and URANS approaches. While CIRA only considered a time-varying profile at the orifice exit to represent the SJ in the crossflow, the rest modeled the cavity in their simulations with a time-dependent and uniform in space vertical velocity (cross-stream) at the bottom of the cavity. Note that in all simulations except those of CIRA, the amplitude of the vertical velocity was adjusted to obtain a reasonable match with the experimental vertical velocity at the orifice exit.

A comparison of the time variation of the jet velocity in the vicinity of the orifice exit, which is seen in Figure 4.13, showed a noticeable difference between different simulations that mainly related to the various unsteady boundary conditions applied for SJ. A reasonable agreement was observed for u velocity in Figure 4.13a between the experiment and the results of NASA and USTO, whereas the modeling of CIRA and ONERA overpredicted and underpredicted this component of velocity, respectively. Although all the CFD simulations obtained zero or nearly zero spanwise velocity near the centerline of orifice exit, the experimental results showed a nonzero value of this quantity during the expulsion phase that might be caused by resonances inside the cavity or asymmetry in experimental setup (Figure 4.13b). The cross-stream component of velocity shown in Figure 4.13c was accurately predicted for all simulations.

The time-averaged streamwise velocity along the centerline was extracted at three stations downstream of the SJ and showed a reasonable agreement with experiment with some dissimilarity within the boundary layer. However, along the domain span a significant discrepancy was observed between different CFD results for time average of cross-stream velocity. It should be pointed out that although two different CFD models (LES and URANS) were applied in this study, both methods produced very similar time- and phase-averaged flow quantities.

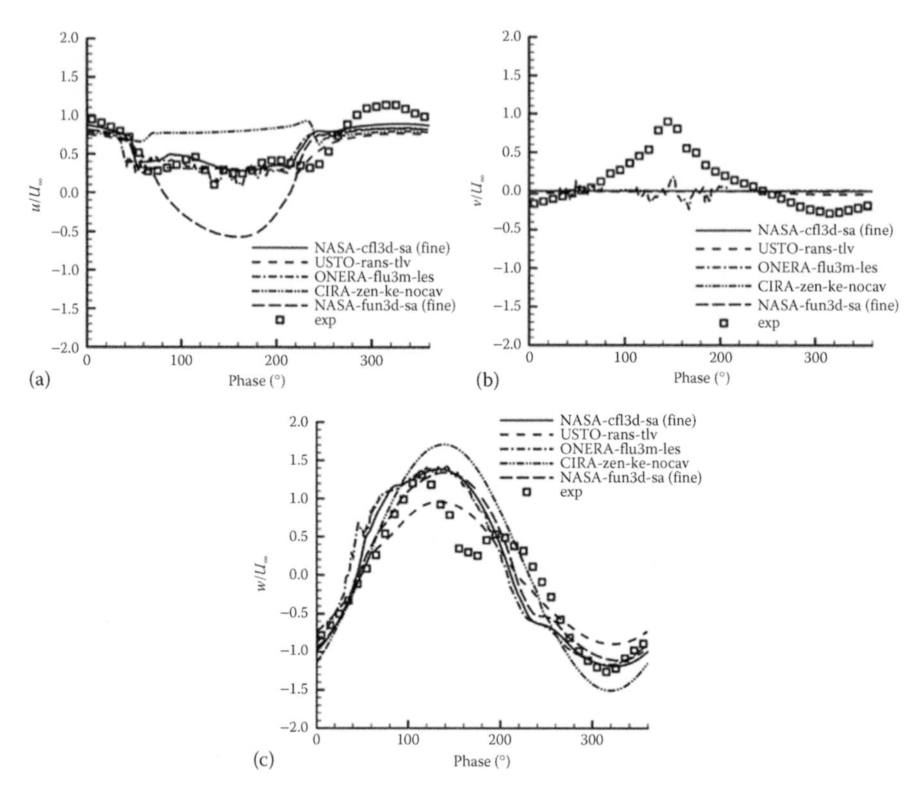

FIGURE 4.13 Time variation of (a) streamwise velocity, (b) spanwise velocity, and (c) cross-stream velocity near the center of orifice exit over one cycle of excitation. The legend provides some information on the CFD code and/or turbulence model employed. (Reproduced with permission from Rumsey, C. L. et al., *AIAA J.*, 44, 194–207, 2006.)

4.4 SJ-Based Flow Control

4.4.1 Separation Control of Flow over an Airfoil

One of the key parameters that determine the effectiveness of the SJ is the momentum coefficient (C_μ). The control authority varies monotonically with the ratio of the jet velocity to the freestream velocity (and therefore with the jet momentum coefficient) up to a point where further increase completely disrupts the boundary layer (Seifert et al. 1996; Seifert and Pack 1999; Glezer and Amitay 2002). On the contrary, the control authority has a highly nonmonotonic variation with dimensionless frequency (F^+). Studies have found a large range of F^+ values that provide effective control (Mittal and Kotapati 2004). Note that F^+ is the equivalent of a Strouhal number when defined as

$$F^+ = \frac{f_J l_c}{U_\infty} = St$$

where:

l_c is some length scale in flow configuration
f_J is the forcing frequency
U_∞ is the freestream velocity

Studies show that for $l_c = c$, optimal values range from 0.55 to 5.5 (Bar-Sever 1988; Seifert et al. 1993; Ravindran 1999; Wygnanski 2000; Darabi and Wygnanski 2002; Gilarranz et al. 2002; Margalit et al. 2002; Funk et al. 2010). For $l_c = X_{TE}$, X_{TE} being the distance from actuator to trailing edge, this range is found to be 0.5 to 2.1 (Greenblatt and Wygnanski 1999, 2001; Seifert and Pack 1999), while for $l_c = L_{sep}$, optimal values range from 0.75 to 2.0 (Seifert et al. 1996; Pack and Seifert 2000; Pack et al. 2002). Overall, $F^+ \sim O(1)$ provides better flow control; however, in certain cases, $F^+ \sim O(10)$ has also shown to result in improved aerodynamics performance (Amitay et al. 2001). Wu et al. (1998) on the contrary have argued that for post-stall cases, the optimal control frequency should be a harmonic of f_{wake}.

Raju et al. (2008) explored SJ-based separation control and its correlation to distinct timescales in the uncontrolled flow for a 2D NACA 4418 section based on the experimental setup of Zaman and Culley (2006) for a chord Reynolds number of 40,000 and angle of attack $\alpha = 18°$. The SJA was modeled as an oscillatory boundary condition [of the form $v_J(t) = V_0 \sin(2\pi f_J^+ t)$, where V_0 is the amplitude and $f_J^+ = fc/U_\infty$ is the forcing frequency] on the airfoil surface, at two different streamwise locations corresponding to $x_J/c = 0.024$ and 0.070. The normalized width of the actuator for these two locations was $w_J/c = 0.012$ and 0.019, respectively. The uncontrolled cases indicated the existence of three distinct frequencies in the flow: a shear layer frequency was identified as $f^+_{shear} \approx 12$ at the upstream end of the separation, and separation bubble and wake frequencies were identified as $f^+_{sep} \approx 2$ and $f^+_{wake} \approx 1$, respectively. Here f^+ represents frequency normalized by U_∞/c. Based on these natural frequencies, 10 separate forcing frequencies were chosen. The first set consisted of $f_J^+ = m \cdot f^+_{sep}$, where m = 1/4, 1/2, 3/4, 1, 3/2, and 2, while the remaining sets were closer to the shear layer frequency (f^+_{shear}).

The response of the separated flow to different forcing frequencies ahead of the separation was put into three categories. The Type I response, which consisted of frequencies lower than f^+_{wake}, had a detrimental effect on the airfoil performance. Type II response consisted of cases where forcing ranges were close to the f^+_{sep}. For these cases, the length of the separation bubble was reduced by nearly 35%–87%, while its height was reduced by 70%–90% under these conditions. Overall, forcing close to the separation bubble frequency yielded approximately a 38% decrease in the drag. Comparison of the lift-to-drag ratio in Figure 4.14d shows an overall improvement in this quantity by nearly 64% for these cases. On the contrary, forcing at higher frequencies (Type III response) close to that of the shear layer increased the size of the recirculation region significantly. This was due to the formation of a number of small discrete vortices in the shear layer, which merged to form a strong clockwise vortex. This forcing also tends to increase the drag by nearly 28%–39% while decreasing the lift-to-drag ratio by nearly 35% (Figure 4.14d). Figure 4.14a–c shows a comparison of the time-averaged streamlines between the baseline case and representative cases from the Type II and III categories. Forcing at separation bubble frequencies also

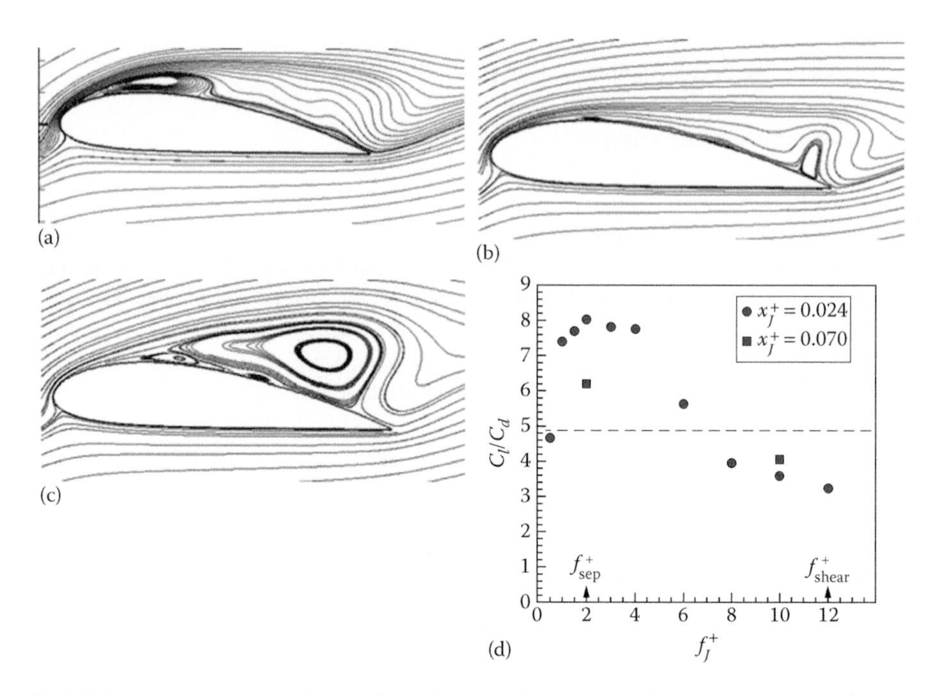

FIGURE 4.14 Time-averaged streamlines showing (a) uncontrolled case, (b) forcing at f_{sep}^+ and (c) f_{shear}^+, and (d) comparison of lift-to-drag ratio for different frequencies. (From Raju, R. et al., *AIAA J.*, 46, 3103–3115, 2008.)

shows a distinct lock-on of the shear layer, and separation bubble frequency to the forcing frequency might explain its suppression mechanism.

Some preliminary investigations on moving the forcing inside the separation bubble ($x_J/c = 0.070$) did not yield a similar improvement in lift-drag ratio. This indicated that flow is most responsive to perturbations ahead of separation where forcing near the natural separation bubble frequency allows for locking onto the local instability of the flow, thereby controlling the leading edge separation.

4.4.2 Control of a Canonical Separated Flow

As discussed in Section 4.4.1, for an SJ of a given geometry, the key operational parameters are the jet frequency (f_J) and the jet velocity (V_J). The former is usually nondimensionalized as

$$F_J^+ = \frac{f_J}{f_n}$$

where:

f_n is a frequency associated with some intrinsic timescale in the uncontrolled flow

The parameter V_J is some characteristic measure of the jet velocity, such as the peak or an average velocity during the discharge phase of the cycle, and is normalized by U_∞. There is generally little leeway for optimizing this device with respect to V_J. On the contrary, as mentioned earlier, control authority has been found to exhibit a highly nonmonotonic variation with F_J^+ (Seifert and Pack 2000; Glezer et al. 2005; Greenblatt and Wygnanski 2003), and this not only suggests the presence of rich flow physics and a highly nonlinear, multimodal system, but also reveals the potential for optimizing the control scheme with respect to this parameter.

As pointed out by Mittal and Kotapati (2004), depending on the flow conditions, there are at least three natural frequencies occurring in a separated airfoil flow. At higher angles of incidence, which are of particular interest in practical applications, separation may occur, often near the leading edge, and the flow may or may not reattach. When the flow does not reattach, there are at least two natural frequencies, f_{wake} and f_{SL}, the latter is due to the shear layer Kelvin–Helmholtz instability. On the contrary, if the flow does reattach (in a time-averaged sense) before the trailing edge, as in a relatively low Reynolds number airfoil (Ol et al. 2005; Raju et al. 2008), a third frequency scale (f_{sep}) corresponding to the separation bubble may also be present. In this case, f_{SL} may lock-on to f_{sep} through subharmonic resonance, and f_{sep} and f_{wake} might also show lock-on type dynamics. Thus, the notion that a separated airfoil flow is a single-frequency system that is dominated by f_{sep} via the dynamics associated with the separation bubble, which is inherent in much of the past work on separation control, is not entirely correct. This lack of understanding has led to a paradigm where forcing schemes are devised mostly through a trial-and-error process, and the nonlinear dynamics of this complex flow system are usually ignored to the detriment of the control effectiveness.

To explore the rich flow physics and nonlinear dynamics of this multiphenomena system in a systematic manner and to establish a firm flow-physics-based grounding for effective forcing schemes, Mittal and Kotapati (2004) devised a novel canonical separated flow configuration (shown in Figure 4.15) that is ideally suited for this purpose. An adverse pressure gradient was introduced on midchord suction side of 2% thick elliptic flat plate at zero angle of attack by implementing a steady suction and blowing on the top boundary of the computational domain. The key aspect of this approach is that the separation bubble can be located in any position along the suction side of air foil with desired size without the complicating influence of curvature. It also provides separate control by prescribing the location and extent of the separated region as well as Reynolds number, which is not possible by variation of angle of attack and/or freestream velocity for an airfoil.

Kotapati et al. (2007) used this configuration to create an aft-chord separation of a 2D 5% thick flat plate with 8:1 elliptic leading edge and blunt trailing edge at a Reynolds number of 6×10^4. Results of the uncontrolled case showed that for this particular case, the entire system locked on to a single frequency. SJ-based perturbation of the shear layer close to this lock-on frequency or its first superharmonic provided the most effective condition of controlling the separation.

Aram et al. (2010) continued the study of Kotapati et al. (2007) to provide an in-depth understanding of the flow physics and nonlinear dynamics of uncontrolled 3D canonical aft-separated flow at chord Reynolds number of 1×10^5 and to use this knowledge to develop effective control strategies that reduce reliance on current trial-and-error methods.

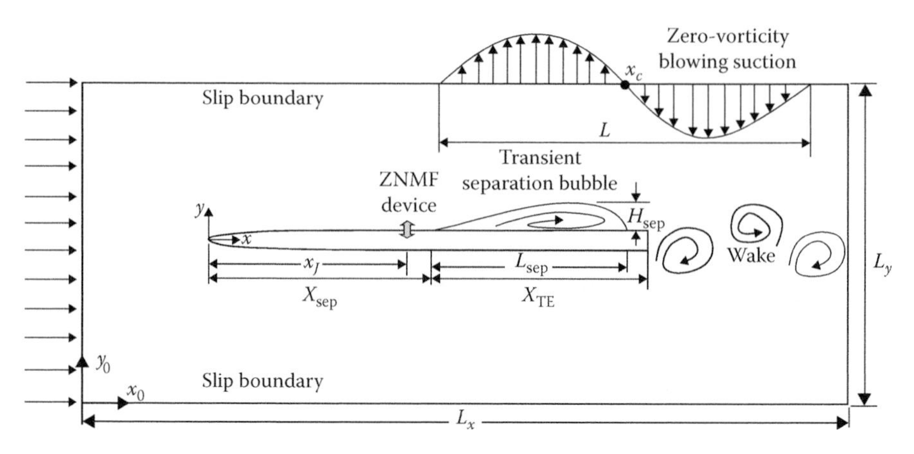

FIGURE 4.15 A schematic of the canonical separated flow configuration (not to scale). (From Mittal, R. and R. Kotapati. Resonant mode interaction in a canonical separated flow. *Proceedings of the 6th IUTAM Symposium on Laminar-Turbulent Transitions*, Bangalore, India, December 13–17, 2004.)

Figure 4.16 shows the computational results of Aram et al. (2010b) for baseline and controlled flow conditions of aft-chord separation case. The results showed that the shear layer and separation bubble were locked on to the same dominant frequency, whereas the wake region has a lower dominant frequency. It was found that forcing the flow at a frequency that was close to the shear layer/separation bubble frequency caused a significant reduction in the size of separation bubble, whereas higher excitation frequency was not effective.

4.4.3 Large-Scale Simulations of SJ-Based Separation Control

An accurate prediction of a SJ in large-scale simulation is a challenging task, since it involves modeling of a wide range of geometrical and flow scales from simulation of unsteady viscous flow within the cavity and slot to predicting a variety of physical phenomena in the large volume of external flow that is being controlled. Even so, many numerical studies have been performed to investigate SJ characteristics and its performance for various applications, mostly by implementing both statistical models such as RANS and scale-resolving approaches including LES and DNS. While the statistical approaches are computationally economical, they are not suitable for unsteady flows, especially in the case where the excitation frequency is in the same order of magnitude as that of turbulent structures (Leschziner and Lardeau 2011). On the contrary, the scale-resolving models need to apply various grid resolutions to account for all length scales.

A number of validation studies of the interaction of SJ with a turbulent crossflow have shown reasonable agreement with corresponding experimental results (see

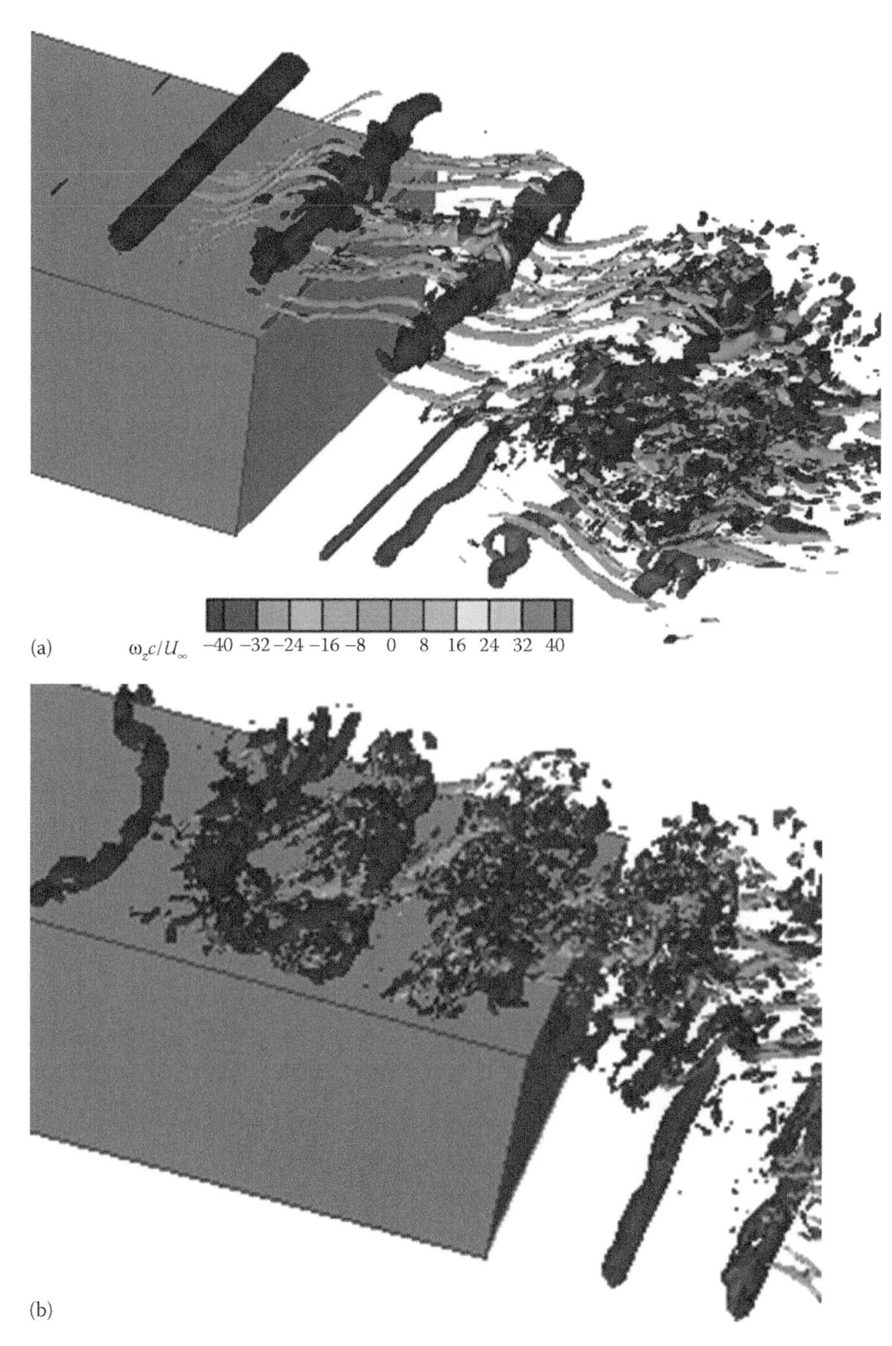

FIGURE 4.16 (See color insert.) Isosurfaces of instantaneous swirl strength ($50\ Qc/U_\infty$) and spanwise vorticity ($\omega_z c/U_\infty$) contours for (a) baseline and (b) controlled aft-chord separated flow. (From Aram, E., R. Mittal, J. Griffin, and L. N. Cattafesta. Toward effective ZNMF jet based control of a canonical separated flow. *The 5th AIAA Flow Control Conference*, Chicago, IL, June 28–July 1, 2010b.)

FIGURE 4.17 Instantaneous spanwise vorticity of flow over a wall-mounted hump.

Section 4.3.4 for more details). Some other numerical studies have attempted to evaluate the effectiveness of SJ for separation control over bluff bodies. Among those is the flow control over a wall-mounted hump that was the third validation case investigated in 2004 CFD validation workshop for SJs. This geometry, which is shown in Figure 4.17, was first proposed by Seifert and Pack (2002) to study the turbulent flow separation and control at high Reynolds number [$O(10^6)$]. They conducted a series of experiments using a cryogenic tunnel for baseline and controlled cases and provided a comprehensive database for CFD validation.

Thirteen groups presented their findings in the workshop for this case, where the results were mainly obtained by employing RANS and URANS methods in 2D or 3D simulations. In some cases, simple boundary conditions were applied on the hump surface to represent the jet in a crossflow, while for other cases, the whole cavity was considered for control purpose. The PIV experimental data and pressure data from wall-mounted pressure taps were used to validate the computed velocity/turbulence profiles and pressure coefficients.

Figure 4.18a shows the comparison between CFD modeling and experimental data for pressure coefficient in baseline uncontrolled case. It is clear that most models overpredicted the suction pressure that this might be related to the effect of sidewalls that has been used in the experiment. It is also observed that the computed pressure coefficient was overpredicted within the separation region near the trailing edge of the hump in most cases. While all simulations were able to predict the separation points with reasonable agreement with the experiment, the predicted reattachment point was significantly downstream of the corresponding experimental result.

Figure 4.18b shows the comparison between CFD modeling and experimental data for pressure coefficient in suction-based separation control case. Similar to the baseline uncontrolled case, the pressure coefficient was underpredicted by the CFD near the suction peak. The separation point was also obtained slightly upstream of the experimental result, and the separation bubble length from CFD was significantly larger than the one from the experiment. A comparison between the experiment and CFD in SJ-based flow control case for pressure coefficient is shown in Figure 4.18c, where the discrepancy between different CFD results is more noticeable than the previous two cases. A significant difference between the experimental and computational results is observed near the trailing edge, which can indicate the dissimilarity between two results for the location of the separation point.

In another study, You and Moin (2008) employed an unstructured grid LES solver to study the active flow control of turbulent separated flow over an NACA 0015 airfoil at chord Reynolds number of 896,000. Oscillatory synthetic jets were introduced into the flow over the suction side of the airfoil through a slot that was connected to a cavity (Figure 4.19a).

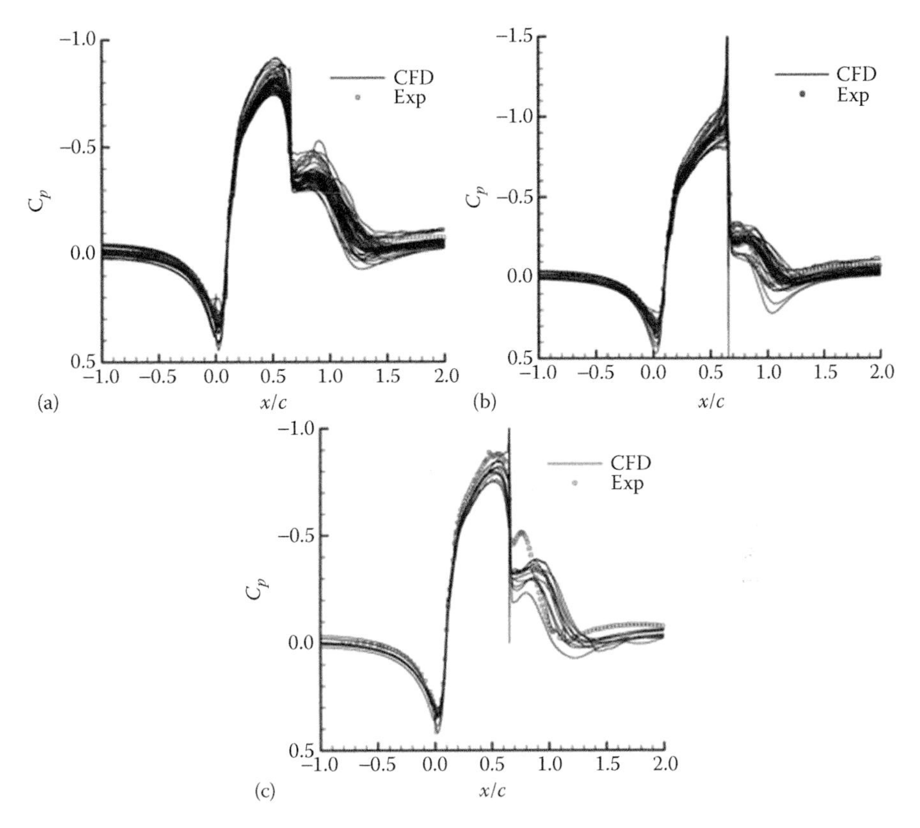

FIGURE 4.18 Comparison of surface pressure coefficient between experimental and CFD for (a) baseline uncontrolled case, (b) suction-based separation control, and (c) SJ-based separation control. (Reproduced with permission from Rumsey, C. L. et al., *AIAA J.*, 44, 194–207, 2006.)

The jet exit was extended over the entire span of the computational domain, consequently eliminating end effects.

Using 15 million grid points for the simulations resulted in a wall-clock time of about 4.8 seconds with 200 CPUs of IBM Power 5. A comparison of time-averaged pressure distribution over the airfoil surface for both uncontrolled and controlled cases with published experimental results (Gilarranz et al. 2005) is shown in Figure 4.19b, which reveals favorable agreement in both cases with experiments. The lift and drag coefficients were also well predicted for both flow conditions.

More recently, Sahni et al. (2011) conducted a complimentary experimental and numerical investigation to examine the 3D structures and interactions of a finite-span rectangular jet in a crossflow over a NACA 4421 airfoil at a chord Reynolds number of 100,000 and zero angle of attack. The numerical setup including the size of the airfoil, orientation, location of the SJ, size and shape of the flow control cavity, and jet blowing ratio matched the physical dimensions and flow parameters. DNS and LES methods were used to solve the

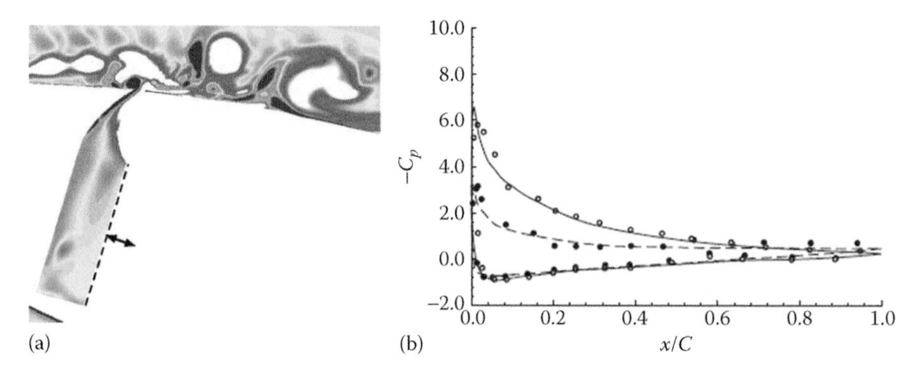

FIGURE 4.19 (a) Instantaneous spanwise vorticity contours inside and around an SJA. (b) Time-averaged pressure distribution over the airfoil surface. Solid line, controlled case; dashed line, uncontrolled case; symbols, experimental data. (Reproduced with permission from You, D. and P. Moin, *J. Fluids Struct.*, 24, 1349–1357, 2008.)

incompressible Navier–Stokes equations on 15 million grid points and 80 million elements. In all cases of study, reasonable agreement was found between experimental and numerical results with some limitations.

4.5 Summary

A broad spectrum of computational studies of SJ flow fields as well as SJ-based flow control conducted over the last twenty years have been reviewed. Computational modeling has been used to gain better insight into the fluid dynamics of these actuators as well as to understand the effectiveness of these actuators for flow control. One key challenge associated with computational modeling of these flows is the large spatial and temporal scale disparity between the SJ and the flow being controlled. Development of more accurate actuator models that can eliminate the need for including the entire actuator in the simulations could help in this regard. The second challenge is associated with the inherent unsteady nature of the SJ perturbation, which makes the application of conventional turbulence models difficult. Further advances in URANS models as well as the application of LES to these flows are expected to help overcome this limitation.

Acknowledgments

We would like to acknowledge the support from AFOSR (grant monitored by Dr. Douglas Smith) and NASA (grant monitored by Dr. Brian Allen). We would also like to acknowledge our collaboration with Professor Louis Cattafesta in much of the research our group has conducted in the area of synthetic jets.

References

Agashe, J., D. P. Arnold, and L. Cattafesta. 2009. Development of compact electrodynamic zero-net mass-flux actuators, *The 47th AIAA Aerospace Sciences Meeting, AIAA Paper* 2009-1308, Orlando, FL, January.

Amitay, M., B. L. Smith and A. Glezer. 1998. Aerodynamic flow control using synthetic jet technology, *The 36th AIAA Aerospace Sciences Meeting and Exhibit, AIAA Paper* 98-0208, Reno, Nevada, January.

Amitay, M., D. Smith, V. Kibens, D. Parekh, and A. Glezer. 2001. Aerodynamics flow control over an unconventional airfoil using synthetic jet actuators. *AIAA Journal*, 39(3): 361–370.

Aram, E., R. Mittal, and L. N. Cattafesta. 2010a. Simple representations of zero-net mass-flux jets in grazing flow for flow-control simulations. *International Journal of Flow Control*, 2(2): 109–125.

Aram, E., R. Mittal, J. Griffin, and L. N. Cattafesta. 2010b. Toward effective ZNMF jet based control of a canonical separated flow. *The 5th AIAA Flow Control Conference*, Chicago, IL, 28 June–1 July.

Bar-Sever, A. 1988. Separation control on an airfoil by periodic forcing. *AIAA Journal*, 27(6): 820–821.

Campbell, J. S., W. Z. Black, A. Glezer, and J. G. Hartley. 1998. Thermal management of a laptop computer with synthetic air microjets. *Intersociety Conference on Thermomechanical Phenomenon in Electronic Systems*, Seattle, WA, May 27–30.

Chatlynne, N., M. Rumigny, M. Amitay, and A. Glezer. 2001. Virtual aero-shaping of a clark-Y airfoil using synthetic jet actuators, *The 39th Aerospace Sciences Meeting and Exhibit, AIAA Paper* 2001-0732, Reno, NV.

Chen, F. and G. B. Beeler. 2002. Virtual shaping of a two-dimensional NACA 0015 airfoil using synthetic jet actuator, *The 40th Aerospace Sciences Meeting and Exhibit, AIAA Paper* 2002-3273, Reno, NV.

Darabi, A. and I. Wygnanski. 2002. On the transient process of flow reattachment by external excitation, *The 40th Aerospace Sciences Meeting and Exhibit, AIAA Paper* 2002-3163, Reno, NV.

Didden, N. 1979. On the formation of vortex rings: Rolling-up and production of circulation. *Journal of Applied Mathematics and Physics*, 30: 101–116.

Filz, C., D. Lee, P. D. Orkwis, and M. G. Turner. 2003. Modeling of two dimensional directed synthetic jets using neural network-based deterministic source terms, *The 33rd AIAA Fluid Dynamics Conference and Exhibit, AIAA Paper* 2003-3456, Orlando, FL, June.

Funk, R., D. Parekh, T. Crittenden, and A. Glezer. 2010. Transient separation control using pulse combustion actuation. *AIAA Journal*, 48(11): 2482–2490.

Gallas, Q., R. Holman, T. Nishida, B. Carroll, M. Sheplak, and L. Cattafesta. 2003a. Lumped element modeling of piezoelectric-driven synthetic jet actuators. *AIAA Journal*, 41(2): 240–247.

Gallas, Q., R. Mittal, M. Sheplak, and L. Cattafesta. 2004. Case 1: Lumped element modeling of a zero-net mass flux actuator issuing into a quiescent medium. *Proceedings of the NASA LaRC Workshop on CFD Validation of Synthetic Jets and Turbulent Separation Control*, Williamsburg, VA, March 29–31.

Gallas, Q., G. Wang, M. Papila, M. Sheplak, and L. Cattafesta. 2003b. Optimization of synthetic jet actuators, *The 41st Aerospace Sciences Meeting and Exhibit, AIAA Paper* 2003-0635, Reno, NV, January 6–9.

Gilarranz, J. L., L. W. Traub, and O. K. Rediniotis. 2002. Characterization of a compact, high-power synthetic jet actuator for flow separation control, *The 40th AIAA Aerospace Sciences Meeting and Exhibit, AIAA Paper* 2002-0127, Reno, NV, January.

Gilarranz, J. L., L. W. Traub, and O. K. Rediniotis. 2005. A new class of synthetic jet actuators—Part II: Application to flow separation control. *Journal of Fluids Engineering*, 127: 377–387.

Glezer, A. and M. Amitay. 2002. Synthetic Jets. *Annual Review of Fluid Mechanics*, 34: 503–529.

Glezer, A., M. Amitay, and A. M. Honohan. 2005. Aspects of low- and high-frequency actuation for aerodynamic flow control. *AIAA Journal*, 43(7): 1501–1511.

Greenblatt, D. and I. Wygnanski. 1999. Parameters affecting dynamic stall control by oscillatory excitation, *The 17th Applied Aerodynamics Conference, AIAA Paper* 99-3121, Norfolk, VA.

Greenblatt, D. and I. Wygnanski. 2001. Use of periodic excitation to enhance airfoil performance at low Reynolds number. *Journal of Aircraft*, 38(1): 190–192.

Greenblatt, D. and I. Wygnanski. 2003. Effect of leading-edge curvature on airfoil separation control. *Journal of Aircraft*, 40: 473–481.

Guarino, J. R. and V. P. Manno. 2001. Characterization of a laminar jet impingement cooling in portable computer Applications. *Proceedings of the 17th Annual Semiconductor Thermal Measurement and Management Symposium*, San Jose, CA, March, pp. 1–11.

Holman, R., Y. Utturkar, R. Mittal, B. L. Smith, and L. Cattafesta. 2005. Formation criterion for synthetic jets. *AIAA Journal*, 43(10): 2110–2116.

Honohan, A. M., M. Amitay, and A. Glezer. 2000. Aerodynamic control using synthetic jets, *The Fluids 2000 Conference and Exhibit, AIAA Paper* 2000-2401, Denver, CO.

James, R. D., J. W. Jacobs, and A. Glezer. 1996. A round turbulent jet produced by an oscillation diaphragm. *Physics of Fluids*, 8(9): 2484–2495.

Kihwan, K., A. Beskok, and S. Jayasuriya. 2005. Nonlinear system identification for the interaction of synthetic jets with a boundary layer. *Proceedings of the American Control Conference*, Portland, OR, June 8–10.

Kotapati, R. B., R. Mittal, and L. Cattafesta. 2007. Numerical study of transitional synthetic jet in quiescent external flow. *Journal of Fluid Mechanics*, 581: 287–321.

Kral, L. D., J. F. Donovan, A. B. Cain, and A. W. Cary. 1997. Numerical simulation of synthetic jet actuators, *The 4th AIAA Shear Flow Control Conference, AIAA Paper* 97-1824, Snowmass, CO.

Lee, C. Y. and D. B. Goldstein. 2002. Two-dimensional synthetic jet simulation. *AIAA Journal*, 40(3): 510–516.

Leschziner, M. and S. Lardeau. 2011. Simulation of slot and round synthetic jets in the context of boundary-layer separation control. *Philosophical Transactions of the Royal Society A*, 369: 1495–1512.

Lockerby, D. A., P. W. Carpenter, and C. Davies. 2002. Numerical simulation of the interaction of microactuators and boundary layer. *AIAA Journal*, 40(1): 67–73.

Margalit, S., D. Greenblatt, A. Seifert, and I. Wygnanski. 2002. Active flow control of a delta wing at high incidence using segmented piezoelectric actuators, *The 1st Flow Control Conference, AIAA Paper* 2002-3270, St. Louis, MO.

Mittal, R. and R. Kotapati. 2004. Resonant mode interaction in a canonical separated flow. *Proceedings of the 6th IUTAM Symposium on Laminar-Turbulent Transitions*, Bangalore, India, December 13–17.

Mittal, R., P. Rampuggoon, and H. S. Udaykumar. 2001. Interaction of a synthetic jet with a flat plat boundary layer, *The 15th AIAA Computational Fluid Dynamics Conference, AIAA Paper* 2001-2773, Anahiem, CA.

Mittal, R. and P. Rampunggoon. 2002. On the virtual aeroshaping effect of synthetic jets. *Physics of Fluids*, 14(4): 1533–1536.

Ol, M., B. R. McAuliffe, E. S. Hanff, U. Scholz, and C. H. Kaehler. 2005. Comparison of laminar separation bubble measurements on a low Reynolds number airfoil in three facilities, *The 35th AIAA Fluid Dynamics Conference and Exhibit, AIAA Paper* 2005-5149, Toronto, ON, Canada.

Pack, L. and A. Seifert. 2000. Dynamics of active separation control at high Reynolds numbers, *The 38th Aerospace Sciences and Exhibit, AIAA Paper* 2000-0409, Reno, NV.

Pack, L. G., N. W. Schaeffler, C.-S. Yao, and A. Seifert. 2002. Active control of flow separation from the slat shoulder of a supercritical airfoil, *The 1st Flow Control Conference, AIAA Paper* 2002-3156, St. Louis, MO, June 24–26.

Raju, R. 2007. Scaling laws and separation control strategies for zero-net mass-flux actuators. DSc thesis, The George Washington University, Washington, DC.

Raju, R., E. Aram, R. Mittal, and L. N. Cattafesta. 2009. Simple models of zero-net mass-flux jets for flow control simulations. *International Journal of Flow Control*, 1(3): 179–197.

Raju, R., R. Mittal, and L. Cattafesta. 2008. Dynamics of airfoil separation control using zero-net mass-flux forcing. *AIAA Journal*, 46(12): 3103–3115.

Rampunggoon, P. 2001. Interaction of a synthetic jet with a flat-plate boundary layer. PhD Thesis, University of Florida, Gainesville, FL.

Rathnasingham, R. and K. S. Breuer. 1997. Coupled fluid-structural characteristics of actuators for flow control. *AIAA Journal*, 35(5): 832–837.

Ravi, B. R. 2007. Numerical study of three dimensional synthetic jets in quiescent and external grazing flows. DSc thesis, The George Washington University, Washington, DC.

Ravindran, S. S. 1999. Active control of flow separation over an airfoil. *NASA Technical Reports*, NASA/TM-209838, Langley Research Centre, Hampton, VA, December.

Rediniotis, O. K., J. Ko, and A. J. Kurdila. 2002. Reduced order nonlinear Navier-Stokes models for synthetic jets. *Journal of Fluid Engineering*, 124(2): 433–443.

Rizzetta, D. P., M. R. Visbal, and M. J. Stanek. 1999. Numerical investigation of synthetic-jet flow fields. *AIAA Journal*, 37(8): 919–927.

Rumsey, C. L. 2009. Successes and challenges for flow control simulations. *International Journal of Flow Control*, 1(1): 1–27.

Rumsey, C. L., T. B. Gatski, W. L. Sellers III, V. N. Vatsa, and S. A. Viken. 2006. Summary of the 2004 computational fluid dynamics validation workshop on synthetic jets. *AIAA Journal*, 44(2): 194–207.

Sahni, O., J. Wood, K. E. Jansen, and M. Amitay. 2011. Three-dimensional interactions between a finite-span synthetic jet and a crossflow. *Journal of Fluid Mechanics*, 671: 254–287.

Seifert, A., T. Bachar, D. Koss, M. Shepshelovich, and I. Wygnanski. 1993. Oscillatory blowing: A tool to delay boundary-layer separation. *AIAA Journal*, 31(11): 2052–2060.

Seifert, A., A. Darabi, and I. Wygnanski. 1996. Delay of airfoil stall by periodic excitation. *Journal of Aircraft*, 33(4): 691–698.

Seifert, A. and L. Pack. 1999. Oscillatory control of separation at high Reynolds numbers. *AIAA Journal*, 37(9): 1062–1071.

Seifert, A. and L. Pack. 2000. Separation control at flight Reynolds numbers: Lessons learned and future directions, *The Fluids 2000 Conference and Exhibit, AIAA Paper* 2000-2542, Denver, CO.

Seifert, A. and L. Pack. 2002. Active flow separation control on wall-mounted hump at high Reynolds numbers. *AIAA Journal*, 40(7): 1363–1372.

Smith, B. L. and A. Glezer. 1997. Vectoring and small-scale motions effected in free shear flows using synthetic jet actuators, *The 35th Aerospace Sciences Meeting and Exhibit, AIAA Paper* 97-0213, Reno, NV.

Smith, B. L. and A. Glezer. 1998. The formation and evolution of synthetic jets. *Physics of Fluids*, 10(9): 2281–2297.

Smith, B. L. and G. W. Swift. 2001. Synthetic jets at large Reynolds number and comparison to continuous jets, *The 35th Aerospace Sciences Meeting and Exhibit, AIAA Paper* 2001-3030, Reno, NV.

Utturkar, Y., R. Holman, R. Mittal, B. Carroll, M. Sheplak, and L. Cattafesta. 2003. A jet formation criterion for synthetic jet actuators, *The 41st Aerospace Sciences Meeting and Exhibit, AIAA Paper* 2003-0636, Reno, NV, January 6–9.

Utturkar, Y., R. Mittal, P. Rampunggoon, and L. Cattafesta. 2002. Sensitivity of synthetic jets to the design of the jet cavity, *The 40th Aerospace Sciences Meeting and Exhibit, AIAA Paper* 2002-0124, Reno, NV.

Wu, J.-Z., X. Lu, A. Denny, M. Fan, and J.-M. Wu. 1998. Post-stall flow control on an airfoil by local unsteady forcing. *Journal of Fluid Mechanics*, 371: 21–58.

Wygnanski, I. 2000. Some new observations affecting the control of separation by periodic forcing, *The Fluids 2000 Conference and Exhibit, AIAA Paper* 2000-2314, Denver, CO.

Yao, C. S., F. J. Chen, D. Neuhart, and J. Harris. 2004. Synthetic jets in quiescent air. *Proceedings of the NASA LARC Workshop on CFD Validation of Synthetic Jets and Turbulent Separation Control*, Williamsburg, VA, March 29–31.

Yamaleev, N. K. and M. H. Carpenter. 2003. A reduced-order model for efficient simulation of synthetic jet actuators. *NASA Technical Reports*, NASA/TM-2003-212664, December.

You, D. and P. Moin. 2008. Active control of flow separation over an airfoil using synthetic jets. *Journal of Fluids and Structures*, 24: 1349–1357.

Zaman, K. and D. Culley. 2006. A study of stall control over an airfoil using "Synthetic Jets," *The 44th AIAA Aerospace Sciences Meeting and Exhibit, AIAA Paper* 2006-98, Reno, NV.

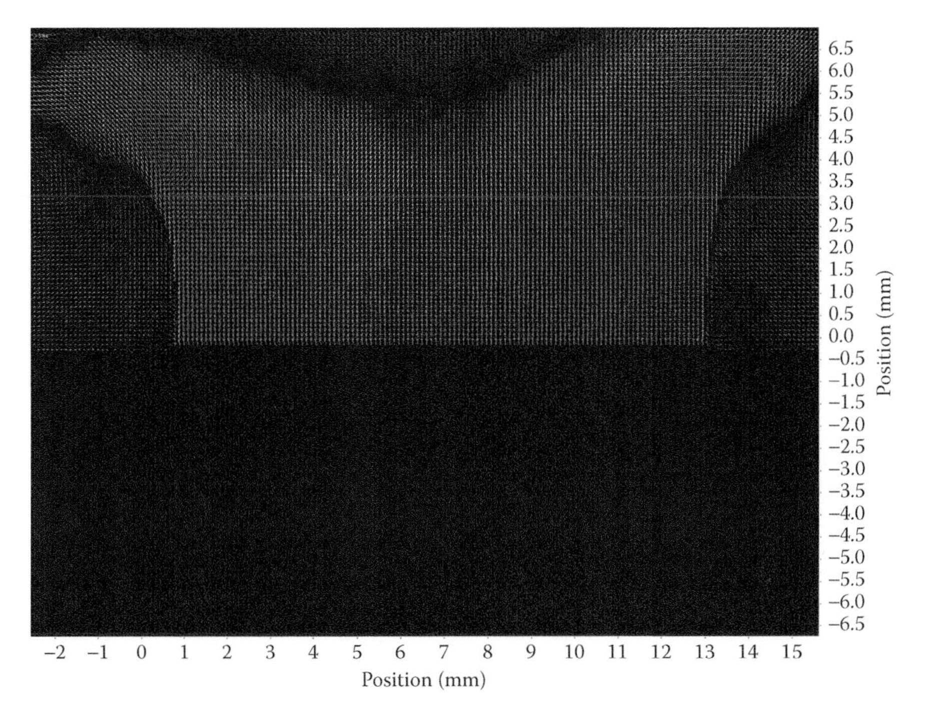

FIGURE 3.13 Example from Nani and Smith (2012) of a phase-averaged vector field used to determine the volume flow rate waveform in an axisymmetric synthetic jet with $L_0/D = 7$ and $Re_\delta = 500$. (From Nani, D. J. and B. L. Smith, *Phys. Fluids*, 24, 115110, 2012. Reprinted with permission.)

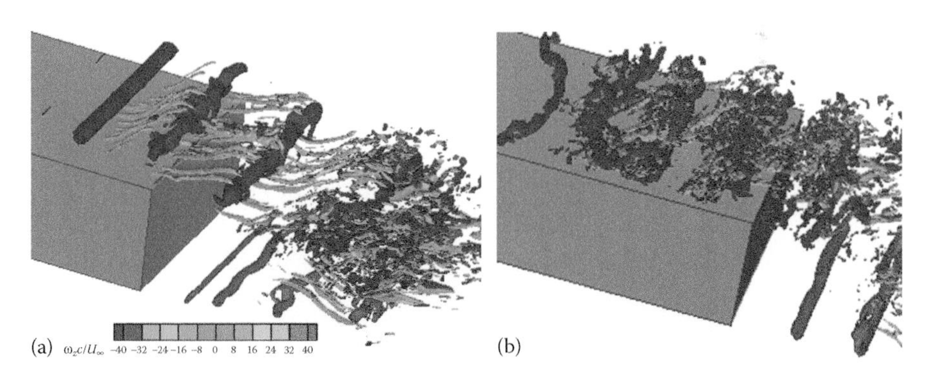

FIGURE 4.16 Isosurfaces of instantaneous swirl strength ($50\ Qc/U_\infty$) and spanwise vorticity ($\omega_z c/U_\infty$) contours for (a) baseline and (b) controlled aft-chord separated flow. (From Aram, E., R. Mittal, J. Griffin, and L. N. Cattafesta. Toward effective ZNMF jet based control of a canonical separated flow. *The 5th AIAA Flow Control Conference*, Chicago, IL, June 28–July 1, 2010b.)

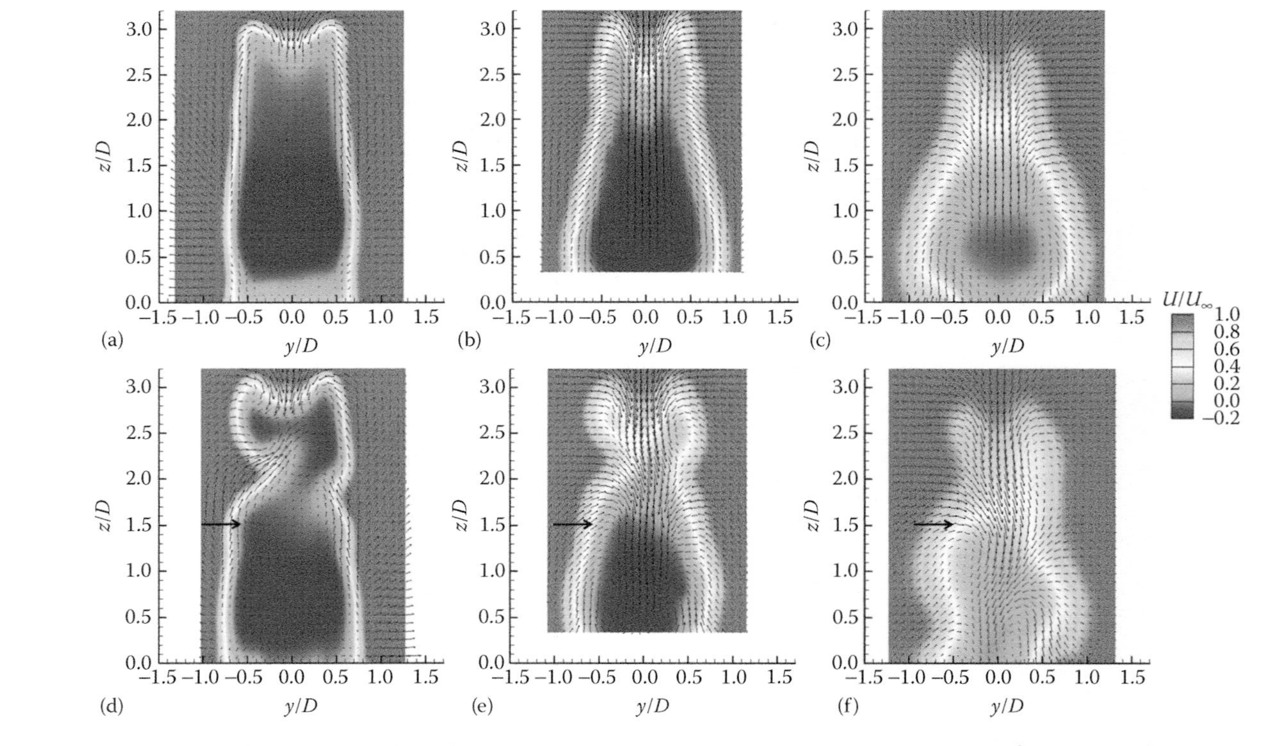

FIGURE 6.4 Color contours of the streamwise (out of plane) velocity with superimposed in-plane vectors, $\theta_{jet} = 113°$ (jet's centerline): (a) unforced, $x/D = 1$; (b) forced, $x/D = 1$; (c) unforced, $x/D = 2$; (d) forced, $x/D = 2$; (e) unforced, $x/D = 3$; and (f) forced, $x/D = 3$. (From DeMauro, E. P. et al., *Exp. Fluids.*, 53, 1969, 2012. With permission.)

5

Reduced-Order Modeling of Synthetic Jets

Nail K. Yamaleev
North Carolina A&T State University

5.1 Introduction

Synthetic jet actuation is a very promising technique that demonstrates the great potential for separation control (Seifert et al. 1996; Crook et al. 1999), lift enhancement (Smith et al. 1998), heat transfer (Mahalingam and Glezer 2005), flow vectoring (Guo and Gary 2001), mixing enhancement (Chen et al. 1999), and transition delay (Rathnasingham and Breuer 1997b). The numerical simulation and optimization of synthetic jet flows have recently received considerable attention. The exact nature of any synthetic jet actuator can be determined (in principle) by simulating the precise three-dimensional (3-D) geometry, including all aspects of diaphragm movement and deformation (see Chapter 5 for detailed discussion of this approach). Note, however, that the full numerical simulation of both the cavity and exterior flowfields is extremely computationally expensive, which makes this approach impractical for analysis and optimization of synthetic jet flows. Indeed, the numerical solution of the cavity flow requires nearly the same number of grid points that are needed to solve the exterior flowfield. Furthermore, the Mach number in the problem varies from $O(10^{-3})$ (near the diaphragm) to $O(1)$ (in the exterior flow). The variation of the flow parameters from fully incompressible to fully compressible regimes and the presence of the moving boundary considerably increase the algorithm complexity. As a result, the computational cost associated with the 3-D unsteady Navier–Stokes model is generally too large, which indicates that reduced-order modeling of synthetic jets is critical, especially for design and optimization studies.

A wide range of simplified models ranging from simple orifice exit boundary conditions to full numerical simulations of the synthetic jet flow have been developed. The most accurate model of a synthetic jet actuator is based on the 3-D unsteady Navier–Stokes simulation of the cavity flow and diaphragm oscillations. The extreme opposite to the full simulation is to impose experimental data at the actuator orifice. If reliable experimental data are available, then this approach is desirable. All general actuator models, however, fall somewhere between these two extremes and attempt to trade fidelity for cost in some measure. A detailed description of the most popular reduced-order models (ROMs) available in the literature is presented in Section 5.2.

5.2 Transpiration Boundary Conditions

5.2.1 Transpiration Boundary Conditions-Based Models for Incompressible Flows

One of the simplest ROMs for simulation of synthetic jets is introduced by Kral et al. (1997). The key idea of this approach is to impose transpiration boundary conditions (TBCs) at the orifice exit, by directly prescribing the harmonic motion generated by the actuator. For incompressible flows, this approximation is a direct consequence of the conservation of mass, which reduces to a simple relation between the membrane and jet velocities:

$$\int_{-d/2}^{d/2} v(x, 0, t)\mathrm{d}x = \int_{-D/2}^{D/2} v(x, -H, t)\mathrm{d}x \tag{5.1}$$

where:

v is the normal component of the velocity vector

H is the actuator depth

d and D are the characteristic lengths of orifice and membrane, respectively, as shown in Figure 5.1

Using the mean-value theorem and assuming that the membrane oscillates sinusoidally, Equation 5.1 can be recast in the following form:

$$\bar{v}(t) = V_a \sin(\omega t) \tag{5.2}$$

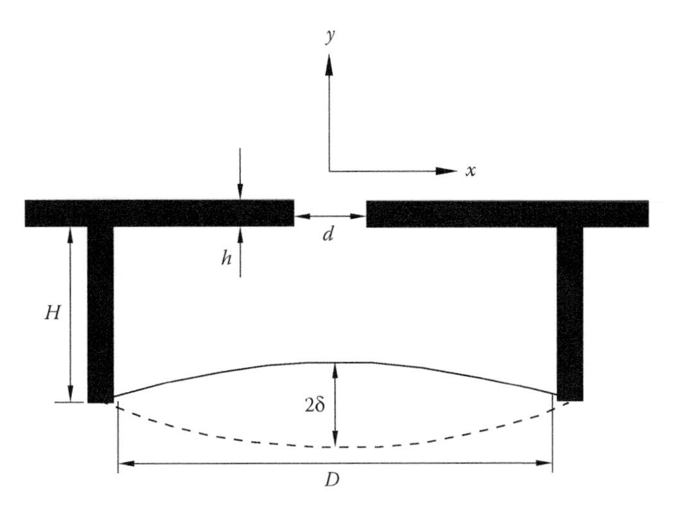

FIGURE 5.1 Schematic of a synthetic jet actuator geometry.

where:
 ω is the diaphragm angular frequency

In Equation 5.2, $\bar{v}$ is the averaged wall-normal component of jet velocity at the orifice exit, whose amplitude can be evaluated as

$$V_a = (D/d)^p v_m \tag{5.3}$$

where:
 v_m is the membrane velocity
 p is either 1 or 2 for slot- or circular-type actuators, respectively

Note that in contrast to the above equation, the jet velocity amplitude V_a in Kral et al. (1997) and Donovan et al. (1998) is adjusted to match the peak jet velocity observed in experiments and assumed to be nonuniform across the orifice. Among various spatial profiles of $V_a(x)$ considered in Kral et al. (1997), the "top-hat" distribution most closely matches the experimental data. The pressure at the orifice exit should be consistent with the time-harmonic velocity perturbations. This is achieved by using the normal momentum equation:

$$\rho \frac{\partial v}{\partial t} = -\frac{\partial p}{\partial y} \tag{5.4}$$

Here, the viscous terms have been neglected. Taking into account the time-harmonic velocity oscillations given by Equation 5.2, the pressure boundary condition becomes

$$\frac{\partial p}{\partial y} = -\rho V_a \omega \cos(\omega t) \tag{5.5}$$

This simplified representation of the synthetic jet eliminates the need of computation of the cavity flow and drastically reduces the computational cost, especially if large arrays of actuators are simulated. The numerical results obtained using the boundary conditions (Equations 5.2 and 5.5) for large orifice diameters show good qualitative agreement with the experiments (Seifert et al. 1993, 1996). A similar approach to modeling the actuator as the blowing/suction-type boundary conditions, which can be fully specified in advance of the computation, is also used in the works by Carlson and Lumley (1996), Hofmann and Herbert (1997), Hassan and JanakiRam (1998), and Lin and Chieng (1999).

5.2.2 Model Based on Localized Forces

To avoid the integration of the incompressible Navier–Stokes equations on a moving grid, an alternative technique is used by Lee and Goldstein (2002). Their method imposes a localized body force along the desired points in the computational mesh to bring the fluid there to a specified velocity so that the force has the same effect as a solid boundary. The desired velocity is incorporated in an iterative feedback loop to determine the appropriate force. For a membrane moving with velocity $v_m(x, t)$, an expression for the body force is given by

$$F(x,t) = \alpha \int_0^t (v - v_m)\mathrm{d}t' + \beta(v - v_m) \tag{5.6}$$

where:

- v is the fluid velocity
- α and β are the user-defined constants that are negative and can be treated as the gain and damping of the force field

Though this approach simulates the moving membrane without using a time-dependent coordinate transformation, it still requires the solution of the cavity flow, thus making the computational cost comparable with that of the full-order model.

5.2.3 TBCs-Based Models for Compressible Flows

The major problem with ROMs based on the TBC is the lack of information of what jet velocity, pressure, and density profiles at the orifice exit should be prescribed to provide the global conservation of mass, momentum, and energy for the entire flow including the exterior and cavity regions. One of the approaches to address this issue is proposed in Rizzetta et al. (1999). The required temporal and spatial velocity distributions at the orifice exit are precomputed by solving the unsteady compressible full Navier–Stokes equations in the external region, cavity itself, and orifice duct on separate grids that are linked with each other through the Chimera methodology. The membrane motion is simulated by varying the position of appropriate boundary points. For low Mach and Reynolds numbers considered in Rizzetta et al. (1999), the internal cavity flow becomes periodic after several cycles of diaphragm oscillation. Therefore, the normal velocity profile across the jet exit at each time step is recorded and used as a boundary condition in subsequent runs involving the external domain only. For computations that consider only the exterior domain, the transverse and spanwise velocity components (orthogonal to jet axis) are set to 0, and the inviscid normal momentum equation

$$\frac{\partial p}{\partial y} = -\rho\left(\frac{\partial v}{\partial t} + v\frac{\partial v}{\partial y}\right) \tag{5.7}$$

is used to establish the pressure. The density at the orifice exit is extrapolated from the interior solution. This approach provides a more accurate description of the flow details at the orifice than the simplified boundary conditions (Equations 5.2 and 5.5). Similar direct numerical simulation of the external and cavity flows is performed by Joslin et al. (1998) and shows good agreement with experimental results. Despite its accuracy, the major drawback of this ROM is the need to solve the entire problem including the cavity flow, which makes this approach impractical for optimization studies.

Another implementation of the TBC (Equations 5.2 and 5.5), which makes this model more accurate, is suggested in Yamaleev et al. (2005) and further modified in Raju et al. (2009) and Aram et al. (2010). In this approach, the blowing/suction boundary conditions are imposed inside the orifice duct rather than at the jet exit. As a result, the compressibility and viscous effects near the jet exit are partially taken into account. Comparison of the pressure integrals obtained using this TBC-based model and the full two-dimensional (2-D) numerical simulation of both the cavity flow and subsonic crossflow ($M_\infty = 0.5$) is shown in Figure 5.2. As evident from this comparison, the simplified TBC for the essentially

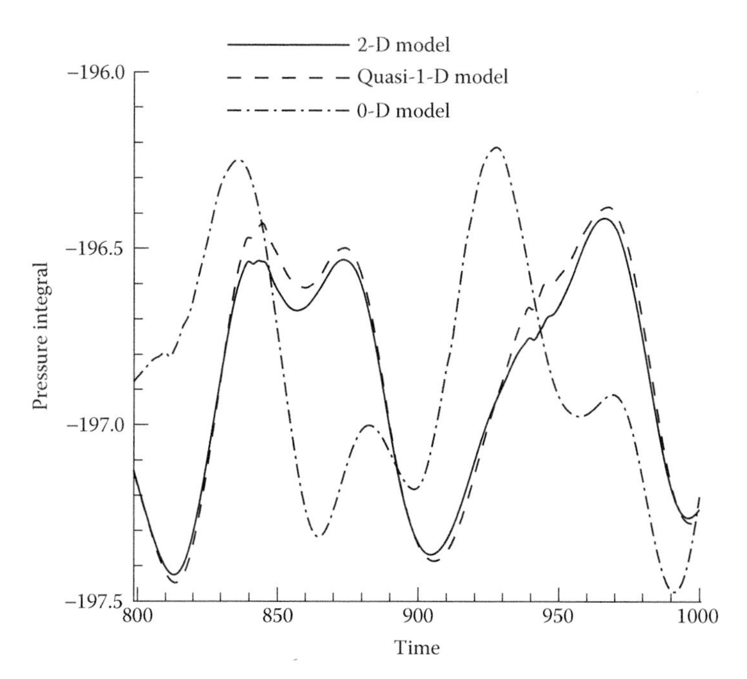

FIGURE 5.2 Pressure integral obtained with the TBC and full-order models. (From Yamaleev, N. K. et al., *AIAA J.*, 43, 357–369, 2005.)

compressible flow cannot even qualitatively predict the time history of the pressure integral.

Despite the simplicity and efficiency, there are serious problems that significantly restrict the area of applications of ROMs based on the TBC. First of all, to close the boundary conditions (Equations 5.2 and 5.5), the jet velocity amplitude V_a is required, which is not readily available and should be obtained either from experimental data or from full numerical simulations of the cavity flow (Rizzetta et al. 1999). Note that the formula (5.3) obtained in the incompressible limit neglects the nonlinear and compressibility effects and therefore fails to predict the jet velocity for realistic orifice diameters (Tang et al. 2007). As shown in Lee and Goldstein (2002), the real streamwise velocity profile and the velocity component in the crossflow direction are far from the analytical expressions of Equation 5.2. Furthermore, the TBC-based models do not satisfy the conservation of momentum and energy. For compressible flows, these models do not also provide the conservation of mass (Yamaleev et al. 2005). Therefore, the TBC-based models are lacking a mechanism to account for changes in the pressure field caused by the interaction of the external boundary layer and the actuator.

5.3 Zero-Dimensional Models

An alternative technique to reduce the computational cost associated with the simulation of synthetic jets is to use some approximation of the volume-averaged conservation laws inside an actuator cavity. This approach leads to a system of ordinary differential equations (ODEs), which is herein referred as a zero-dimensional (0-D) model.

5.3.1 Model Based on the Unsteady Bernoulli Equation

A 0-D model based on a system of ODEs describing a resonant synthetic jet actuator is developed in Rathnasingham and Breuer (1997a). In this approach, the actuator is simulated as a closed cavity of height H (Figure 5.1). The actuator membrane is represented as an oscillating piston of diameter D. The piston displacement normalized by the cavity height $\eta = \delta/H$ represents an averaged true membrane deflection described by its mode shape. Assuming that the flow is isothermal and combining structural and fluid effects, a set of the following four coupled nonlinear first-order ODEs is derived:

$$\begin{aligned}
m\eta'' &= -k\eta - c\eta' - \frac{\pi D^2}{4H}(p - p_\infty) + \frac{F}{H}\sin(\omega t) \\
\rho' &= \frac{\rho(D^2 H\eta' - ud^2)}{D^2 H(1-\eta)} \\
p' &= RT\rho' \\
u' &= \frac{p - p_\infty}{\rho s} - \frac{u|u|}{2s}
\end{aligned} \tag{5.8}$$

where:

m is the membrane mass
k is the membrane stiffness
c is the structural damping coefficient
p and p_∞ are pressure values inside the cavity and in the ambient atmosphere
F and ω are the forcing amplitude and frequency
d is the orifice diameter
ρ and T are the air density and temperature inside the cavity
s is the length of an accelerating streamline through the orifice

In Equation 5.8, the membrane motion is described by the first equation whose right-hand side includes the structural stiffness, damping, fluid, and forcing terms. The second equation represents the conservation of mass in the cavity, while the third equation is the equation of state differentiated with respect to time. The last equation is obtained from the unsteady Bernoulli equation that provides a relation between the jet velocity and the cavity pressure.

The second and third equations of the above-mentioned reduced-order model take into account some nonlinear and compressibility effects. If dominant, these effects may lead to generation of high-frequency harmonics, modulation of the fundamental amplitude, and a shift in the natural frequency, which is a well-known result in the theory of nonlinear oscillators (Nayfeh 1979). Note, however, that Equation 5.8 neglects the viscous losses associated with the flow through the orifice. Indeed, in the limit of an incompressible fluid, the time derivative of density in the second equation vanishes, thus leading to the following relation between the jet membrane velocities:

$$u = \frac{D^2 H \eta'}{d^2} \tag{5.9}$$

As follows from this equation, the jet velocity becomes infinitely large as $d/D \to 0$. Note that in reality, the viscous effects become dominant as d approaches 0, which restricts the mass flow through the orifice. To account for the viscous effects for small Stokes numbers ($St = \sqrt{\omega d^2/\nu}$), Rathnasingham and Breuer (1997a) suggested to use the Hagen–Poiseuille formula:

$$u(r,t) \approx \frac{p_y(d^2/4 - r^2)}{4\nu}$$

where:

r is the radial coordinate
p_y is the pressure gradient
ν is the fluid kinematic viscosity

Matching the velocities at the orifice center for the inviscid and fully viscous regimes and assuming the harmonic solution for the membrane deflection, an optimal Stokes number that provides the maximum jet velocity has been derived (Rathnasingham and

Breuer 1997a). This ROM successfully predicts trends in actuator jet velocities observed in the experiments for low Stokes numbers.

5.3.2 Static Compressible Model

A similar approach called a static compressible (SC) flow model, which takes into account both the orifice and cavity regions, is developed in the work by Tang et al. 2007. Similar to the previous ROM, a fully developed pipe flow model is used in the orifice duct region. This implies a direct application of a constant pressure gradient and, therefore, a linear variation of pressure along the orifice duct. In addition to the assumption that the flow is incompressible in the orifice region, it is also assumed that density and velocity variations through the orifice are small and approximately linear. Under the above assumptions, the Navier–Stokes equations for the circular orifice can be approximated as follows:

$$\begin{gathered} u_o = \frac{\rho_i u_i}{\rho_o} - \frac{h}{2\rho_o}\frac{\partial \rho_i}{\partial t} \\ \frac{\partial u_i}{\partial t} + \left(\frac{\rho_i}{\rho_o} - 1\right)\frac{u_i^2}{h} - \frac{u_i}{2\rho_o}\frac{\partial \rho_i}{\partial t} = \frac{p_i - p_o}{\rho_i h} + \frac{\mu}{\rho_i}\frac{\partial^2 u_i}{\partial r^2} \end{gathered} \tag{5.10}$$

where:

h is the orifice depth

subscripts i and o denote the values at inlet and outlet of the orifice duct, respectively

The SC model neglects the unsteady effects inside the cavity and assumes that the pressure in the actuator is uniform. Note that the outlet quantities are taken from the solution of the exterior problem in which this actuator model is embedded in, while p_i and ρ_i are obtained from the mass conservation in the cavity region:

$$\frac{d}{dt}\left\{\rho_i \frac{\pi D^2}{4}\left[H - \frac{\delta}{8}\sin(\omega t)\right]\right\} = -2\pi\rho_i \int_0^{d/2} u_i(r)dr$$

and the assumption that the cavity flow is isothermal. Since u_i is assumed to be parabolic in the radial direction, Equation (5.10) becomes a system of ODEs with the only unknown, the jet velocity at the orifice exit. As has been reported in Tang et al. 2007, the SC model overpredicts the maximum peak velocity. This overprediction is caused by the inability of the SC model to accurately model the flow dynamics through the orifice duct, which is essentially nonlinear in contrast to the linear approximation used in Equation (5.10).

5.3.3 Models Based on the Unsteady Pipe-Flow Theory

Another technique based on the unsteady pipe-flow theory, which is applicable to synthetic jet actuators whose orifice length-to-diameter ratio is greater than unity, is proposed in Carpenter et al. (2002). For this type of actuators, an adequate assumption is that the streamlines in the orifice exit are nearly parallel to its axis, thus leading to the following governing equation describing the orifice flow:

$$\rho\frac{\partial v}{\partial t} + \frac{\rho v_O |v_O|}{2h} = -\frac{\partial p}{\partial y} + \mu r\frac{\partial}{\partial r}\left(\frac{1}{r}\frac{\partial v}{\partial r}\right) \tag{5.11}$$

where:

y and r are the axial and radial coordinates, respectively
v is the axial velocity
h is the orifice length
v_O is the velocity at the orifice exit

Though compressibility effects in Equation 5.11 are neglected, the density (ρ) is assumed to be a function of time and taken to be the instantaneous mean of the cavity and external densities. The convective term $\rho v v_y$ is approximated by the second term on the left-hand side of Equation 5.11. Note that this approach is similar to that of the SC model given by Equation (5.10). The flow dynamics in the cavity is ignored, and the pressure there is calculated by means of the perfect gas law. The cavity and external boundary layer flowfields are linked by requiring continuity of velocity and pressure at the orifice exit. Though, no comparisons of the reduced- and full-order models are presented in Carpenter et al. (2002), the proposed ROM is used for predicting the Helmholtz resonance frequency which turns out to be close to the frequency of oscillation of the actuator jet. Parametric studies performed using this ROM have shown that actuators optimized for maximum jet velocity are susceptible to the Helmholtz resonance, which could cause structural problems. Furthermore, the inactive actuator can generate and amplify Tollmien–Schlichting waves, thus triggering transition to turbulence.

5.3.4 Lumped Element Models

Another group of models uses the equivalent circuit representation of the piezoelectric-driven synthetic jet flow (Gallas et al. 2003; Tang et al. 2007). Similar to the other 0-D ROMs outlined in this section, in the lumped element model (LEM) approach, the governing multidimensional flow equations are averaged or "lumped" into a set of coupled ODEs. This averaging is based on the assumption that the wavelength of the actuator diaphragm vibration (λ) is much larger than the actuator characteristic length (L_a). If this assumption holds, then at each instant in time the deviation of spatial distributions of the pressure and velocity fields inside the actuator cavity from their averaged values is proportional to L_a/λ and can therefore be neglected. In LEM, the individual components of the actuator are modeled as elements of an electrical circuit by using conjugate power variables. The linear composite plate theory is used to calculate the net volume velocity of the diaphragm and the distributed vertical velocity. The acoustic mass and resistance in the orifice duct are obtained by using the incompressible fully developed laminar pipe-flow theory. The result is that the lumped element model neglects the compressibility effects inside the orifice duct. For further details on the lumped element modeling, we refer the reader to Chapter 2 of this book, which presents various LEMs and discusses their assumptions and limitations.

5.4 Low-Dimensional Models

A low-dimensional model of realistic synthetic jet actuators, which combines the accuracy and conservation properties of the full numerical simulation methods with the efficiency of the simplified blowing/suction-type boundary conditions, has been introduced and developed in Yamaleev et al. (2005) and Yamaleev and Carpenter (2006).

This class of ROMs is based on the following three observations. First, the size of a realistic actuator is much less than the wavelength of diaphragm vibration. Indeed, the actuator size L_a is determined by the size of a control surface (e.g., wing and blade) and usually of the order of $O(10^{-2})$ m. The wavelength of diaphragm oscillations $\lambda = c_\infty/\omega$ (where ω is the diaphragm frequency, c_∞ is the speed of sound) is of the order of $O(1)$ m for diaphragm frequencies of $O(1)$ kHz. Hence, the characteristic size of realistic actuators is much less than the wavelength of diaphragm oscillations. All changes in the pressure and velocity fields inside the actuator cavity occur on the scale of the wavelength of diaphragm vibration and are, therefore, by a factor of L_a/λ less than the pressure and velocity themselves. Another conclusion that can be drawn from this consideration is that the cavity shape has no significant effect on the actuator performance if $L_a \ll \lambda$. It should be noted, however, that the jet momentum strongly depends on the actuator volume and the orifice duct length (Yamaleev and Carpenter 2006).

The second observation is that vortices ingested inside the actuator cavity have no perceptible impact on the actuator characteristics. One can speculate that because the exterior flow is essentially turbulent, vortices generated in the turbulent boundary layer can be ingested into the actuator cavity, which makes the cavity flowfield substantially multidimensional. Note, however, that the vortex energy and the total energy of gas inside the cavity are proportional to their volumes (see Yamaleev and Carpenter, 2006, for details), that is,

$$\frac{E_{\text{vortex}}}{E_{\text{cavity}}} \approx \frac{d^3}{L_a^3} \ll 1 \tag{5.12}$$

where:

d is the orifice diameter
L_a is the actuator characteristic size

In the above equation, it has been taken into account that the maximum vortex size is bounded by the orifice diameter d. As follows from Equation 5.12, even if the vortex is ingested inside the cavity, its energy is much smaller than the total energy of the system and can therefore be neglected. Assuming that the problem under consideration is well-posed, one can conclude that the negligibly small perturbations in the energy of the system caused by the vortex ingestion result in negligibly small perturbations in the actuator solution.

The third important observation is that the actuator acoustic resonance frequency is nearly independent of the cavity shape. Indeed, because of design constraints, the diaphragm deflection is much less than the actuator characteristic size. Therefore, the volume variation due to diaphragm oscillations is negligibly small as compared with the cavity volume, and the Helmholtz resonance theory can be applied to estimate the actuator resonance frequency, thus leading to

$$\omega_r \sim c_\infty \sqrt{\frac{S_O}{Vh}} \tag{5.13}$$

where:

V is the actuator volume
h is the characteristic orifice duct length
S_O is the orifice cross-sectional area
c_∞ is the speed of sound

From the above equation, it is observed that the acoustic resonance frequency does not depend on the actuator shape, but does depend on the actuator volume V, orifice duct length h, and orifice size d. Equation 5.13 indicates that to quantitatively predict the resonance characteristics of a 3-D actuator, the actuator volume, orifice size, and duct length should be retained in a ROM.

The above-mentioned considerations suggest that the flow inside a realistic 3-D synthetic jet actuator can be accurately described by conservation law equations whose dimensionality is less than the dimensionality of the flow governing equations used for modeling the exterior flowfield. In Yamaleev et al. (2005) and Yamaleev and Carpenter (2006), the quasi-one-dimensional (1-D) conservation law equations are used to simulate the compressible flow inside the actuator cavity, while the diaphragm is treated as a moving boundary. The quasi-1-D unsteady Euler equations in the time-dependent coordinate frame (τ, η) are written in the following conservation law form:

$$\frac{\partial U}{\partial \tau} + \frac{\partial E}{\partial \eta} + G = 0 \tag{5.14}$$

$$U = \frac{A}{J}\begin{bmatrix} \rho \\ \rho v \\ \rho e \end{bmatrix}, \quad E = \frac{A}{J}\begin{bmatrix} \eta_t \rho + \eta_y \rho v \\ \eta_t \rho v + \eta_y(\rho v^2 + p) \\ \eta_t \rho e + \eta_y v(\rho e + p) \end{bmatrix}, \quad G = -\frac{1}{J}\begin{bmatrix} 0 \\ p\dfrac{\partial A}{\partial y} \\ 0 \end{bmatrix}$$

$$y(\eta, \tau) = (\eta - 1)\,[H + \delta \sin(\omega \tau)], J = \frac{1}{y_\eta}$$

where:

A is the cross-sectional area of the quasi-1-D actuator, which is assumed to be independent on time
δ and ω are the diaphragm amplitude and frequency, respectively
H is the mean depth of the quasi-1-D actuator

Since the η coordinate depends on time, a moving mesh technique is applied to solve the quasi-1-D Euler equations. Diaphragm oscillations are forced by varying the position of the diaphragm $y(0, \tau)$ where the impermeable wall boundary condition is imposed. The diaphragm velocity is calculated by differentiating y with respect to time to give $v(0, \tau) = -\delta\omega \cos(\omega\tau)$.

A region near the jet exit is characterized by strong flow separation that cannot be described by the quasi-1-D Euler equations. In Yamaleev et al. (2005) and Yamaleev and Carpenter (2006), this small region near the actuator orifice is modeled by solving the 2-D compressible unsteady Navier–Stokes equations, whereas the flow inside the actuator cavity is described by the quasi-1-D Euler equations (5.14), as shown in Figure 5.3. This approach accurately predicts the interaction of the synthetic jet with the external flow and vorticity generation in the vicinity of the actuator orifice, while drastically reducing the computational cost as compared with the full 2-D Navier–Stokes simulation of the cavity flow.

The low-dimensional actuator model possesses several advantages that make this approach very attractive. First, the quasi-1-D model is fully conservative and provides conservation of mass, momentum, and energy. Second, in contrast to the TBC, the quasi-1-D model does not require any information about the velocity amplitude and profile at the jet exit. Furthermore, the ROM is computationally much more efficient as compared with the 2-D or 3-D Navier–Stokes simulation of the cavity flow. Another important property of the quasi-1-D ROM is its ability to account for the nonlinear, compressibility, and resonance effects inherent in actuator devices, which can be efficiently used for quantitative prediction and optimization of the synthetic jet flows.

In Yamaleev et al. (2005) and Yamaleev and Carpenter (2006), the quasi-1-D reduced-order model has been validated and tested on various realistic synthetic jet actuator

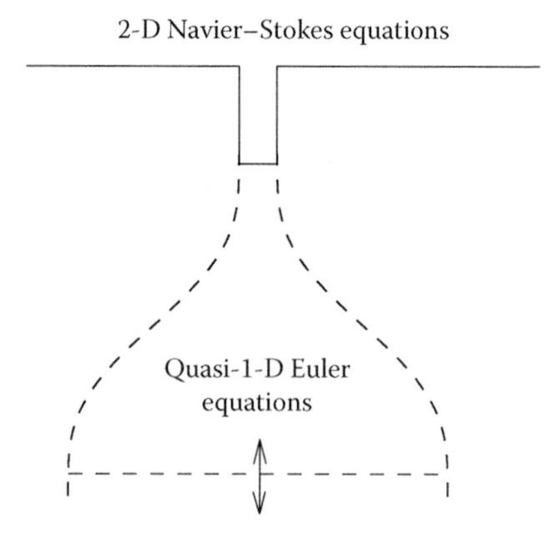

FIGURE 5.3 Schematic of the quasi-1-D ROM. (From Yamaleev, N. K. et al., *AIAA J.*, 43, 357–369, 2005.)

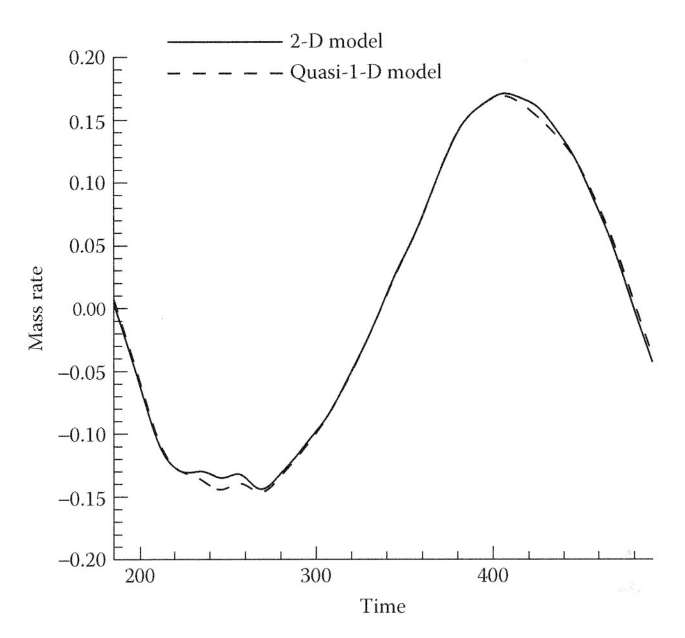

FIGURE 5.4 Mass rate time histories obtained with the quasi-1-D and 2-D Navier–Stokes models for the box actuator. (From Yamaleev, N. K. et al., *AIAA J.*, 43, 357–369, 2005.)

geometries ranging from short-orifice-duct box-type devices to long-orifice-duct narrow-cavity-type actuators. Time histories of the mass rate at the orifice exit, as predicted by the quasi-1-D model and the full 2-D Navier–Stokes simulation, for a $21d \times 20d$ box actuator with a short $1d$ orifice duct are shown in Figure 5.4. Although the quasi-1-D Euler equations do not describe the complex behavior of vortex structures inside the actuator cavity, the agreement between the reduced-order and full 2-D Navier–Stokes models is excellent.

The quasi-1-D ROM has also demonstrated a very good quantitative agreement with the experimental data (Yao et al. 2004). In this experiment performed at NASA Langley Research Center for the computational fluid dynamics (CFD) validation workshop, a synthetic jet in quiescent air is generated by a cavity-pumping actuator that consists of a slot of 0.05 inches wide and 1.4 inches long and a narrow cavity. The actuator diaphragm, 2 inches in diameter, mounted on the side of the cavity was driven at a constant frequency of 444.7 Hz.

Figure 5.5 shows phase-averaged time histories of the vertical velocity near the slot exit, obtained with the quasi-1-D model and measured using the particle image velocimetry (PIV) technique (Yao et al. 2004). As follows from this comparison, the quasi-1-D ROM quantitatively predicts the major local characteristics of realistic 3-D synthetic jet flows.

The accuracy of the quasi-1-D ROM is quite remarkable, because it is capable of predicting not only integral and local quantities, but also the entire exterior flowfield.

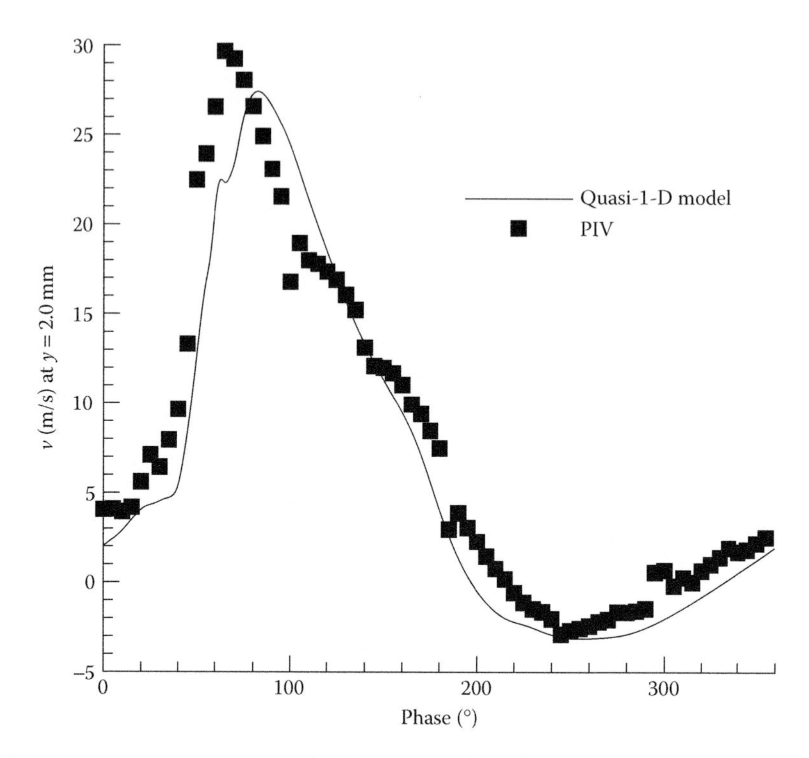

FIGURE 5.5 Comparison of the quasi-1-D model and the PIV experimental data. (From Yao, C. S. et al., Synthetic jet in quiescent air. *Proceedings of Langley Research Center Workshop on CFD Validation of Synthetic Jets and Turbulent Separation Control*, Williamsburg, VA, March 29–31, 2004; Yamaleev, N. K. and M. H. Carpenter, *AIAA J.*, 44, 208–216, 2006.)

Figure 5.6 shows phase-averaged v-velocity contours predicted by the full 2-D Navier–Stokes equations and the quasi-1-D model for the interaction of the same synthetic jet actuator used in the CFD validation workshop experiment and a crossflow (Yamaleev and Carpenter 2006).

Although the quasi-1-D Euler equations do not take into account the viscous losses and flow separation inside the actuator cavity, the ROM provides very good accuracy in the exterior flowfield. The high accuracy of the quasi-1-D model can be explained by its conservation properties and the detailed simulation of the multidimensional nonlinear and viscous effects near the slot exit.

Though the viscous terms have not been included in Equation 5.14, they can easily be incorporated into the quasi-1-D ROM by properly averaging the 2-D/3-D Navier–Stokes equations. Another obvious extension of this ROM is to use the 2-D unsteady Navier–Stokes equations for the prediction of the 3-D flow inside the actuator cavity. Similar to the quasi-1-D counterpart, the computational cost of the 2-D ROM is negligibly small as compared with that of the full 3-D Navier–Stokes simulation of the cavity flow. This approach can be used at final stages of the design process, when resolution of more detailed flow features inside the actuator cavity is desirable.

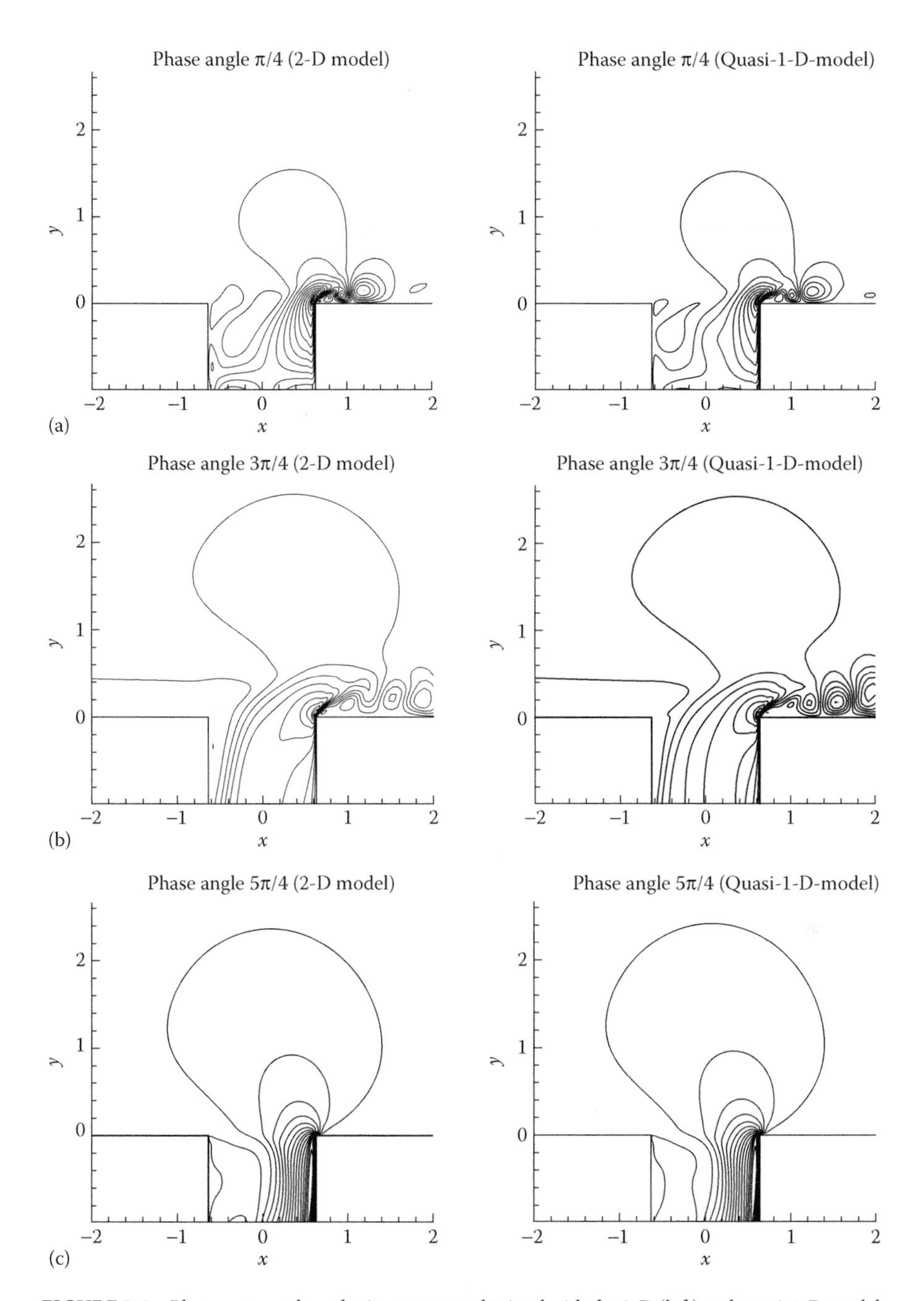

FIGURE 5.6 Phase-averaged v-velocity contours obtained with the 2-D (left) and quasi-1-D models at phase angles (a) $\pi/4$, (b) $3\pi/4$, and (c) $5\pi/4$. (From Yamaleev, N. K. and M. H. Carpenter, *AIAA J.*, 44, 208–216, 2006.)

5.5 Input/Output-Based Models

5.5.1 ROM Based on Proper Orthogonal Decomposition

The proper orthogonal decomposition (POD) is a popular order reduction technique for various unsteady turbulent flows (Berkooz et al. 1993). In Addington et al. (2002) and Rediniotis et al. (2002), the POD methodology is used for derivation of a ROM of 2-D incompressible synthetic jet flows. The POD basis functions associated with the velocity vector field are constructed using the method of snapshots (Sirovich 1987), which can be formulated as follows. For a given collection of M snapshots of the velocity field $\{v^1, \ldots, v^M\}$, find a subspace of fixed, much smaller dimension, which is optimal in the sense that the error in the projection onto the subspace is minimized in the L_2 sense. It can be shown (Berkooz et al. 1993; Sirovich 1987) that this constrained optimization problem is reduced to the following eigenvalue problem:

$$C\phi = \lambda\phi$$

where:
C is a correlation matrix defined as

$$c_{ij} = \frac{1}{M}\int_{\Omega}(v^i - v)^T(v^j - v)d\Omega;\ i, j = 1, \ldots, M$$

The mean velocity vector v is given by

$$v = \frac{1}{T}\int_0^T v dt$$

where:
T is the period of diaphragm oscillation

The correlation matrix C is the positive semidefinite real matrix. Therefore, it has a complete set of orthogonal eigenvectors $\{\phi_1, \ldots, \phi_M\}$ with the corresponding eigenvalues $\lambda_1 \geq \ldots \geq \lambda_M \geq 0$ that are enumerated in decreasing order. The POD basis functions are obtained as a linear combination of the snapshot basis:

$$\Phi_k = \sum_{k=1}^{M} \phi_i^k \left(v^i - v\right) \qquad (5.15)$$

where:
ϕ_i^k is the ith element of the kth eigenvector ϕ^k corresponding to eigenvalue λ_k

As follows from Equation 5.15, the POD basis functions are divergence free, because the snapshots of the velocity field satisfy the continuity equation $\nabla \cdot v = 0$. For synthetic jet flows considered in Addington et al. (2002) and Rediniotis et al. (2002), the eigenvalues decay very rapidly, and most of the flow energy is captured in the first m POD modes, that is,

$$\sum_{i=1}^{m} \lambda_i \approx \sum_{i=1}^{M} \lambda_i$$

where:
$m \ll M$

As a result, order reduction can be achieved by approximating the velocity field using only the first m POD modes:

$$v(x, y, t) \approx v(x, y) + \sum_{i=1}^{m} a_i(t)\Phi_i(x, y) \tag{5.16}$$

Note that the same time-dependent amplitude $a_i(t)$ is used for both components of the velocity vector. Substituting Equation 5.16 in the momentum equations of the 2-D incompressible Navier–Stokes equations and using the Galerkin projection of these equations onto the POD basis, the ODEs for the time-varying amplitudes $a_i(t)$, $i = 1, \ldots, m$ can be derived as follows:

$$St\frac{da_i}{dt} = B_i + \sum_{j=1}^{m} F_{ij}a_j + \sum_{j=1}^{m}\sum_{k=1}^{m} D_{ijk}a_ja_k \tag{5.17}$$

where:
St is the Strouhal number

The coefficients B_i, F_{ij}, and D_{ijk} are given by

$$B_i = -[\Phi_i, (v \cdot \nabla)v] - \frac{1}{Re}(\nabla\Phi_i, \nabla v) + \frac{1}{Re}(\Phi_i, \nabla v)$$

$$F_{ij} = -[\Phi_i, (v \cdot \nabla)\Phi_j] - [\Phi_i, (\Phi_j \cdot \nabla)v] - \frac{1}{Re}\left(\nabla\Phi_i, \nabla\Phi_j\right) + \frac{1}{Re}\left(\Phi_i \nabla\Phi_j\right)$$

$$D_{ijk} = -\left[\Phi_i, (\Phi_j \cdot \nabla)\Phi_k\right]$$

In Equation 5.17, there is no contribution from the pressure gradient terms, since the POD basis functions given by Equation 5.16 are divergence free and homogeneous at the boundaries. Note that the approximation of the velocity vector field constructed this way does not necessarily satisfy the energy equation. The spatial operators in Equation 5.17 have to be approximated numerically, and some care must be exercised in discretizing these terms. It is well known that the governing ODEs arising in POD-based models may become unstable and require additional stabilization. This problem can be overcome by constructing a POD-based ROM that approximates the spatial derivative terms by utilizing the same numerical scheme used for discretization of the original Navier–Stokes equations (Pathak and Yamaleev 2011).

Equation 5.17 represents a POD-based ROM of the discretized 2-D incompressible Navier–Stokes equations. The original system consisting of $O(4N_g)$ equations is reduced to a system of $2m$-coupled ODEs, where the number of POD modes m is much smaller than the total number of grid points N_g. In Rediniotis et al. (2002), this POD-based ROM has been validated for a 2-D synthetic jet in the quiescent flow at the Reynolds number of $Re_d = 228$ and the Strouhal number of $St = 0.175$, which are based on the orifice diameter and the peak jet velocity. For low frequencies of 10 Hz considered in Rediniotis et al. (2002), the first five POD modes capture over 99% of the entire energy in the system. This can be explained by the fact that for this relatively low Strouhal number, the majority of the flow energy is contained in the two counterrotating vortices and the jet flow, which is exactly the flow pattern that is observed in the first two POD eigenmodes. The numerical results presented in Rediniotis et al. (2002) suggest that this class of synthetic jet flows is amenable to the POD order reduction technique, which provides good qualitative prediction of the velocity field in the neighborhood of the jet exit.

5.5.2 ROM Based on the Volterra Series

In Hofmann and Herbert (1998), a ROM of a membrane actuator is constructed using the Volterra series formulation. To reproduce the linear flow responses to disturbances generated by a membrane actuator in the incompressible laminar boundary layer on a flat plate, the corresponding flow governing equations are considered as a dynamical system. The system input is an excitation introduced by the actuator, and the output is unsteady streamwise velocity disturbances that develop in response to the actuation. Based on the Volterra theory, the linear input–output relation of the crossflow/actuator system is evaluated using the classical Duhamel convolution integral:

$$u(x,t) = F(0)u_0(x,t) + \int_0^t u_0(x,t-\tau)\frac{dF(\tau)}{d\tau}d\tau \qquad (5.18)$$

where:

$u_0(x,x)$ is the streamwise velocity obtained in response to a unit stepwise forcing
$F(t)$ is a prescribed time-dependent forcing function

The flow responses are determined by using an incompressible Navier–Stokes code developed in Liu et al. (1993), which simulates the actuator disturbances as nonhomogeneous boundary conditions similar to Equations 5.2 and 5.5.

According to the Volterra kernel identification procedure, $u_0(x, t)$ in Equation 5.18 is a flow response generated by step function excitation. Since the instantaneous step function motion of the actuator membrane cannot be numerically simulated, the membrane deflection is linearly ramped up over two time steps to its maximum value and thereafter kept unchanged. To compute $u_0(x, t)$, the discretized Navier–Stokes equations are integrated using a time step associated with the timescale of the most unstable mode generated by the actuation. The convolution integral given by Equation 5.18 is approximated as

$$u(x,t) = F(0)u_0(x,t) + \sum_{n=1}^{N} u_0(x, t - \tau_n)\,[F(\tau_n) - F(\tau_{n-1})] \tag{5.19}$$

where:

τ_n represents nondimensional time at the nth step

The above ROM provides significant reduction in the computational time, because the flow response to an arbitrary diaphragm deflection $F(t)$ can be reproduced using Equation 5.19, which requires a single computation of the unit step function response $[u_0(x,t)]$. Note that Equation 5.19 is linear with respect to the forcing function $[F(t)]$. Therefore, the amplitude of actuator diaphragm deflection should be sufficiently small, so that velocity disturbances generated by the actuator are much smaller than the mean flow velocity, and the nonlinear effects can be neglected.

Numerical experiments presented in Hofmann and Herbert (1998) have shown that the above ROM can accurately reproduce the streamwise velocity disturbances generated by small-amplitude membrane deflections. The interaction of a membrane actuator with a crossflow would normally generate streamwise and normal velocity disturbances. However, only the streamwise velocity disturbances are considered in Hofmann and Herbert (1998). In principle, the quadratic and higher-order terms can be retained in the Volterra series to simulate the nonlinear effects inherent in synthetic jet flows. Note, however, that the kernel identification procedure in this case becomes much more complicated and computationally expensive.

5.5.3 Nonlinear System Identification Model

An input/output-based ROM that is capable of taking into account the nonlinear effects inherent in synthetic jet flows has been developed by Kim et al. (2005). This ROM is derived for the interaction of a synthetic jet with a crossflow by using a polynomial nonlinear autoregressive moving average model with exogenous inputs (NARMAX). The incompressible boundary layer flow on a flat plate is solved using an *hp* spectral element method. The synthetic jet in the full-order model is simulated as the TBC (Equations 5.2 and 5.5) with the velocity profile given by $V(x) = \sin[\pi(1 - x/d)]$, where d is the actuator slot width.

To reduce the computational cost associated with feedback control of flow separation via synthetic jet actuation, the jet/crossflow interaction in the neighborhood of the orifice exit is represented as a dynamic model. The inner region of the boundary layer on the control surface near the actuator slot is treated as a domain for the system identification. The polynomial NARMAX method is applied to the numerical results obtained using the full Navier–Stokes simulation to identify the input/output ROM. In this model, the actuation frequency is taken as an input, and pressure values at two sections located $1d$ upstream and $6d$ downstream of the orifice exit are used as system outputs. Since for the synthetic jet–crossflow interaction problem, the freestream velocity is the most significant variable affecting the model uncertainty; each model parameter vector (Θ_i) in the NARMAX model is computed via CFD simulation and system identification for different fixed freestream velocities (u_i) and gathered into a set $\{\Theta\}$. The discrete nonlinear model is then constructed as follows:

$$p(k) = \sum_{i=1}^{n_\theta} \hat{\theta}_i r_i(k) + \xi(k) = R^T(k)\Theta + \xi(k) \tag{5.20}$$

where:

p is the measured output
r_i is the regressor term
$\hat{\theta}$ is the model coefficient
$R = \{r_1(k), \ldots, r_{n_\theta}(k)\}$
$\Theta = \{\theta_1, \ldots, \theta_{n_\theta}\}$
ξ is the residual vector

The regressor term is given in the following polynomial form:

$$\begin{aligned} &r_i(k) = \prod_{j=1}^{s} p(k - n_{p_j}) \prod_{l=1}^{q} \omega(k - n_{\omega_l}) \\ &r_1(k) = 1 \\ &i = 2, \ldots, n, s, q \geq 0, 1 \leq p + q \leq M \\ &1 \leq n_{p_j} \leq n_p, 1 \leq n_{\omega_l} \leq n_\omega \end{aligned} \tag{5.21}$$

where:

ω denotes the input actuation frequency

The permutations of input/output pairings in Equation 5.21 generate a very large number of possible regressor terms. To identify the most significant terms in Equation 5.20 and evaluate the corresponding model parameter vector (Θ), the orthogonal least-square algorithm is used, which gives the following auxiliary model:

$$p(k) = \hat{p}(k) + \xi(k) = \sum_{i=1}^{n_\theta} \hat{g}_i v_i(k) + \xi(k) \tag{5.22}$$

where:

$\hat{g}_1, \ldots, \hat{g}_{n_\Theta}$ are orthogonal and calculated by the Gram–Schmidt method

The decision whether or not to retain each regressor term in Equation 5.22 is made based on the error reduction ratio (ε), which is given by

$$\varepsilon_i = \frac{\hat{g}_i^2 \sum_{k=1}^{N} v_i^2(k)}{\sum_{k=1}^{N} p_i^2(k)} \tag{5.23}$$

The ith regressor term is retained if the corresponding ε_i value is greater than a user-defined threshold. The above-mentioned nonlinear model captures the flow response to the synthetic jet actuation as well as the influence of uncertain flow conditions. For different freestream velocities, the model parameters are computed while keeping the regressors fixed. As a result, the effect of synthetic jets under a varying freestream velocity is identified as the NARMAX model with parameter uncertainty. The numerical results show that for small variations of the freestream velocity (within 25%), the pressure at the given locations near the orifice exit can be predicted by the NARMAX model with accuracy of 7%–8%.

5.5.4 Neural Network-Based Model

Another input/output approach based on lumped deterministic source terms (LDSTs) trained by a neural network is developed for simulation of synthetic jets in Pes et al. (2002) and Filz et al. (2003). The key idea of this technique is to model synthetic jet flows in a time-averaged sense by adding source terms to the right-hand side of the Navier–Stokes equations instead of resolving the actual unsteadiness. The source terms are derived from unsteady simulations of the synthetic jet geometry by using the LDST technique.

The LDST method is based on the observation that the solution of the steady-state Navier–Stokes equations is not the same as a time average of the corresponding unsteady solution. Indeed, the unsteady flow solution ($\mathbf{Q}$) can be represented as

$$\mathbf{Q} = \bar{\mathbf{Q}} + \tilde{\mathbf{Q}} + \mathbf{Q}' \tag{5.24}$$

where:

$\bar{\mathbf{Q}}$ is the time-averaged value of $\mathbf{Q}$ over a large time interval T

$\tilde{\mathbf{Q}}$ is the phase-averaged periodic fluctuations

$\mathbf{Q}'$ is the stochastic fluctuations

Substituting Equation 5.24 into the unsteady governing equations and averaging all the terms in time yield

$$\boldsymbol{R}(\mathbf{Q}) + \boldsymbol{S} = 0 \tag{5.25}$$

where:

$\boldsymbol{R}(\mathbf{Q})$ is the spatial residual operator

$\boldsymbol{S}$ is the LDST that takes into account the unsteady effects

Note that in Equation 5.25, the stochastic fluctuations $\mathbf{Q}'$ are modeled via the Reynolds stress terms and the linear $\tilde{\mathbf{Q}}$ terms vanish, so that only the steady-state residual operator acting on the time-averaged solution plus the time average of all the higher-order perturbation terms are left. The presence of the LDST term in Equation 5.25 indicates that the solution of the steady-state equations $\boldsymbol{R}(\mathbf{Q}) = 0$ and the time-averaged solution $\bar{\mathbf{Q}}$ are not equal to each other. Another interesting observation is that the LDST term corresponding to the continuity equation is identically equal to zero, that is, $\boldsymbol{S}_1 = 0$, because this equation is linear in the conservative variables. It implies that there is no net mass injection into the system, which is an inherent feature of synthetic jets.

For optimal flow control problems, the solution of the unsteady Navier–Stokes equations and consequently the LDST terms have to be updated at each optimization iteration. If the LDST terms are continuous functions of the input parameters, then neural networks can be used to reduce the computational cost associated with evaluating the source terms. In Pes et al. (2002) and Filz et al. (2003), for given amplitude and frequency of diaphragm oscillations, a neural network is constructed using the Levenberg–Marquard algorithm (Levenberg 1944; Marquardt 1963). This neural network can provide nonlinear mappings from input parameters (the geometrical location, diaphragm frequency and amplitude, and Mach number) to output parameters (the momentum coefficient, momentum thickness, displacement thickness, and shape factor) when the network has a sufficiently large database of prior examples from which to learn. Data consisting of each individual source term at each grid node for all the test cases are stored in a series of vectors. These vectors are then randomly reordered to achieve efficient training and then divided into training data and testing data. The latter is used to set a stopping criterion, so that the network does not waste the computational time by continued training after the testing error reached its minimum. Once the training process is stopped, the weights corresponding to the minimum testing error are used to generate the LDST terms.

Unsteady simulations in Pes et al. (2002) and Filz et al. (2003) are performed using the commercial CFD package CFD++ to characterize the behavior of an isolated jet for different diaphragm frequencies and amplitudes corresponding to the jet velocities in the range of 0.01–20 m/s as well as for several freestream Mach numbers varying from 0.15 to 0.6. It has been shown that the neural network-based source terms quite accurately reproduce the time-averaged physics of the synthetic jet and global boundary layer quantities such as the time-averaged momentum thickness.

The main drawback of all input/output-based ROMs outlined earlier is that they require the output quantities that can be obtained either by solving the multidimensional unsteady Navier–Stokes equations or by measuring them in an experiment. In either case, the full-order model solution has to be available over the entire time interval of interest to construct the input/output-based ROM. In principle, this class of ROMs can be used for a broader range of input parameters over which they have been constructed. Note, however, that the ROM accuracy may rapidly deteriorate for those values of the input parameters that are outside of the trust region. Of course, one can "retrain" the ROM for the extended range of the input parameters, which again requires solutions of the full-order model equations, thus increasing the computational cost. An area of applications, for which the input/output-based ROMs are particularly attractive, is optimization of synthetic jet flows. If a ROM contains enough information to accurately approximate the synthetic jet dynamics encountered throughout the entire optimization process, so that the number of ROM updates is much less than the total number of optimization iterations, the substantial computational cost savings are possible, and the ROM can be used as an efficient tool for optimization studies.

5.6 Conclusions

Reduced-order modeling of synthetic jets is a very challenging problem, mainly due to the complex physics involved and lack of confidence in computational methods for such time-dependent, multiscale flows. Designers typically model actuators with different levels of fidelity depending on the acceptable level of error in each circumstance. If crude properties of the actuator (e.g., peak mass rate and frequency) are sufficient for some preliminary designs, the simplified models based on the TBC or 0-D models are desirable because of their efficiency. If more detailed information (e.g., Helmholtz and structural resonances, nonlinear and compressibility effects, flow separation in jet exit region, and multiple actuator interactions) is needed for design and optimization studies, more sophisticated ROMs such as the low-dimensional and input/output-based models are preferred over the simplified counterparts. The decision on which reduced-order model is appropriate in each particular case should be based on the ratio of the magnitude of error committed by an actuator model to the computational cost involved. As follows from the literature review presented in this chapter, reduced-order modeling of synthetic jet flows is a very active area of research which opens new avenues for maximizing the performance of actuator devices and designing efficient active flow control systems.

References

Addington, G., J. Hall, and J. Myatt. 2002. Reduced order modeling applied to reactive flow control, *The AIAA Atmospheric Flight Mechanics Conference and Exhibit, AIAA Paper* 2002-4807, Monterey, CA, August 5–8.

Aram, E., R. Mittal, and L. Cattafesta. 2010. Simple representations of zero-net mass-flux jets in grazing flow for flow-control simulations. *International Journal of Flow Control*, 2(2): 109–125.

Berkooz, G., P. Holmes, J. L. Lumley, L. Sirovich. 1993. The proper orthogonal decomposition in the analysis of turbulent flows. *Annual Review of Fluid Mechanics*, 25: 539–575.

Carlson, H. A. and J. L. Lumley. 1996. Flow over an obstacle emerging from the wall of a channel. *AIAA Journal*, 34: 924–931.

Carpenter, P. W., D. A. Lockerby, and C. Davies. 2002. Numerical simulation of the interaction of Microactuators and boundary layers. *AIAA Journal*, 40(1): 67–73.

Chen, Y., S. Liang, K. Aung, A. Glezer, and J. Jagoda. 1999. Enhanced mixing in a simulated combustor using synthetic jet actuators, *The 37th AIAA Aerospace Sciences Meeting and Exhibit, AIAA Paper* 99-0449, Reno, NV, January 11–14.

Crook, A., A. M. Sadri, and N. J. Wood. 1999. The development and implementation of synthetic jets for the control of separated flow, *The 17th Applied Aerodynamics Conference, AIAA Paper* 99-3173, Norfolk, VA, June 28–July 1.

Donovan, J. F., L. D. Kral, and A. W. Cary. 1998. Active flow control applied to an airfoil, *The 36th AIAA Aerospace Sciences Meeting and Exhibit, AIAA Paper* 98-0210, Reno, NV, January 12–15.

Filz, C., D. Lee, P. Orkwis, and M. Turner. 2003. Modeling of two dimensional directed synthetic jets using neural network-based deterministic source terms. *The 33rd AIAA Fluid Dynamics Conference and Exhibit*, Orlando, FL, June 23–26.

Gallas, Q., J. Mathew, A. Kaysap, R. Holman, T. Nishida, B. Carrol, M. Sheplak, and L. Cattafesta. 2003. Lumped element modeling of piezoelectric-driven synthetic jet actuators. *AIAA Journal*, 41(2): 240–247.

Guo, D. and A. W. Gary. 2001. Vectoring control of a primary jet with synthetic jets, *The 39th Aerospace Sciences Meeting and Exhibit, AIAA Paper* 2001-738, Reno, NV, January 8–11.

Hassan, A. W. and R. D. JanakiRam. 1998. Effects of zero-mass "synthetic" jets on the aerodynamics of the NACA-0012 airfoil. *Journal of the American Helicopter Society*, 43: 303–311.

Hofmann, L. M. and T. Herbert. 1997. Disturbances produced by motion of an actuator. *Physics of Fluids*, 9: 3727–3732.

Hofmann, L. M. and T. Herbert. 1998. Reproducing the flow response to actuator motion. *Journal of Computational Physics*, 142: 264–268.

Joslin, R. D., J. T. Lachowicz, and C. S. Yao. 1998. DNS of flow induced by a multi-flow actuator. *Proceedings of the ASME Fluids Engineering Conference, Forum on Control of Transitional and Turbulent Flows*, Washington, DC, June 21–25.

Kim, K., A. L. Beskok, and S. Jayasuriya. 2005. Nonlinear system identification for the interaction of synthetic jets with a boundary layer. *Proceedings of the 2005 American Control Conference*, June 8–10, Portland, OR: American Automatic Control Council (AACC), pp. 1313–1318.

Kral, L. D., J. F. Donovan, A. B. Cain, and A. W. Cary. 1997. Numerical simulation of synthetic jet actuators, *The 4th Shear Flow Control Conference, AIAA Paper* 97-1824, Snowmass, CO, June 29–July 2.

Lee, C. Y. and D. B. Goldstein. 2002. Two-dimensional synthetic jet simulation. *AIAA Journal*, 40(3): 510–516.

Levenberg, K. 1994. A method for the solution of certain non-linear problems in least squares. *Quarterly Applied Mathematics*, 2: 164–168.

Lin, H. and C. C. Chieng. 1999. Computations of compressible synthetic jet flows using multigrid/dual time stepping algorithm, *The 17th Applied Aerodynamics Conference, AIAA Paper* 99-3114, Norfolk, VA, June 28–30.

Liu, C., Z. Liu, and S. McCormick. 1993. Multigrid methods for flow transition in three-dimensional boundary layers with surface roughness. NASA Contractor Report 4540, National Aeronautics and Space Administration, Las Cruces, NM.

Mahalingam, R. and A. Glezer. 2005. Design and thermal characteristics of a synthetic jet ejector heat sink. *Journal of Electronic Packaging*, 127(2): 172–177.

Marquardt, D. W. 1963. An algorithm for least squares estimation of nonlinear parameters. *SIAM Journal on Applied Mathematics*, 11(2): 431–441.

Nayfeh, A. H. 1979. *Nonlinear Oscillations*. New York: John Wiley & Sons.

Pathak, K. and N. K. Yamaleev. 2011. POD-based reduced-order model for arbitrary Mach number flows. *The 6th Theoretical Fluid Mechanics AIAA Conference*, Hawaii, June 27–30.

Pes, M., B. Lukovic, P. Orkwis, and M. Turner. 2002. Modeling of two dimensional synthetic jet unsteadiness using neural network-based deterministic source terms. *The 32nd AIAA Fluid Dynamics Conference and Exhibit*, St. Louis, MS, June 24–26.

Raju, R., E. Aram, R. Mittal, and L. Cattafesta. 2009. Simple models of zero-net mass-flux jets for flow control simulations. *International Journal of Flow Control*, 1(3): 179–196.

Rathnasingham, R. and K. S. Breuer. 1997a. Coupled fluid-structural characteristics of actuators for flow control. *AIAA Journal*, 35(5): 832–837.

Rathnasingham, R. and K. S. Breuer. 1997b. System identification and control of a turbulent boundary layer. *Physics of Fluids*, 9(7): 1867–1869.

Rediniotis, O. K., J. Ko, and A. J. Kurdila. 2002. Reduced order nonlinear Navier–Stokes models for synthetic jets. *Journal of Fluids Engineering*, 124: 433–443.

Rizzetta, D. P., M. P. Visbal, and M. J. Stanek. 1999. Numerical investigation of synthetic jet flowfields. *AIAA Journal*, 37(8): 919–927.

Seifert, A., A. Darabi, and I. Wygnanski. 1996. Delay of airfoil stall by periodic excitation. *AIAA Journal*, 33(4): 691–707.

Seifert, A., T. Bachar, D. Koss, M. Shepshelovich, and I. Wyganski. 1993. Oscillatory blowing: A toll to delay boundary-layer separation. *AIAA Journal*, 31(11): 2052–2060.

Sirovich, L. 1987. Turbulence and the dynamics of coherent structures. *Quarterly of Applied Mathematics*, 45(3): 561–590.

Smith, D., M. Amitay, V. Kibens, D. Parekh, and A. Glezer. 1998. Modification of lifting body aerodynamics using synthetic jet actuators, *The 36th AIAA Aerospace Sciences Meeting and Exhibit, AIAA Paper* 98-0209, Reno, NV, January 12–15.

Tang, H., S. Zhong, M. Jabbal, L. Garcillan, F. Guo, N. Wood, and C. Warsop. 2007. Towards the design of synthetic-jet actuators for full-scale flight conditions—Part 2: Low-dimensional performance prediction models and actuator design method. *Flow Turbulence Combustion*, 78: 309–329.

Yamaleev, N. K. and M. H. Carpenter. 2006. Quasi-one-dimensional model for realistic three-dimensional synthetic jet actuators. *AIAA Journal*, 44(2): 208–216.

Yamaleev, N. K., M. H. Carpenter, and F. Ferguson. 2005. Reduced-order model for efficient simulation of synthetic jet actuators. *AIAA Journal*, 43(2): 357–369.

Yao, C. S., F. J. Chen, D. Neuhart, and J. Harris. 2004. Synthetic jet in quiescent air. *Proceedings of Langley Research Center Workshop on CFD Validation of Synthetic Jets and Turbulent Separation Control*, Williamsburg, VA, March 29–31.

Applications

6

Separation Control

Michael Amitay
Rensselaer Polytechnic Institute

John Farnsworth
Rensselaer Polytechnic Institute

6.1 Introduction

The topic of separation control has been widely investigated throughout the flow control community over the years, because it has provided one of the most clear and profound examples of where fluidic actuation can have a dramatic effect on altering the natural (i.e., unforced) aerodynamics of a flow field. This is in part due to the nature of separated flows, whose naturally occurring instabilities make them very receptive to small disturbances, and thus an ideal test bed for implementing fluidic control. In the last decade, several studies on synthetic jet applications have demonstrated that flow separation can be mitigated or even suppressed altogether (Amitay et al. 1999; Crook et al. 1999; He et al. 2001; Amitay and Glezer 2002b). Synthetic jets have also been used for separation control in inlet ducts (Amitay et al. 2002). It has also been demonstrated on unmanned aerial vehicle (UAV) (Amitay et al. 2003; Parekh et al. 2003), jet vectoring (Smith and Glezer 2002) as well as for flight control on scaled models (Ciuryla et al. 2007). More recently, it has been used for vibration suppression in wind turbines by Maldonado et al. (2009).

Clearly, with the broad application of synthetic jet-based separation control, it is a topic that deserves much attention. This chapter is not intended to be a review of all these application areas, but rather a discussion of the broader implementation of synthetic jet-based separation control. As a result, an extensive discussion on the operational considerations of synthetic jet actuators for separation control is first discussed in Section 6.2. Specifically, the role that the synthetic jet actuation frequency plays and the proper quantification and calibration of the synthetic jet amplitude are both thoroughly discussed. Section 6.3 then presents a collection of experimental examples of synthetic jet-based separation control. The examples are presented in a progressive manner, first starting with conventional steady two-dimensional (2D) geometries before incorporating unsteady or three-dimensional (3D) effects. In the examples, separation control on two widely studied canonical aerodynamic bodies is presented: a circular cylinder and an airfoil wing surface. The final example progresses out of the study of the fundamental flow physics and into an engineering application where synthetic jet-based separation control is demonstrated on a wind tunnel scale model of a conventional Cessna 182 Aircraft for stall control. A summary of this chapter is presented in Section 6.4, which includes a discussion on the future direction that synthetic jet-based separation control is headed.

6.2 Active Flow Control Techniques

The major goal in aerodynamics is to improve vehicle performance over a wide range of operating conditions. This can be achieved by optimizing the shape or by using passive and/or active flow control techniques. Active flow control techniques based on fluidic actuators have been widely studied and used due to their dynamic applicability over a broad range of flow conditions along with their efficiency (e.g., separation mitigation at high angles of attack and virtual aeroshaping at low angles of attack). However, to be truly successful and efficient, one must understand the pertinent parameters associated with controlling a particular flow field.

In controlling separated flows, it has been widely shown that the location of actuation is drastically important. Specifically to control or mitigate flow separation, the actuation needs to be located at or just upstream from the point where the flow naturally separates to be optimal. This dependence of location is true with all separation control devices, passive and active. The primary added benefit of active control techniques is that their operational frequency and amplitude can be tailored for a particular flow situation, meaning that they can have a broader range of operation, reducing the dependence on actuation location. In addition even when they are optimally located, they can produce dramatically larger effects compared with passive devices, due to the fact that their amplitude can be incrementally adjusted and their frequency of operation can be tuned. As a result, the importance and quantification of the operational frequency and amplitude are thoroughly discussed for active flow control techniques and more specifically for synthetic jets in Sections 6.2.1 through 6.2.3.

6.2.1 Effect of Actuation Frequency

A significant benefit of active flow control techniques, and specifically synthetic jets, is that they can be operated at various frequencies. As a result, by taking advantage of the natural flow field's receptivity to particular harmonic frequencies, unsteady active devices can dramatically alter the flow as compared to steady devices. One approach is to couple the actuation frequency to instabilities inherent to separated flows and thus alter the global flow field by modifying the large-scale vortical structures (Oster and Wygnanski 1982; Ho and Huerre 1984; Roberts 1985; Wygnanski 2000). In this approach, the time period of actuation scales with the advection time through the length of the flow domain downstream of the separation (as measured by the reduced or nondimensional frequency, F^+). In such an approach, control input is effective within a limited spatial domain immediately upstream of separation; however, when the flow is not separated (e.g., at low angles of attack), the efficacy of this approach is negligible (Seifert et al. 1993).

A different approach can be utilized where fluidic actuators are driven at much larger frequencies than characteristic flow frequencies (such as synthetic jets; Glezer and Amitay 2002). This approach (Amitay and Glezer 2002a; Glezer et al. 2005) allows more control through the modification of the apparent aerodynamic shape of the lifting surfaces (Amitay et al. 1997, 2001b). This approach does not necessarily rely on coupling the actuation frequency to global flow instabilities and therefore can be applied at various spatial locations and over a broader range of flow conditions (Amitay et al. 2001a; Glezer and Amitay 2002). Furthermore, it can accommodate a broader band of control algorithms where more complex actuation waveforms can be used (the pulse modulation technique; Amitay and Glezer 2006). The application of excitation at both low and high frequencies, and combinations thereof, is more thoroughly discussed in Sections 6.2.1.1 through 6.2.1.3.

6.2.1.1 Low Frequency (Actuating at the Frequency of the Separating Shear Layer)

Active manipulation of separated flows over lifting surfaces at moderate and high angles of attack, to achieve complete or partial flow reattachment with the objective of improving the aerodynamic performance and extending the flight envelope, which relies on the excitation of naturally unstable modes, has been the focus of a number of investigations since the early eighties. Coandă-like reattachment is normally effected by exploiting the receptivity of the separating shear layer to external excitation, which affects the evolution of the ensuing vortical structures and their interactions with the flow boundary. Active flow control schemes that rely on the instability of the separating shear layer were demonstrated in a number of earlier investigations of separated flow (Ahuja and Burrin 1984; Huang et al. 1987; Hsiao et al. 1990; Williams et al. 1991; Seifert et al. 1996).

Separation control by means of internal acoustic excitation (Huang et al. 1987) has typically employed an acoustically driven cavity within the airfoil in which (normally time-harmonic) acoustic excitation is applied through a spanwise slot upstream of separation (typically near the leading edge of the airfoil). The work of Chang et al. (1992) confirmed the earlier results of Hsiao et al. (1990): at low excitation levels, excitation at or near the unstable frequency of the separating shear layer ($St = f_{act} \cdot c/U_\infty = 2$, where f_{act} is the actuation frequency, c is the chord, and U_∞ is the freestream velocity) can lead to a 50% increase in post-stall lift. A similar approach for the coupling of internal forcing to the predominant instabilities of the separating shear layer was also employed in the experiments of Seifert et al. (1996). These authors used unsteady jet blowing over several airfoil models to achieve various degrees of separation control by employing dimensionless actuation frequencies $St \sim O(1)$ (instead of St, these authors chose to denote the dimensionless actuation frequency as F^+; in their work, the characteristic length scale x_s is the streamwise extent of the separated flow domain). However, an important contribution in the work of Chang et al. (1992) was the demonstration that the application of acoustic excitation at levels that are somewhat higher than those of their baseline experiments resulted in effective control of separation over a broad range of excitation frequencies (up to $St = 20$) that far exceed the unstable frequency of the separating shear layer.

6.2.1.2 High Frequency (An Order of Magnitude above the Natural Shedding Frequency)

A different approach for separation control using synthetic jets is used when these fluidic actuators are driven at much larger frequencies than the characteristic flow frequencies. Smith et al. (1998) and Amitay et al. (2001b) demonstrated the utility of synthetic jet actuators for the suppression of separation over an unconventional airfoil at moderate Reynolds numbers (up to 10^6), resulting in a dramatic increase in lift and decrease in pressure drag. The jets were operated at dimensionless frequencies that are an order of magnitude higher than the shedding frequency of the airfoil [i.e., $F^+ \sim O(10)$], and because they are zero net mass flux in nature, their interaction with the crossflow leads to local modification of the apparent shape of the flow surface. Full or partial reattachment including the controlled formation of a quasi-steady recirculating region can be controlled by the streamwise location and the strength of the jets. The excitation is effective over a broad streamwise domain that extends well upstream of where the flow separates in the absence of actuation and even downstream of the front stagnation point on the pressure side of the airfoil. Furthermore, the work of Amitay and Glezer (1999) showed that while the circulation of the attached flow when the actuation is applied at $St \sim O(10)$ is nominally time-invariant, the shedding of organized vortical structures when the actuation is applied at $St \sim O(1)$ leads to time-periodic variations in the circulation (and therefore in the lift).

The effect of the actuation frequency on the manipulation of the global aerodynamic forces on lifting surfaces using surface-mounted fluidic actuators based on synthetic jet technology was demonstrated in wind tunnel experiments by Amitay and Glezer (2002a). The effect of the actuation was investigated at two ranges of (dimensionless) jet formation frequencies on the order of or well above the natural shedding frequency. The vortical structures within the separated flow region varied substantially, while the dimensionless actuation frequency (F^+) varied between $O(1)$ and $O(10)$. When F^+ was $O(1)$, the reattachment was characterized by the formation of large vortical structures at the driving frequency that persisted well beyond the trailing edge of the airfoil. The formation and shedding of these vortices lead to unsteady attachment and consequently to a time-periodic variation in vorticity flux and in circulation. Actuation at F^+ of $O(10)$ lead to a complete flow reattachment that was marked by the absence of organized vortical structures along the flow surface. This suggests that when the actuation frequency is high enough, the Coandă-like attachment of the separated shear layer to the top (suction) surface of the airfoil can be replaced by completely attached flow for which separation may be bypassed altogether.

6.2.1.3 Frequency Modulation (Combining the Low and High Frequencies)

Another actuation technique (frequency modulation), which was shown to improve the control efficiency in some cases, was introduced by Amitay and Glezer in 2002. This actuation technique can be used to achieve an improvement in the efficiency of the jet actuators by using a pulse-modulated excitation input. The pulse modulation of the actuation frequency produces successive bursts of the driving signal which help to "capture" the vorticity produced during the initial stages of the separation process on the suction side of the airfoil and thus to increase the lift force. The actuator resonance waveform (nominally at $F^+ = 10$) is pulse modulated such that the period and duty cycle of the modulating pulse train are independently controlled. In their work, Amitay and Glezer (2002) restricted the duty cycle to 25% and varied the modulating frequency (F_m^+) between 0.27 and 5.0. When the modulation frequency was $F_m^+ = 0.27$ (corresponding to the "natural" passage frequency of the vortices during the initial transient stages of the reattachment process), the circulation exhibited oscillations that are similar to the transient stages of the reattachment with shedding of similar vortical structures. When F_m^+ was increased to 1.1, the large oscillations in the circulation were substantially attenuated. A further increase in F_m^+ to 3.3 resulted in a circulation (and consequently lift coefficient) that was similar to the magnitude of the piecewise-averaged circulation at $F_m^+ = 1.1$, but with the absence of oscillations. When the modulating frequency was increased further to $F_m^+ = 5$, the effectiveness of the modulation was minimal and the lift coefficient returns to the same levels obtained with a continuous (high frequency) pulse train.

Given these results, it might be argued that a lift coefficient increment similar to $F_m^+ = 3.3$ may be obtained by simply operating the jet actuators time-harmonically at $F^+ = F_m^+$, thus bypassing the need for pulse modulation altogether. This argument was tested by comparing the effect of both actuation approaches—time-harmonic excitation at $F^+ = 3.3$ and pulse modulation at $F_m^+ = 3.3$—while maintaining the same jet momentum coefficient. This comparison showed that time-harmonic actuation does not yield the same levels of lift coefficient as pulse-modulated excitation.

The pulse modulation technique was implemented successfully on a UAV with 50° leading edge sweep (Amitay et al. 2003, 2004; Parekh et al. 2003). In this case, synthetic jets were mounted in the leading edges of the vehicle for separation control. Active flow control was used to enhance vehicle control at moderate and high angles of attack for take-off and landing activities or gust load alleviation. The surface-mounted synthetic jet actuators were operated at various waveforms where the carrier frequency was at least an order of magnitude higher than the characteristic shedding frequency of the flow. Actuation yielded a suction peak near the leading edge; however, while excitation with a sinusoidal waveform resulted in a sharp suction peak near the leading edge, pulse modulation yielded a larger and wider suction peak, which, in turn, resulted in larger moments and forces, especially for angles of attack larger than 15°. Perhaps the most significant finding from the work of Parekh et al. was the degree to which one can vary the control moments simply by varying modulation frequency. Not only the magnitude of the change but also the direction of the change is affected. The overall integrated effect on the forces and moments is strongly dependent on the nature of the local interaction between the actuators and the embedding flow. The same actuator parameters that can improve local suction in a nominally separated flow can reduce local suction in a nominally attached flow. These types of interactions not only provide a rich set of data to be mined but also suggest intriguing new approaches for vehicle aerodynamic control.

6.2.2 Effect of Actuation Amplitude

In addition to the actuation frequency, the actuation amplitude is a significant factor in the efficacy of a flow control technique. In synthetic jet actuators, the maximum achievable jet amplitude is dependent upon several design parameters, which will not be discussed in detail here. Instead, for a chosen actuator design, the amplitude then becomes primarily dependent upon only two operating conditions: (1) the actuation frequency (or carrier frequency in modulated waveforms) and (2) the synthetic jet driver stroke length.

In typical designs, a piezoelectric diaphragm is used as the synthetic jet driver, where the strain in the diaphragm (and thus the driver's stroke length) is controlled via a voltage potential across the piezoelectric material. Then, when the actuation frequency is appropriately chosen, near the resonance of the device, the fluidic amplitude becomes solely dependent upon the input voltage to the piezoelectric material. In optimal designs, the fluidic amplitude of the device is proportional to this input voltage, but will need to be calibrated to quantify the effectiveness of an individual actuator. This is typically accomplished through measuring the phase-averaged jet velocity at the centerline of the orifice of the synthetic jet with a hotwire anemometer.

To be thoroughly rigorous in quantifying the fluidic amplitude of a synthetic jet actuator, the exit velocity profile should be completely mapped (Smith and Swift 2001; Holman et al. 2005). However, for the purposes of only a consistent calibration of the devices, an assumed plug-flow exit profile with the amplitude equivalent to the measured centerline velocity is valid (Shuster and Smith 2007). With this information, the two primary actuator effectiveness benchmarks can be produced for a given flow control application, specifically the blowing coefficient (C_b) and the momentum coefficient (C_μ) each of which are discussed in detail in Sections 6.2.2.1 and 6.2.2.2.

6.2.2.1 Blowing Coefficient

The blowing coefficient (C_b) provides a simple useful metric for comparing and scaling the synthetic jet amplitude with the freestream conditions for a given flow control application and is defined as follows:

$$C_b = \frac{U_0}{U_\infty}$$

where:

U_∞ is the freestream velocity
U_0 is the average jet exit velocity over the blowing portion of the synthetic jet cycle

Due to the oscillating nature of the mass flux through the synthetic jet orifice, the average jet exit velocity is not equivalent to the peak centerline velocity. The transport of the fluid in the flow field is primarily due to the vortical structures that are formed during the expulsion portion of actuation cycle. As a result, the average centerline exit velocity is calculated as the integral of the centerline velocity across only the expulsion half of the actuation cycle, averaged over the entire period of the actuation cycle (T), which is given as follows:

$$U_0 = \frac{1}{T}\int_0^{T/2} u_0(t)\ dt$$

When the synthetic jet is driven by a sinusoidal carrier waveform (as is typical), the exit velocity as a function of time can be represented by

$$u_0(t) = u_p \sin\left(2\pi f_{\text{act}} t\right)$$

where:

u_p is the peak centerline exit velocity
f_{act} is the actuation frequency ($f_{\text{act}} = 1/T$)

As a result, for sinusoidal actuation, the time-averaged blowing coefficient reduces to

$$C_b = \frac{u_p}{\pi U_\infty}$$

The blowing coefficient, though an important metric, is not sufficient on its own in quantifying the performance of a synthetic jet actuator. In fact, blowing coefficient can be increased or decreased for a similar actuator design by scaling the actuator exit area appropriately. This alters the time-varying exit velocity while conserving the time-varying mass flux through the synthetic jet orifice. Clearly, as a result, a metric accounting for the orifice cross-sectional area is also needed to realistically compare the efficiencies of similar actuators.

6.2.2.2 Momentum Coefficient

In controlling boundary layers and reattaching separated flows, the velocity ratio of the control device to the freestream is important; however, it can be argued that a measure of the local momentum addition is more pertinent. As a result, a time-averaged momentum coefficient (C_μ) is commonly used and is defined as follows:

$$C_\mu = \frac{\sum_{i=1}^{n} I_{0,i}}{(1/2)\rho_\infty U_\infty^2 A_\infty} \tag{6.1}$$

Here, the momentum added to the flow is given by the summation of the momentum of individual jets (I_0) for each of the n actuators that are activated in a given time. This added momentum is then normalized by a freestream momentum, typically the product of the dynamic pressure [$(1/2)\rho_\infty U_\infty^2$] with a reference area, A_∞ (for wing surfaces, the planform area [S] is commonly used as the reference area).

Similar to the time-averaged velocity in the blowing coefficient, the time-averaged momentum for individual actuator can be calculated as the integral of the momentum added during the expulsion portion of the synthetic jet cycle averaged over the entire period of the synthetic jet cycle:

$$I_0 = \frac{1}{T}\int_0^{T/2} \rho_{jet} A_{jet} u_0^2(t)\,dt$$

where:

A_{jet} is the cross-sectional area of the synthetic jet orifice
ρ_{jet} is the density at the synthetic jet exit (assuming incompressible flow: $\rho_{jet} = \rho_\infty$)

Again, similar to the time-averaged velocity when a sinusoidal actuation waveform is utilized, the time-averaged momentum reduces to

$$I_0 = \frac{\rho_{jet} A_{jet} u_p^2}{4}$$

Thus, the momentum coefficient becomes

$$C_\mu = \frac{\sum_{i=1}^{n}\left(A_{jet} u_p^2\right)_i}{2U_\infty^2 A_\infty}$$

This common formulation of the total time-averaged momentum coefficient is utilized for 3D applications. Additionally, a momentum coefficient per unit span can also be defined for 2D applications, where the freestream reference area and the synthetic jet orifice cross-sectional area are replaced by a streamwise reference length (for airfoil applications a chord length) and the synthetic jet orifice thickness (in the streamwise direction), respectively. In this chapter, each of the proceeding examples employs the 3D formulation of the time-averaged momentum coefficient, unless otherwise stated.

6.3 Examples

With a growing interest in active flow control techniques in the past decade, synthetic jet-based separation control has evolved significantly. Much of the initial mass of work has focused on understanding the fluid interactions in controlled 2D and static model environments. With a progression closer toward industrial application, current research interests have advanced to better understand the interactions of synthetic jets in 3D and/or dynamically changing separated flows. A selection of examples displaying these current research interests is presented in Sections 6.3.1 through 6.3.4.

6.3.1 Circular Cylinder

6.3.1.1 Two-Dimensional (Infinite) Cylinder

Due to the wealth of information related to their bluff body separation, 2D circular cylinders provide the most logical starting point for a discussion on separation control.

Amitay et al. (1997) used a pair of adjacent plane synthetic jet actuators with spatially oriented rectangular orifices normal from the circular cylinder's surface to alter the global flow field about a 2D cylinder. The effects of the jets on the flow at a number of azimuthal positions are presented. Through smoke visualization, they demonstrated that the synthetic jets were capable of altering the streamlines in the flow field. As the azimuthal angle was increased, from $\alpha = 0^\circ$ to 180°, a reduction in separation (and flow reattachment) was observed on both sides of the cylinder (the actuated and nonactuated sides). In conjunction, they also mapped the surface pressure distribution around the circular cylinder at a higher Reynolds number. This showed that the synthetic jets were capable of altering the flow separation through locally accelerating the flow field in the immediate vicinity of the jets (and on the other side of the circular cylinder) and that they were not simply triggering the onset of turbulence.

6.3.1.2 Three-Dimensional (Finite Length and Small Aspect Ratio) Cylinder

The flow about a circular cylinder has been studied by many researchers as a representative bluff body for testing different methods of active flow control, due to the wealth of knowledge regarding its flow field.* In general, mainly the quasi-2D (i.e., large span-to-diameter ratio) cylinder has been of interest, neglecting the possible consequence that three-dimensionalities such as free ends or low aspect ratio might have on the efficacy of flow control. This is particularly important with regard to real-world objects, where such three-dimensionalities can rarely be avoided. A bluff body that better represents the flow about such real-world objects would be a finite-span cylinder with a single free end, whose flow field is significantly different from a corresponding 2D cylinder, and thus, flow control is not expected to have the same effect.

The flow field of a finite-span cylinder is complex and depends largely on the aspect ratio of the cylinder, which in turn determines the degree to which downwash from the free end affects the cylinder wake (Norberg 1993). With a finite-span cylinder, flow over the free end is drawn to the rear of the cylinder due to the lower base pressure. This interaction results in an increase in base pressure, compared to its 2D counterpart, along with a decrease in suction peak, which becomes more dramatic with decreasing aspect ratio (Kawamura et al. 1984). For aspect ratios above a critical value, the downwash does not interact with the wake along the entire span of the cylinder, and there is a small 2D region away from the cylinder's ends, relatively unaffected by the downwash from the top (Farivar 1981; Uematsu and Yamada 1993). The exact value of the critical aspect ratio varies greatly in the literature (Afgan et al. 2007); below this critical value, the effect of the downwash extends along the entire span of the cylinder, and the flow becomes increasingly 3D (Kawamura et al. 1984).

In the current study, the interactions of a single synthetic jet actuator with a finite-span cylinder are experimentally investigated. The experiments were conducted in the open-return low-speed wind tunnel at Rensselaer Polytechnic Institute. The wind tunnel has a test section with a 0.8×0.8 m cross section and is 5 m long with a maximum speed

* The results presented in this section have been previously published in *Experiments in Fluids* (DeMauro et al. 2012).

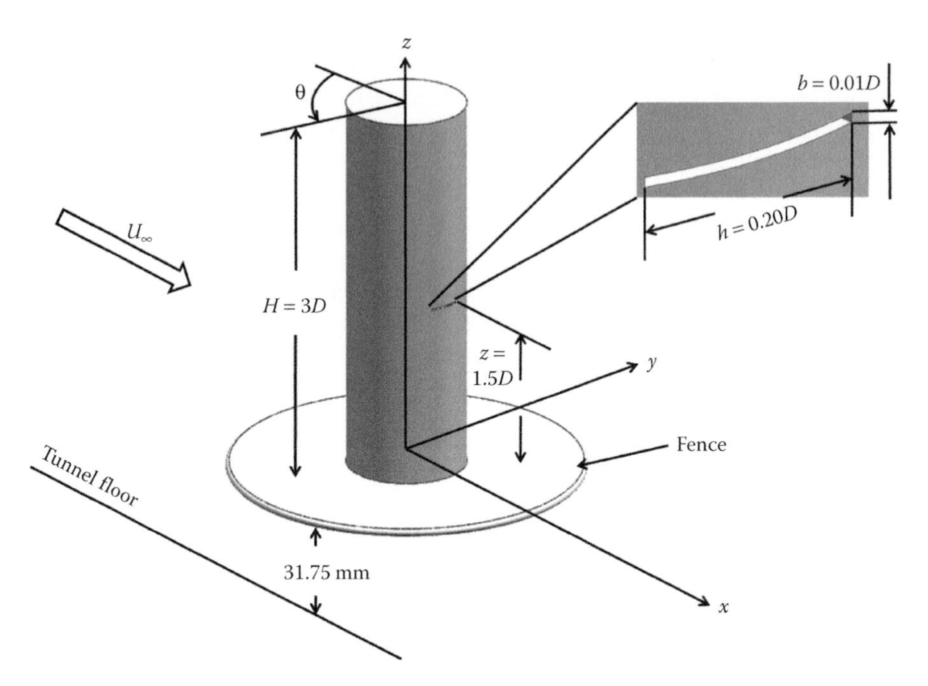

FIGURE 6.1 Schematic of the circular cylinder and axes orientation. The jet is at an arbitrary azimuthal angle. (From DeMauro, E. P. et al., *Exp. Fluids.*, 53, 1965, 2012. With permission.)

of 50 m/s and a turbulence level of less than 0.2%. A finite-span circular cylinder with an outer diameter, $D = 101.6$ mm, was used for the experiments. The cylinder had a height-to-diameter ratio of 3 ($AR = 3$), creating a blockage in the tunnel of 5%. The coordinate system of the cylinder is defined in Figure 6.1, where x, y, and z are the streamwise, cross-stream, and spanwise directions, respectively. The cylinder was instrumented with 112 surface pressure taps arranged in 7 rows along the span of the cylinder ($z/D = 1.31, 1.37, 1.44, 1.50, 1.56, 1.63$, and 1.69). Each row contained 16 pressure taps distributed azimuthally about the circumference of the cylinder. The cylinder was instrumented with a single synthetic jet located at $z/D = 1.5$. The synthetic jet used in this study had a rectangular exit orifice of length $b = 20.32$ mm and width $h = 1$ mm and was aligned such that its long axis was parallel to the freestream, in order to create maximum three-dimensionality by introducing streamwise vortices into the flow (Figure 6.1). Therefore, the orifice of the jet swept 22° along the cylinder surface ($\pm 11^\circ$ from the centerline of the orifice). The azimuthal positioning of the synthetic jet (with respect to the freestream) was controlled by a motor attached to the base of the cylinder. Lastly, the synthetic jet was driven with a sinusoidal waveform at a frequency ($f_{jet} = 2.6$ kHz), resulting in a Strouhal number ($St_{jet} = 14.7$) two orders of magnitude greater than the Strouhal number of the cylinder shedding frequency of $St = 0.16$ (Sumner et al. 2004). The Strouhal number is defined as follows:

$$St = \frac{fD}{U_\infty}$$

where:

f is the frequency
U_∞ is the freestream velocity
D is the cylinder diameter

The flow field in the wake of the cylinder was measured using stereoscopic particle image velocimetry (SPIV) system. The velocity components (U, V, and W) were computed using the stereoscopic cross-correlation of pairs of successive images with 50% overlap between the interrogation domains. The images were processed using an advanced multi-pass method where the initial and final correlation passes were 64×64 and 32×32 pixels, respectively. About 500 image pairs were acquired to produce a time-averaged vector field for each plane.

The effect of the synthetic jet on the pressure distribution around the finite-span cylinder and on the wake at a cylinder diameter-based Reynolds number of 116,000 is presented. First, the surface pressure results are discussed before the velocity field measurements in the cylinder's wake are presented. Surface pressure distributions for the unforced and forced cases are shown in Figure 6.2 when the centerline of the synthetic jet is rotated into six different azimuthal locations with respect to the freestream velocity (θ_{jet} at the orifice = 53°, 68°, 98°, 113°, 128°, and 143°, Figure 6.2a–f, respectively). The unforced azimuthal surface pressure distribution at $z/D = 1.5$ (midspan) resembles the data shown in the works of Kawamura et al. (1984). The base pressure has a relatively constant value of $C_p \approx -0.5$, which is greater than the base pressure measured by Amitay et al. (1997) on a quasi-2D cylinder, due to the mixing of the downwash from the free end with the wake deficit. Likewise, the unforced azimuthal pressure distribution shows that separation occurs at ~80°, indicating a laminar boundary layer. When the synthetic jet is actuated, the suction peak closest to the physical location of the synthetic jet (marked by the dashed line) is increased. The change in the suction peak on the opposite (unactuated) side of the cylinder indicates a change in circulation about the cylinder (a global effect). Figure 6.2 also shows that the change in the pressure distribution has a greater dependency on the azimuthal location of the jet with respect to the freestream velocity than to the jet's blowing ratio. In all plots shown in Figure 6.2, separation is delayed to angles greater than 80°, which is associated with a narrowing of the wake and a decrease in the sectional pressure drag.

The asymmetry shown in Figure 6.2d (for $\theta_{jet} = 113°$) differs from the results of Amitay et al. (1997), where a symmetric pressure distribution was obtained for $\theta_{jet} = 110°$ for a quasi-2D cylinder. In both studies, the value of C_μ is comparable. Matejka et al. (2009) obtained a similar asymmetric pressure distribution to the current study, with synthetic jets whose orifices were oriented normal to the freestream direction, mounted on a cylinder of $AR = 1.7$. In comparison with the current study, the asymmetry of the azimuthal surface pressure distribution is independent of the orientation of the synthetic jet orifice with respect to the freestream and therefore must be attributed to the presence of the downwash. Note that when the jet is located at either $\theta_{jet} = 128°$ or $\theta_{jet} = 143°$, the pressure distribution around the cylinder is symmetric. At these azimuthal locations, the synthetic jet is able to better entrain equal amounts of flow from both the actuated and unactuated sides of the cylinder.

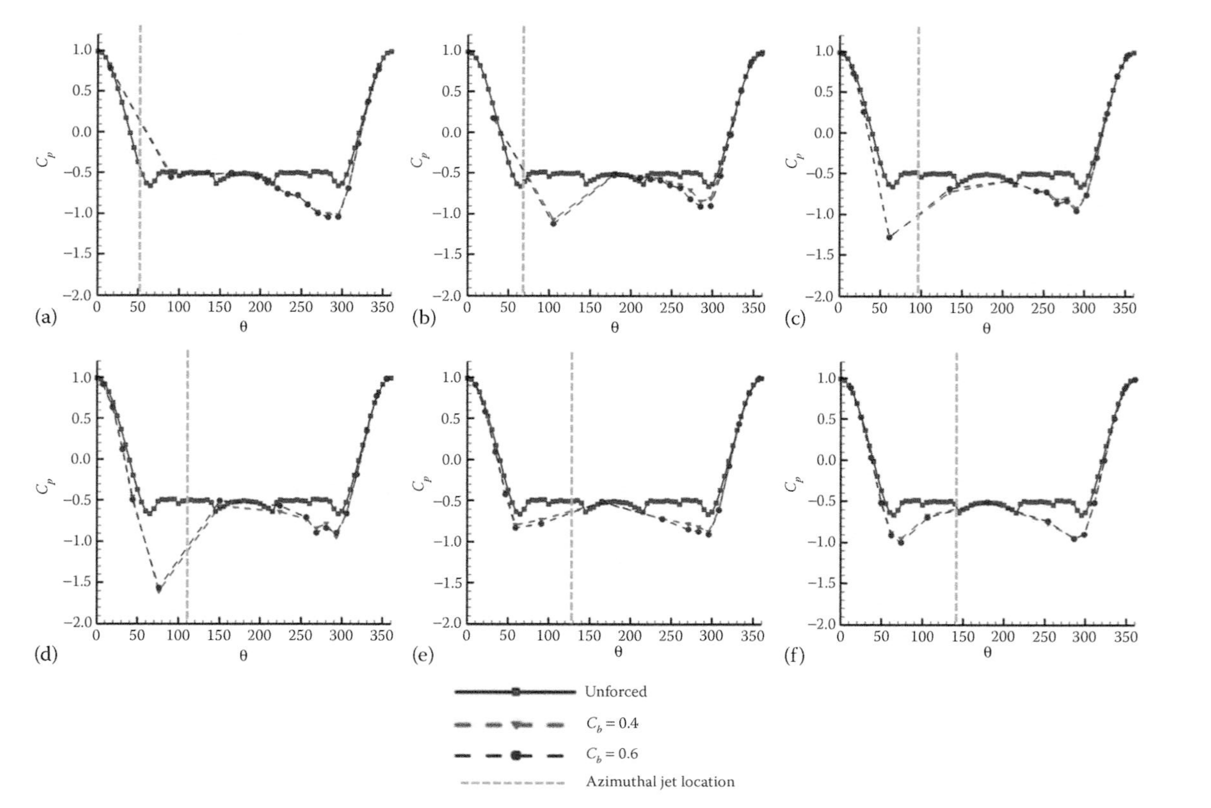

FIGURE 6.2 Azimuthal surface pressure distributions at midspan ($z/D = 1.5$). The centerline of the synthetic jet is at (a) $\theta_{jet} = 53°$ ($42° \leq \theta_{jet} \leq 64°$), (b) $\theta_{jet} = 68°$ ($57° \leq \theta_{jet} \leq 79°$), (c) $\theta_{jet} = 98°$ ($87° \leq \theta_{jet} \leq 109°$), (d) $\theta_{jet} = 113°$ ($102° \leq \theta_{jet} \leq 124°$), (e) $\theta_{jet} = 128°$ ($117° \leq \theta_{jet} \leq 139°$), and (f) $\theta_{jet} = 143°$ ($132° \leq \theta_{jet} \leq 154°$). (From DeMauro, E. P. et al., *Exp. Fluids.*, 53, 1967, 2012. With permission.)

Figure 6.3a–f presents the distributions of the pressure coefficient along the span at the azimuthal location (θ_{model}) at or near the location of the suction peak (Figure 6.2) for $\theta_{jet} = 53°, 68°, 98°, 113°, 128°$, and $143°$, respectively. For all cases, the synthetic jet induces a spanwise effect much greater than the physical dimensions of its orifice. Given the small orifice width (1 mm) compared to the span of the cylinder (304.8 mm), it was initially assumed that the spanwise effect of the jet would be localized around the orifice itself. Figure 6.3 shows that the spanwise change in the surface pressure extends even beyond the measurement domain of the pressure ports. This large spanwise extent of the jet's influence is due to the fact that the synthetic jet is synthesized by drawing air from the surrounding. In the current orifice orientation, the synthetic jet induces spanwise velocity (toward its orifice) across the span of the cylinder, resulting in the large spanwise effect seen in Figure 6.3.

Based on the data shown in Figures 6.2 and 6.3, a large azimuthal and spanwise effect was measured when the centerline of the synthetic jet was located at $\theta_{jet} = 113°$, in comparison with the other measured jet angles. Thus, the interaction of the downwash with the synthetic jet, when the centerline of the synthetic jet is located at $\theta_{jet} = 113°$, was investigated using SPIV measurements in the near wake of the cylinder for both unforced and forced cases.

Figure 6.4 shows the time-averaged color contours of streamwise velocity (i.e., out of the plane velocity component), superimposed with in-plane velocity vectors for both the unforced and forced cases at three streamwise locations ($x/D = 1, 2$, and 3). The unforced wake at $x/D = 1$ (Figure 6.4a) has a relatively constant cross-stream extent along the span of the cylinder, and the downwash from the cylinder's free end (at $|y/D| < 0.5$) is small in magnitude. On both sides of the cylinder ($|y/D| > 1$), the spanwise velocity is positive (i.e., upward motion), which is due to recirculation of the downwash in the near wake. The forced wake at $x/D = 1$ (Figure 6.4b) shows localized regions (at $z/D \approx 2$) of inwardly directed flow on both sides of the cylinder, indicating that the synthetic jet imparts a global effect on the flow field. Note that the synthetic jet orifice is located at $z/D = 1.5$, while the effect of the jet is near $z/D \approx 2$, which is due to the upward motion on both sides of the cylinder.

At $x/D = 2$ (Figure 6.4c), the magnitude of the downwash is increased. Due to entrainment of fluid from the freestream, the velocity deficit near the free end is decreased (relative to $x/D = 1$) and the wake is wider near the base of the cylinder. The forced wake at $x/D = 2$ (Figure 6.4d) shows that the effect of the synthetic jet has evolved and has both narrowed and vectored the wake, while also redirecting the downwash. At $x/D = 3$, the unforced wake (Figure 6.4e) has a slightly decreased magnitude of downwash in comparison with the downwash at $x/D = 2$ (Figure 6.4c). Furthermore, the wake is wider and its spanwise extent is somewhat reduced. The global shape of the forced wake, shown in Figure 6.4f, is affected by the redirected path of the downwash at this location. Note that the effect of the synthetic jet extends throughout the length of the cylinder, which was also evident from the surface pressure data shown in Figure 6.3.

As shown in this section, a finite-span synthetic jet was able to alter the circulation about the cylinder and to induce a large spanwise change in the surface pressure, much

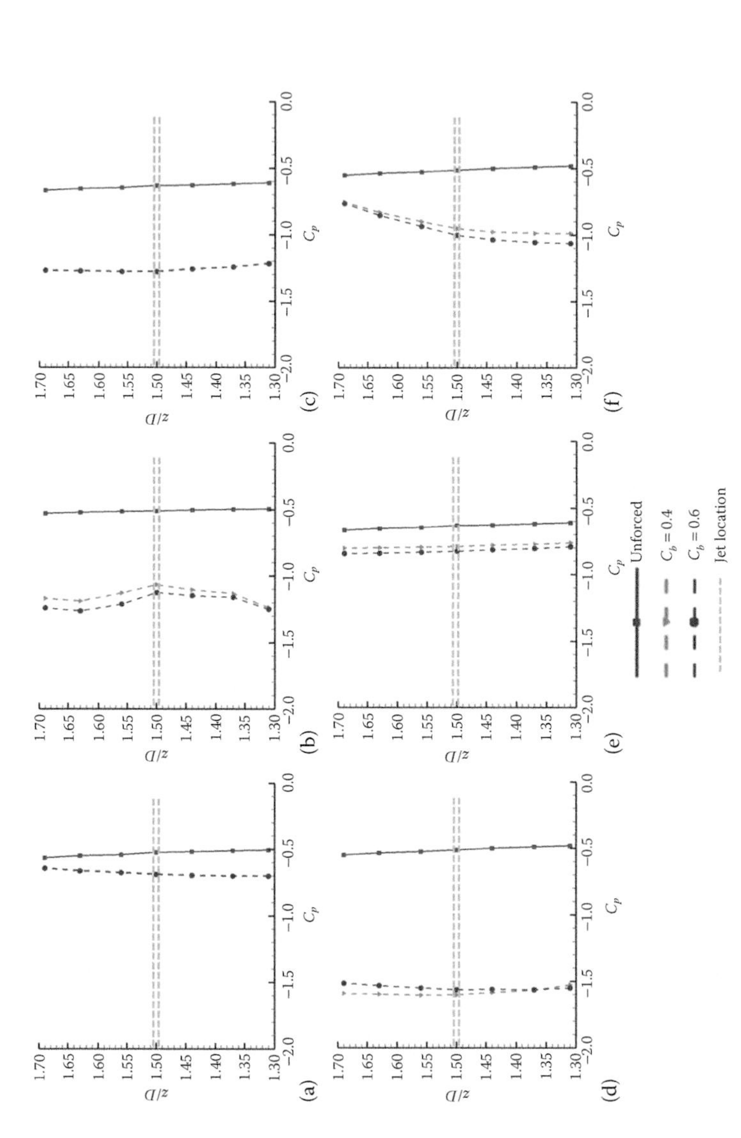

FIGURE 6.3 Spanwise surface pressure distributions. The centerline of the synthetic jet is at various azimuthal locations with respect to the freestream velocity, θ_{jet}. The spanwise distributions are taken at azimuthal location of suction peak, θ_{model}: forcing at (a) $\theta_{jet} = 53°$ and $\theta_{model} = 307°$, (b) $\theta_{jet} = 68°$ and $\theta_{model} = 105°$, (c) $\theta_{jet} = 98°$ and $\theta_{model} = 61°$, (d) $\theta_{jet} = 113°$ and $\theta_{model} = 76°$, (e) $\theta_{jet} = 128°$ and $\theta_{model} = 59°$, and (f) $\theta_{jet} = 143°$ and $\theta_{model} = 74°$. (From DeMauro, E. P. et al., *Exp. Fluids.*, 53, 1968, 2012. With permission.)

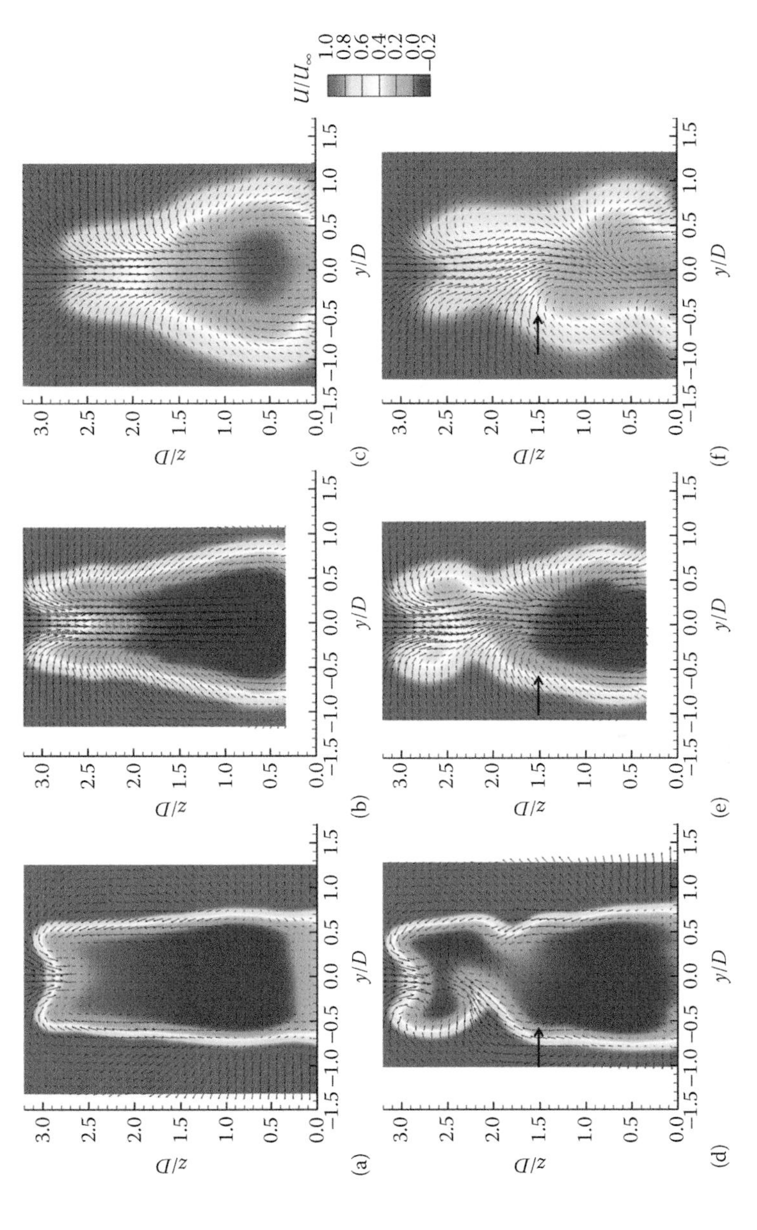

FIGURE 6.4 **(See color insert.)** Color contours of the streamwise (out of plane) velocity with superimposed in-plane vectors, $\theta_{jet} = 113°$ (jet's centerline): (a) unforced, $x/D = 1$; (b) forced, $x/D = 1$; (c) unforced, $x/D = 2$; (d) forced, $x/D = 2$; (e) unforced, $x/D = 3$; and (f) forced, $x/D = 3$. (From DeMauro, E. P. et al., *Exp. Fluids.*, 53, 1969, 2012. With permission.)

greater than the dimensions of its orifice. Moreover, the synthetic jet can enhance mixing of the downwash from the cylinder free end with the wake deficit, vectoring and narrowing the wake.

6.3.2 Two-Dimensional (Infinite) Airfoil

Separation control on 2D airfoils has been the topic of multiple investigations over the last few decades.* An example of such research is the control of flow separation on an unconventional symmetric airfoil using synthetic jet actuators (Amitay et al. 2001a), which is discussed in this section. A 2D symmetric airfoil was used, which comprised of the aft portion of a National Advisory Committee for Aeronautics (NACA) four-digit series airfoil and a leading edge section that is one-half of a round cylinder (Figure 6.5). The experiments were conducted over a range of Reynolds numbers between 310,000 and 725,000. In this range, the flow separated near the leading edge at angles of attack exceeding 5°. When synthetic jet control was applied near the leading edge, upstream of the separation point, the separated flow reattaches completely for angles of attack up to 17.5° and partially for higher angles of attack (Figure 6.6).

The effect of the actuation frequency, actuator location, and momentum coefficient was also investigated for different angles of attack. Distributions of the pressure coefficient about the airfoil for F^+= 0.95, 2.05, and 3.3 and F^+= 10, 14.7, and 20 are shown in Figure 6.7a and b, respectively. The pressure distributions for the (baseline) stalled flow are also shown for reference (solid line). The location of the actuator on the upper surface is marked by an arrow on each plot. Clearly, without control the flow is completely separated from the upper surface of the airfoil (as indicated by the constant pressure coefficient). Actuation at low reduced frequencies (Figure 6.7a) results in a sharp suction peak around x/c = 0.075, which corresponds to the location of separation and the onset of the separated shear layer in the baseline flow. While the streamwise decrease in static pressure for $x/c < 0.15$ appears to be independent of F^+, the degree of pressure recovery toward the trailing edge of the airfoil decreases with increasing F^+ leading to a reduction in lift and a slight increase in pressure drag. The corresponding lift and drag coefficients (as well as the lift to [pressure] drag ratios) were calculated (not shown here), which indicated that when the actuation frequency is of the same order as the shedding frequency, the lift-to-pressure

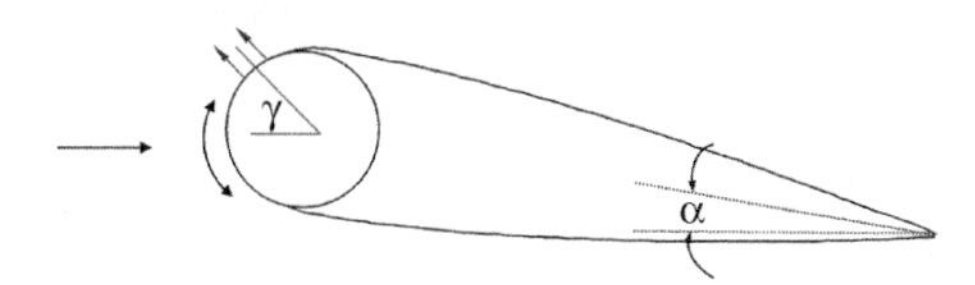

FIGURE 6.5 Airfoil model. (From Amitay, M. et al., *AIAA J.*, 39, 24, 2001b. With permission.)

* The work in this section has been previously published in Amitay et al. (2001a) and Amitay and Glezer (2002a, 2002b). This work was supported by Air Force Office of Scientific Research (AFOSR).

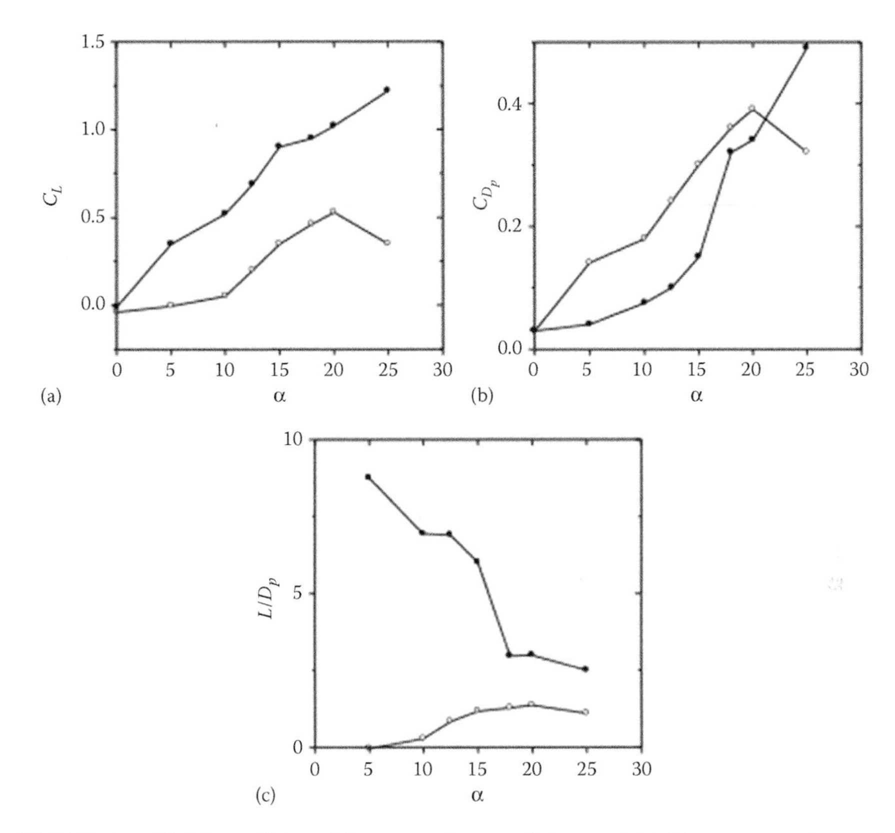

FIGURE 6.6 (a) Lift coefficient, (b) pressure drag coefficient, and (c) lift-to-pressure drag ratio versus angle of attack, α for $\gamma = 60°$. Forced (—•—) and baseline (—○—). (From Amitay, M. et al., *AIAA J.*, 39, 26, 2001. With permission.)

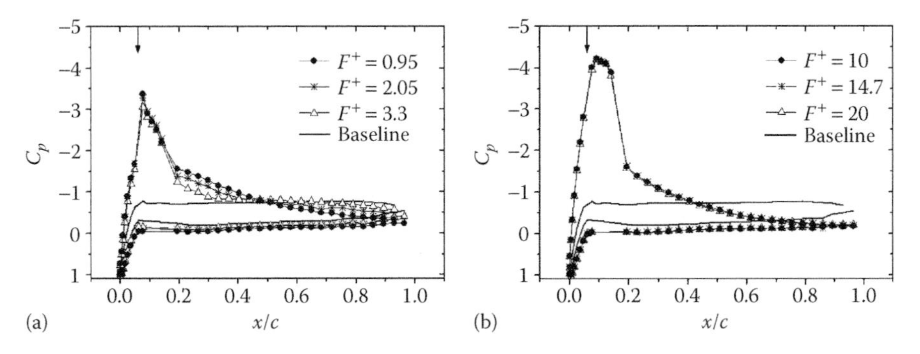

FIGURE 6.7 Variation of the pressure coefficient with the dimensionless forcing frequency (a) $F^+ \sim O(1)$ and (b) $F^+ \sim O(10)$. The distribution for the unforced flow is shown by a solid line. The arrow represents the streamwise location of the synthetic jet. (From Amitay, M. et al., *AIAA J.*, 39, 44, 2001. With permission.)

drag ratio L/D_P decreases with increasing actuation frequency (as would be expected from the reduced receptivity of the separated shear layer to the increased actuation frequency).

The corresponding distributions of pressure coefficient for actuation at high reduced frequencies ($F^+ \geq 10$, Figure 6.7b) exhibit a larger and wider suction peak than at low F^+ (Figure 6.7a), and consequently, the actuation results in a larger increment in lift (compared to $F^+ \sim O[1]$). Unlike the data in Figure 6.7a, the pressure recovery downstream of the suction peak in Figure 6.7b appears to be essentially independent of the actuation frequency suggesting that for sufficiently high frequency, the details of the reattachment become frequency independent. Furthermore, as shown in Figure 6.7b, the pressure difference between opposite stations on the suction and pressure surfaces of the airfoil for $x/c > 0.5$ is smaller than that at corresponding streamwise stations for the low-frequency actuation, resulting in smaller contributions to the pressure drag. Similar to the data in Figure 6.7a, the lift and pressure drag coefficients for the three actuation frequencies were also calculated (not presented here), and at least within the present range of high F^+, the lift to pressure drag ratio appears to be invariant with actuation frequency suggesting that the mechanism that leads to the suppression of separation is not associated with the stability of the separated shear layer. It is conjectured that for a given actuation momentum coefficient, there is a domain between the two frequency ranges where the effectiveness of the actuation has a local minimum (which might be thought of as a threshold) above which the actuation effectiveness increases monotonically and reaches a fixed level where it becomes more or less frequency invariant.

6.3.3 Finite Wing

6.3.3.1 Flow Control on a Static Finite-Span Wing

Many aerodynamic surfaces encounter highly 3D flows due to the shape, finite span, and sweep of typical wing surfaces.* To better understand the effect and predict the performance of synthetic jets in these flows, a complete understanding of their interactions with complex crossflows needs to be developed. This section addresses the interaction of a single finite-span synthetic jet with a crossflow over a 30° swept-back finite-span wing configuration at a Reynolds number of 100,000 and at two angles of attack of 9° and 15.5°. These angles of attack correspond to different pressure gradient conditions present in the baseline flow field in the vicinity of the synthetic jet, which are of interest for understanding the interaction of the synthetic jet. At the angle of attack of $\alpha = 9°$, the flow in the vicinity of the synthetic jet is attached, and at $\alpha = 15.5°$, the flow is separated, and thus the focus of the work is to capture and analyze the secondary flow structures in the vicinity of the synthetic jet in the presence or absence of a local flow separation.

The model used in this study was a finite and swept-back (30° sweep) wing having a cross-sectional shape of the NACA 4421 airfoil, a constant chord of 0.2 m, and a semispan

---

* The results presented in this section are thanks to the experimental work by Dr. Y. Elimelech, Mr. J. Vasile, and Professor M. Amitay and have been published in *Physics of Fluids* (Elimelech et al. 2011). This work has been supported by AFOSR (grant number FA9550-08-1-0233, monitored by Dr. Douglas Smith).

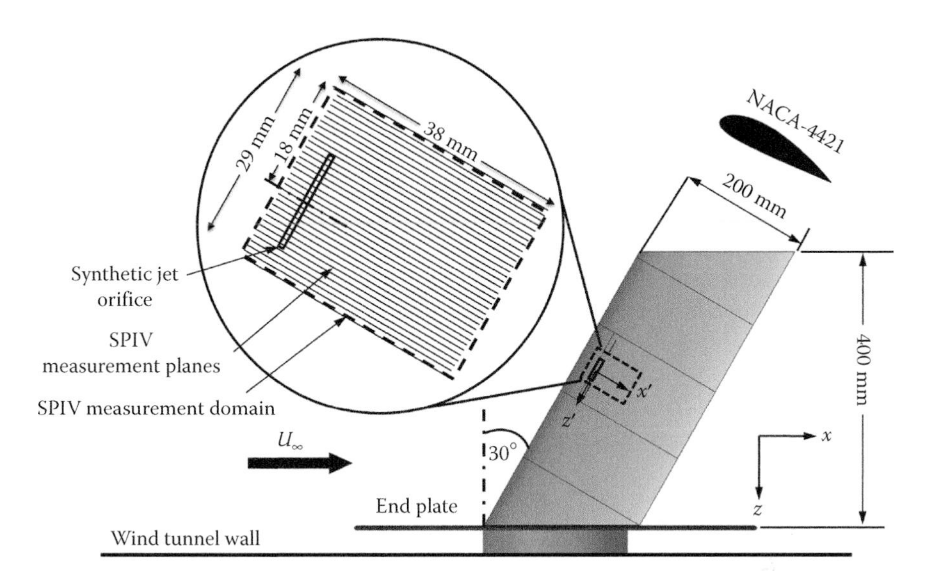

FIGURE 6.8 Experimental model and measurement domain. (Reprinted with permission from Elimelech, Y. et al., *Phys. Fluids*, 23, 094104, 3, 2011. Copyright 2011, American Institute of Physics.)

of 0.4 m. The model was mounted on a circular end-plate at the model's centerline (Figure 6.8). The model consisted of three spanwise sections, where in the present work only one synthetic jet, located at the center of the middle section and at 17% of the chord downstream from the leading edge, was activated. The synthetic jet orifice was oriented perpendicular to the surface and had a spanwise length and width of $l = 18$ mm and $h = 0.625$ mm, respectively. In the present work, all dimensionless quantities were normalized by the orifice width, h, and the freestream velocity, U_∞.

The experiments presented utilized a SPIV system to acquire multiple planes of 3D velocity data. The 3D flow fields were reconstructed from these measurement planes in a technique described in the work by Sahni et al. (2011). To analyze the 3D boundary layer in the vicinity of the synthetic jet orifice, the velocity field was projected to a body-fitted mesh, which yielded the local tangent and normal to the wall velocities, together with the spanwise velocity component. Using this terminology, all chordwise locations are given along the model surface, referenced to the synthetic jet orifice chordwise location (where $x' = 0$).

The synthetic jet actuator was driven for a continuous sinusoidal waveform where the actuation frequency was $f_{act} = 1750$ Hz. The synthetic jet actuation frequency was determined by taking into account the several characteristic timescales associated with the present model: (1) the characteristic frequency of the main flow (or the global timescale), defined as the reciprocal of the time of flight ($f_{char} = U_\infty/c$), which in the present work is approximately 37 Hz; (2) the local timescale associated with the most amplified Tollmien–Schlichting waves of the incoming boundary layer where the baseline boundary layer is attached in the vicinity of the synthetic jet ($\alpha = 9^\circ$); or alternatively, (3) the most amplified

Kelvin–Helmholtz waves that characterized the baseline separated flow field ($\alpha = 15.5^\circ$). At $\alpha = 9^\circ$, the Reynolds number based on the boundary layer thickness of the baseline flow field in the vicinity of the synthetic jet orifice is approximately 855, and thus, the frequency of the most amplified mode is ~420 Hz (assuming a flat plate and no pressure gradient). At $\alpha = 15.5^\circ$, the most amplified disturbance in the vicinity of the synthetic jet orifice has a frequency of ~470 Hz, derived from the local mixing layer thickness and the velocity difference across the mixing layer. Therefore, the actuation frequency is considerably larger than the frequencies that characterize the baseline flow. As this actuation frequency is more than an order of magnitude higher than the global timescale, it essentially results in a quasi-steady "virtual aeroshaping" of the surface.

Before discussing the interaction mechanisms between the synthetic jet and the crossflow, the time-averaged baseline flow field obtained from the SPIV measurements at the two angles of attack is presented in Figure 6.9. The figure presents the boundary layer velocity profiles of the tangent (Figure 6.9a–d) and the normal to the surface velocity component (Figure 6.9e–h). These velocity profiles are measured along the synthetic jet centerline ($z'/h = 0$; Figure 6.8) at four normalized downstream locations of $x'/h = 5$ (Figure 6.9a and e), $x'/h = 15$ (Figure 6.9b and f), $x'/h = 25$ (Figure 6.9c and g), and $x'/h = 40$ (Figure 6.9d and h). The two lines in each plot represent the velocity profiles at angle of attack of $\alpha = 9^\circ$ (black) and 15.5° (gray). At $x'/h = 5$ (just downstream of the synthetic jet orifice) for $\alpha = 9^\circ$, the flow is attached, as indicated by the velocity profiles that show positive tangential velocity and essentially zero normal velocity (i.e., minimal spreading). Farther downstream, the flow is attached up to $x'/h = 40$, where separation begins, as suggested by the negative tangential velocity near the wall (Figure 6.9d) and the increase in the positive normal velocity (Figure 6.9h). At $\alpha = 15.5^\circ$, the cross-stream distributions of the velocity components suggest that flow is separated at all four chordwise locations. Once the baseline flow was established, the interaction of the synthetic jet with the crossflow was studied and is presented in the subsequent paragraphs.

With the primary interest, here is the separation control by the synthetic jet, only the interactions at $\alpha = 15.5^\circ$ are discussed. Figure 6.10a–d and e–h presents the cross-stream distributions of the normalized time-averaged tangential velocity component (U/U_∞) and the normal velocity component (V/U_∞) at $x'/h = 5$, 15, 25, and 40, respectively (Figure 6.10a–h). Note that the baseline velocity profiles were discussed previously and are shown here for reference. Two jet blowing ratios are presented, corresponding to $C_b = 0.8$ and 1.2. Here, the flow along the centerline is fully attached when the jet is activated at both blowing ratios. Figure 6.10 shows that at all four chordwise locations, the tangential velocity near the wall increases significantly due to the additional momentum added by the synthetic jet. Furthermore, the activation of the synthetic jet yields reduced spreading of the flow (denoted by the normal velocity) as well as thinning of the boundary layer thickness.

The 3D geometries of the finite-span synthetic jet and the swept-back wing are likely to introduce 3D flow structures. To depict this interaction, isosurfaces of the time-averaged total velocity were calculated and are presented in Figure 6.11 for $\alpha = 9^\circ$ (Figure 6.11a–c) and $\alpha = 15.5^\circ$ (Figure 6.11d–f). Here, the volumetric data were collected over a spanwise strip of $\Delta z'/h = 46$. This figure clearly shows that in the time-averaged sense, secondary flow structures are formed at both angles of attack.

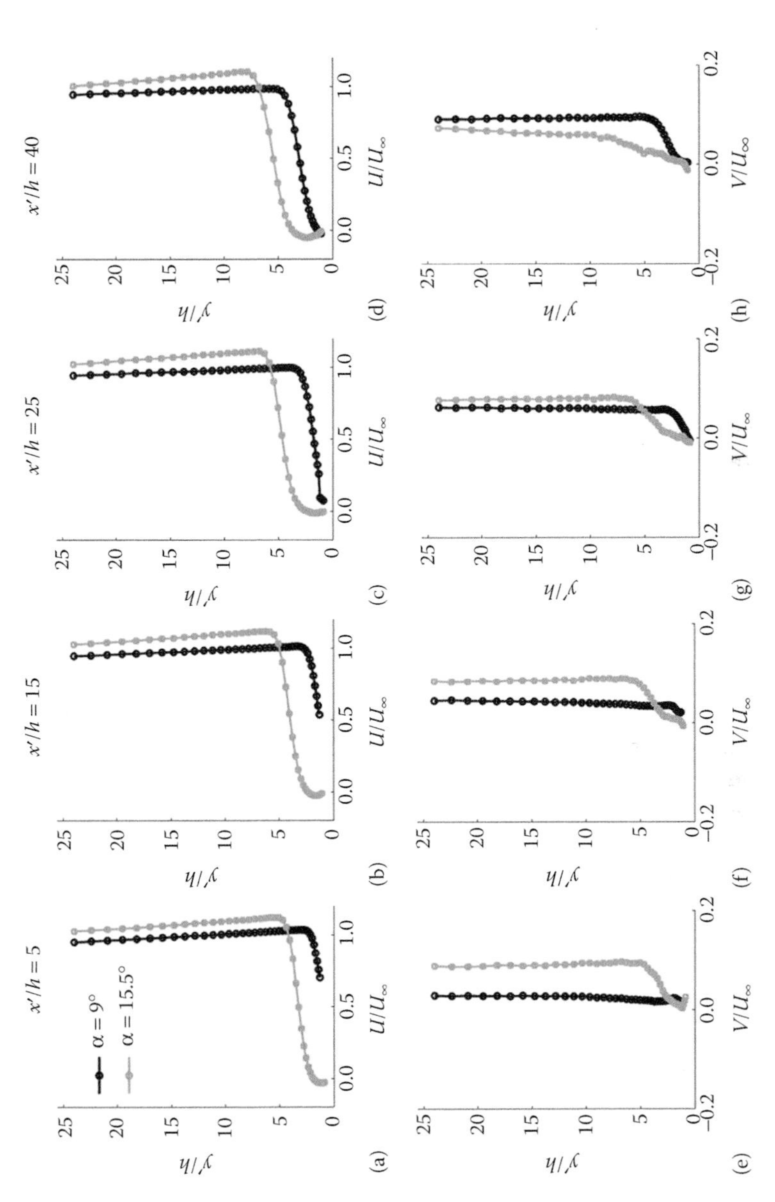

FIGURE 6.9 Cross-stream distributions of the time-averaged velocity at $z' = 0$; U/U_∞ (a–d) and V/U_∞ (e–h) at $x'/h = 5$ (a, e), 15 (b, f), 25 (c, g), and 40 (d, h). $\alpha = 9°$ (black) and 15.5° (gray) indicate baseline flow. (Reprinted with permission from Elimelech, Y. et al., *Phys. Fluids*, 23, 094104, 5, 2011. Copyright 2011, American Institute of Physics.)

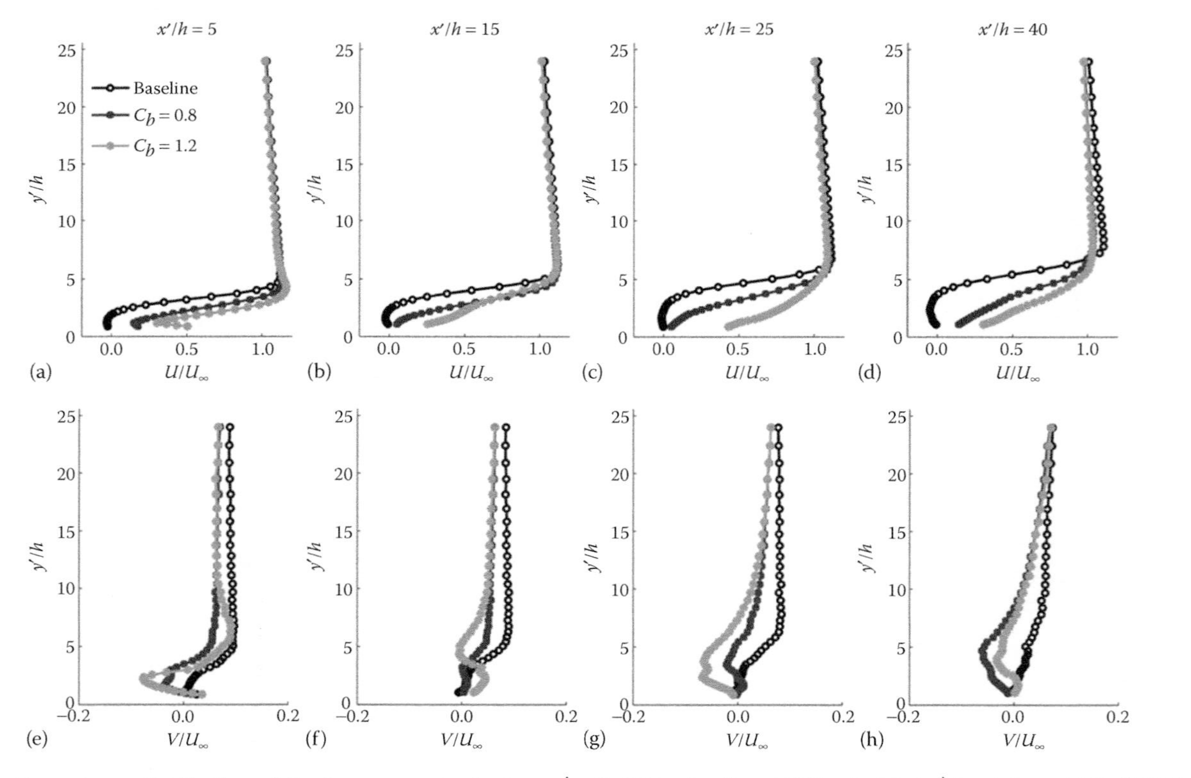

FIGURE 6.10 Cross-stream distributions of the time-averaged velocity at $z' = 0$; U/U_∞ (a–d) and V/U_∞ (e–h) at $x'/h = 5$ (a, e), 15 (b, f), 25 (c, g), and 40 (d, h). $\alpha = 15.5°$. (Reprinted with permission from Elimelech, Y. et al., *Phys. Fluids*, 23, 094104, 6, 2011. Copyright 2011, American Institute of Physics.)

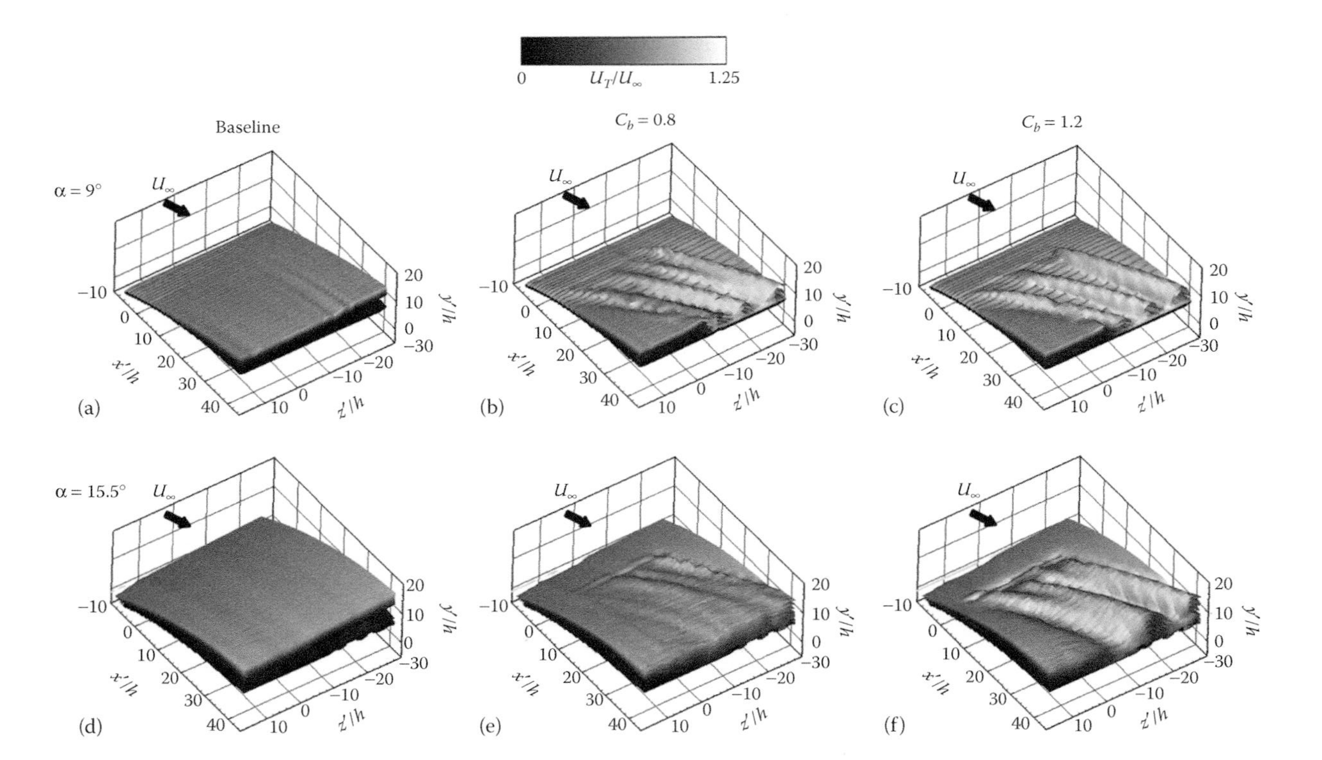

FIGURE 6.11 Iso-total velocity surfaces at $\alpha = 9°$ (a–c) and 15.5° (d–f); baseline (a, d), and forced with $C_b = 0.8$ (b, e) and 1.2 (c, f). (Reprinted with permission from Elimelech, Y. et al., *Phys. Fluids*, 23, 094104, 9, 2011. Copyright 2011, American Institute of Physics.)

The baseline flow fields for both angles of attack are shown here for reference, where the isosurfaces suggest that the flow is separated for $\alpha = 15.5^\circ$ (total velocity surfaces cannot represent separation; however, the presence of a large region near the surface where the total velocity is 0 suggests that separation is present). Moreover, this figure shows that the normal (to the wall) extent of the low-velocity region increases toward the tip (negative z'/h). Activation of the synthetic jet results in an alteration of the flow field, where a higher blowing ratio causes deeper penetration of the jet into the crossflow and the appearance of secondary flow structures closer to the slit. These results are in agreement with the recent findings by Sahni et al. (2011) that explored the interaction of a finite-span synthetic jet with an unswept wing configuration at $\alpha = 0^\circ$. There are a couple of apparent differences between $\alpha = 9^\circ$ and $\alpha = 15.5^\circ$ cases: (1) for the higher angle of attack, the effect of the synthetic jet is felt throughout the entire spanwise measurement domain, and (2) the coalescence of secondary structures is more pronounced at the higher angle of attack. It is noteworthy to mention that the presence of secondary flow structures in a finite-span synthetic jet was previously shown by Amitay and Cannelle (2006) for a finite-span synthetic jet issued into a quiescent flow (in their work, the strength of the synthetic jet was determined using the stroke length, L_0, due to the absence of a freestream). Moreover, Amitay and Cannelle (2006) showed that the strength of the secondary flow structures and their spanwise wavelength vary with the aspect ratio of the synthetic jet slit as well as the stroke length.

To conclude, the flow structures' evolution scenario is described in Figure 6.12. The flow field over a swept configuration (Figure 6.12a) is a superposition of a chordwise boundary layer (denoted by the black boundary layer) and a spanwise boundary layer (denoted by the gray boundary layer). Pairs of spanwise vorticity rollers at high frequency are generated and shed away from the synthetic jet orifice. In addition, chordwise (or streamwise) vorticity is being generated at both edges of the synthetic jet orifice, in opposite directions. As these two vorticity fields do not share the same characteristic timescale, a superposition was measured. This scale separation also explains why the secondary structures are largely seen in the time-averaged results while the coherence of the synthetic jet spanwise rollers is being averaged. By that, the chordwise structures perturb the coherence of the spanwise rollers, tilt and warp them, and eventually break them into vorticity concentrations (Figure 6.12b). In the time-averaged sense (Figure 6.12c), the periodic vorticity field along the synthetic jet orifice creates the secondary flow structures observed.

6.3.3.2 Flow Control on a Dynamically Pitched Finite-Span Wind Turbine Blade

Wind energy is the fastest growing source of electrical energy in the world.* Associated with this growth, industrial interests have focused on producing larger wind turbines

* The results presented in this section are thanks to the experimental work by Mr. K. Taylor, Dr. C. M. Leong, and Professor M. Amitay. Funded by the National Science Foundation (Award Number: 0964989, monitored by Dr. Gregory Rorrer) and New York State (NYS) Energy Research and Development Authority.

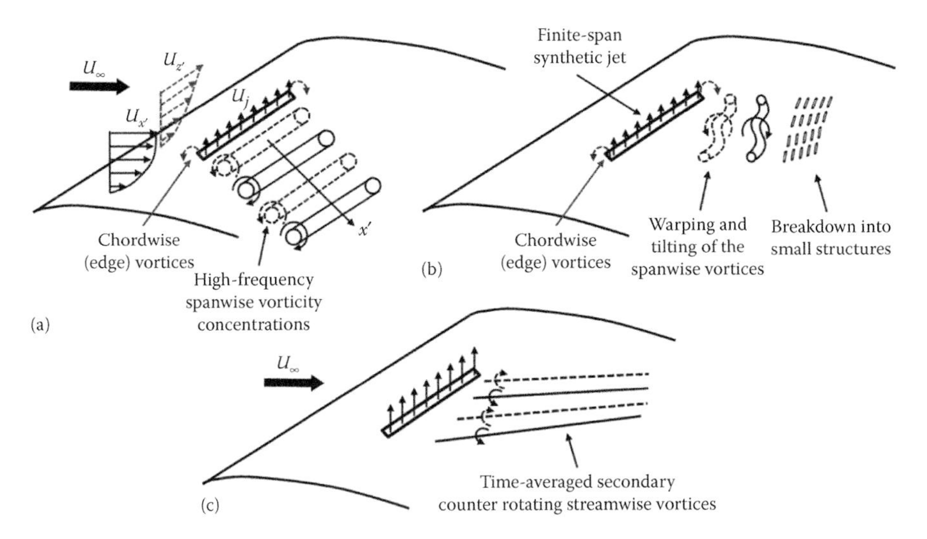

FIGURE 6.12 Schematics of vorticity generation due to the interaction of the finite-span synthetic jet and the crossflow; (a) no interaction, (b) phase-averaged interaction, and (c) time-averaged interaction. (Reprinted with permission from Elimelech, Y. et al., *Phys. Fluids*, 23, 094104, 12, 2011. Copyright 2011, American Institute of Physics.)

than are commonly used today to more efficiently capture the wind energy. However, large turbines also come with significant technical challenges. The structural vibration of the turbine blades could compromise wind energy capture, generate undesirable acoustic emission, and cause blade fatigue and premature failure. Through the application of synthetic jet-based separation control on the blade surface, the structural vibrations present during the dynamic motions of the blade can be significantly reduced. To investigate this, a finite-span S809 airfoil model (displayed in Figure 6.13) was constructed incorporating an array of synthetic jet actuators (with spanwise-oriented rectangular orifices) at $x/C = 0.1$ on the suction surface of the blade. The moments and forces exerted on the blade were measured using a six-component load cell located at the fixed end of the model span. Both static and dynamic loads were measured for changes in the blades' angle of attack to quantify the effect that the synthetic jets had on reducing the static and dynamic stall of blade. The synthetic jets were operated at a continuous sinusoidal actuation frequency of $f_{act} = 2.6$ kHz providing $F^+ = 12.4$. All the measurements were collected for chord-based freestream Reynolds number of 220,000.

The variation of the lift coefficient (C_L), drag coefficient (C_D), and pitching moment coefficient (C_m), with angle of attack (α), for static blade conditions is presented in Figure 6.14a–c, respectively. Without synthetic jet actuation, sharp leading edge stall occurs at $\alpha \approx 18°$. When the synthetic jets were activated with $C_\mu = 3.6 \times 10^{-4}$, the stall characteristics of the S809 airfoil are softened, which, as will be shown later, yields a reduction in blade's vibrations and a significant reduction in hysteresis during dynamic pitch. As C_μ is increased to 1.4×10^{-3}, which corresponds to a doubling of the velocity of the jets, an increase in the lift coefficient is seen with a further smoothing of the transition into stall.

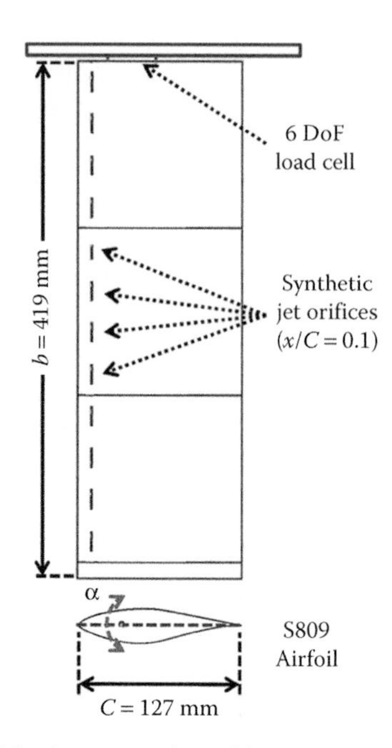

FIGURE 6.13 Schematic of the dynamic pitch model.

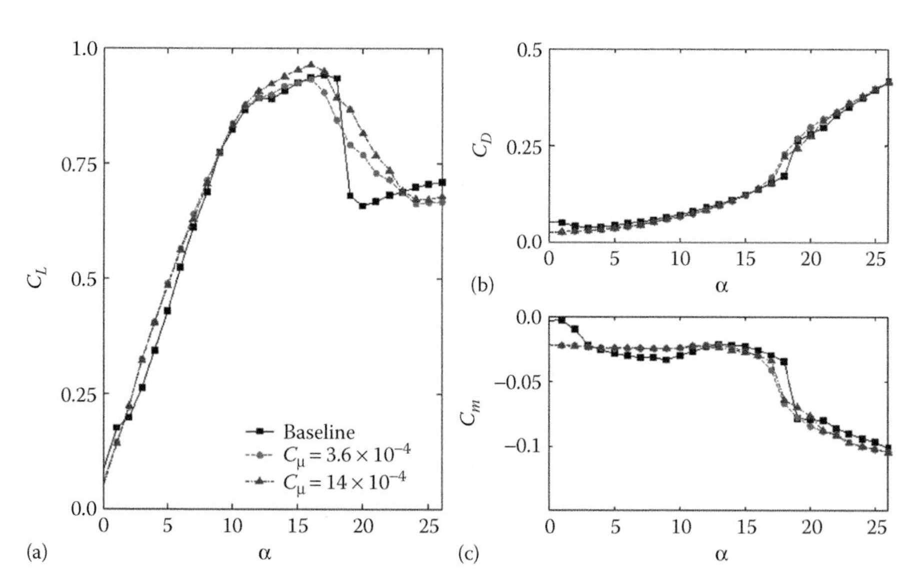

FIGURE 6.14 (a) Lift, (b) drag, and (c) pitching moment coefficient versus angle of attack during static blade conditions.

Next, the effect of synthetic jet actuation during dynamic pitching was examined. To simulate dynamic pitching, the angle of attack was varied sinusoidally as follows:

$$\alpha = \bar{\alpha} + \alpha_A \cdot \sin\left(f_P \cdot t\right)$$

Here, $\bar{\alpha}$ is the mean angle about which the periodic motion is performed, α_A is the amplitude of the motion, and f_P is the frequency of the pitching motion. A wide range of these parameters was investigated; however, for brevity only the following cases are presented: $\bar{\alpha} = 14°$, $\alpha_A = 5.5°$, and $f_P = 0.1$ and 1 Hz. These frequencies correspond to nondimensional frequencies, $k_f = 4.8 \times 10^{-4}$ and 4.8×10^{-3}, respectively (note that $k_f = 1$ is the nondimensional frequency associated with the time of flight). The load cell measurements were collected for each cycle, and all cycles were averaged to produce phase-averaged lift, drag, and pitching moment coefficients for the dynamic pitch cycle.

The variation in the aerodynamic force and moment coefficients versus angle of attack during dynamic pitch is presented in Figure 6.15 for the slower pitching frequency of $f_P = 0.1$ Hz ($k_f = 4.8 \times 10^{-4}$). Without flow control, the dynamic pitch yields a higher lift past $\alpha = 18°$ as the blade pitches up, while the lift coefficient is much lower during the pitch down. This dynamic motion creates a hysteresis in the lift, drag, and pitching moment coefficients as the blade cycles into and out of the stall. When the synthetic jets are activated, the higher lift is maintained through the dynamic stalling, and there is no hysteresis. This demonstrates the potential for active flow control, particularly synthetic jets, to stabilize aerodynamic loads and moments during changing wind conditions. When

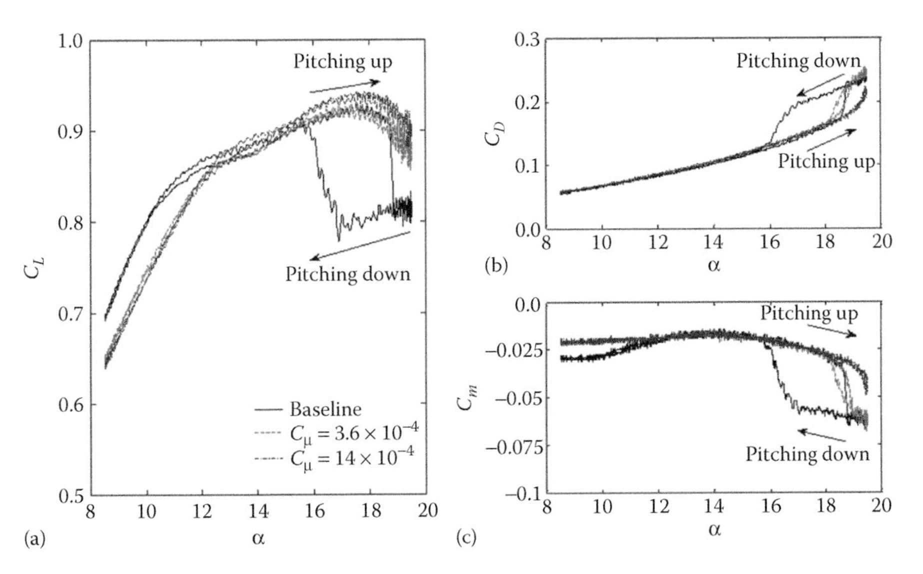

FIGURE 6.15 (a) Lift, (b) drag, and (c) pitching moment coefficient versus angle of attack during the dynamic blade motion where $f_P = 0.1$ Hz ($k_f = 4.8 \times 10^{-4}$).

the jets' momentum coefficient is increased, data do not show a large difference in lift between the two momentum coefficients; however, this is a target of future work to quantify how much momentum is needed to eliminate this hysteresis. Moreover, this shows that it may be advantageous to activate the flow control during only a portion of the cycle, as the use of the jets at the lower angles of attack actually reduces the lift produced. The critical effect illustrated here is the elimination of the hysteresis in lift, drag, and pitching moment coefficients during dynamic motion. The elimination of this hysteresis has the effect of evening out the loading on the blade, reducing stress, which can lead to a longer lifetime for wind turbine blades.

The effect of synthetic jet actuation at an order of magnitude higher than the pitching frequency of $f_p = 1$ Hz ($k_f = 4.8 \times 10^{-4}$) is presented in Figure 6.16. At this higher pitching rate, the severity of the structural variations and the amplitude of the hysteresis are significantly increased. The frequency of the coherent fluctuations during the hysteresis loop matches the natural frequency of the blade model, $\omega_n = 13$ Hz, which was measured separately using a laser vibrometer. Furthermore, the increase in the pitching frequency increases the range of angles of attack that encounters hysteresis. When the jets are activated at the lowest momentum coefficient, there is a reduction in the amplitude of the variance for all coefficients. In addition, the range of hysteresis for all three coefficients is reduced. Increasing the momentum coefficient results in a larger reduction in the variance and the hysteresis during dynamic stall, trending toward elimination. Thus, the use of synthetic jets for dynamic separation control demonstrates the potential to reduce vibrations, increase lift, and reduce drag during dynamic stall conditions. This further suggests that synthetic jet-based separation control could significantly enhance the wind turbine blade life, further improving the wind energy system performance.

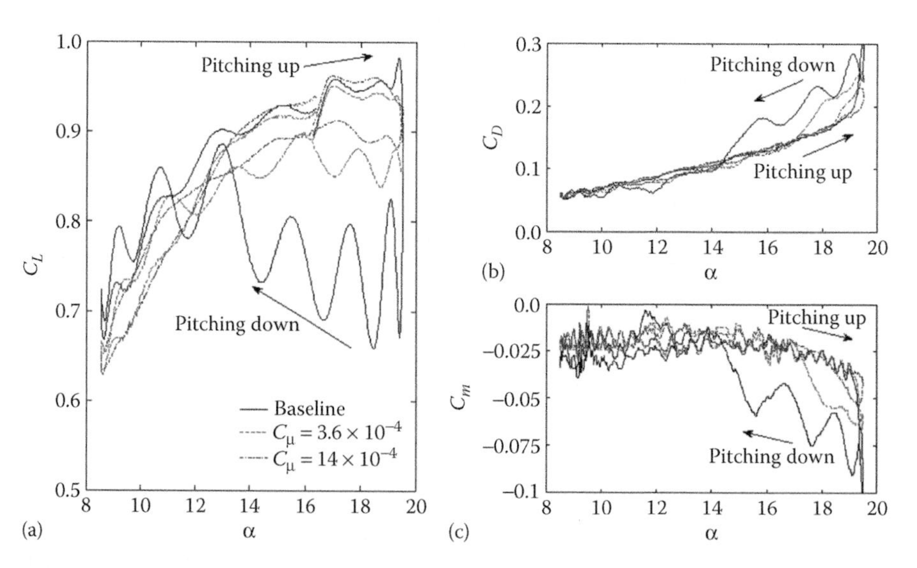

FIGURE 6.16 (a) Lift, (b) drag, and (c) pitching moment coefficient versus angle of attack during the dynamic blade motion where $f_p = 1$ Hz ($k_f = 4.8 \times 10^{-3}$).

6.3.4 Vehicle Flight Control through Separation Control (Modified Cessna 182 Model)

Optimum aerodynamic performance that avoids flow separation on wing surfaces has been traditionally achieved by appropriate aerodynamic design of airfoil sections.* However, when the wing design is driven by nonaerodynamic constraints (stealth, payload, etc.), the forces and moments of the resulting unconventional airfoil shape may be much smaller than a conventional airfoil. Therefore, either active or passive flow control can be used to maintain aerodynamic performance throughout the normal flight envelope. The present section has two objectives: (1) to explore, experimentally, the feasibility of using active flow control, via synthetic jet actuators, as a roll control mechanism instead of conventional ailerons for the small-scale Cessna 182 model; and (2) to develop a closed-loop stall suppression system on a wind tunnel model of the Cessna 182.

The experiments were conducted in a closed-return low-speed wind tunnel, having a test section measuring 60.8 × 60.8 cm, which is equipped with a sting-mounted six-component balance to measure the aerodynamic forces and moments. All forces and moments reported are about the quarter-chord of the wing along the model centerline. The experiments were conducted at chord-based Reynolds numbers of 67,300 and 134,600.

A 1/24th Cessna 182 scaled model (45.7 cm span) was constructed from stereolithography, where the main wing consists of a NACA 2412 section, and a NACA 0005 was used for both the horizontal and vertical tails. The main wing was designed such that different wingtips could be used (Figure 6.17) where aileron deflections from 0° to 18° in 3° increments could be achieved. Note that each aileron's streamwise length is 20% of the root chord on the outer span portions (25% of the span for each aileron) of the wing.

In addition to the aileron deflection wingtips, several wingtips with embedded synthetic jets were also designed and built (Figure 6.17b). The synthetic jet orifice was at $x/C = 0.25$ (near the separation point for high angles of attack). The synthetic jet momentum coefficient (C_μ) was varied between 1.5×10^{-4} and 8.7×10^{-3} (when both wingtips are actuated) and between 7.4×10^{-5} and 4.3×10^{-3} (when a single wingtip is actuated). The synthetic jet wingtips were designed without any taper (C = 67 mm), but with the same planform area and overall span as the aileron wingtips. The synthetic jet-instrumented wingtips have an 18% thick Clark-Y airfoil section (the effect of flow control, via synthetic jet actuators, on a 2D Clark-Y airfoil was previously investigated by Amitay et al. 2001a). A thicker airfoil (than the baseline wingtip) was chosen to provide enough room for the synthetic jet cavities. Moreover, a stall fence was added to the synthetic jet wingtips to prevent any tip effects from imposing on the inner span of the wing. Each wingtip consists of three synthetic jets (at the 25% chord location and 7.5 mm apart), each having a 0.5-mm-wide slit and extending 25.4 mm along the span. Each synthetic jet actuator was driven at an actuation frequency of $f_{act} = 750$ Hz. In addition, a hot-film shear stress sensor was placed downstream the synthetic jet exit (at 35% chord).

* The work in this section has been previously published in Ciuryla et al. (2007). This work was supported as part of Small Business Innovation Research (SBIR) phase I (Contract Number FA8650-05-M-3539, monitored by John Casey and James Myatt).

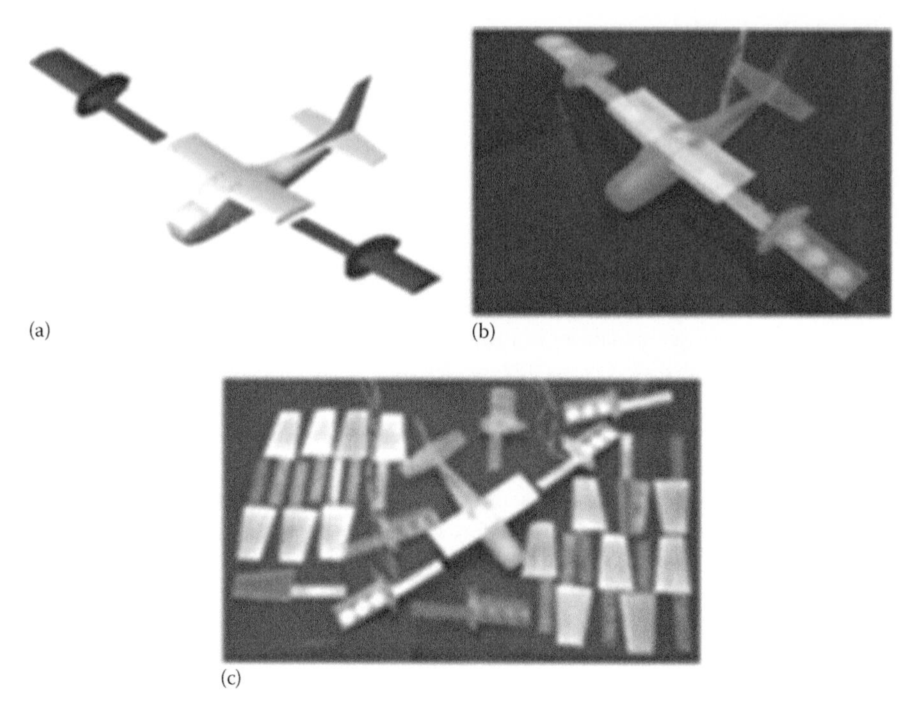

FIGURE 6.17 (a)–(c) Cessna wind tunnel model. (From Ciuryla, M. et al., *J. Aircr.*, 44, 643, 2007. With permission.)

As mentioned earlier, the airfoil used for the synthetic jet wingtips was an 18% thick Clark-Y, which does not have the same aerodynamic performance as the baseline airfoil (NACA 2412). Thus, it is expected that the lift coefficient of the Cessna will be smaller when the synthetic jet-instrumented wingtips are used (without flow control). Figure 6.18b presents a comparison of the lift coefficient (for different angles of attack) for the baseline (aileron) configuration and the synthetic jet configuration (with and without activation of the synthetic jets). Without activation of the jets, the lift coefficient of the synthetic jet configuration has lower performance than the baseline configuration for $\alpha > 2^\circ$. When the synthetic jets are activated (with $C_\mu = 8.6 \times 10^{-3}$), the lift coefficient is recovered (to the baseline values) for $\alpha < 6^\circ$, whereas for $\alpha > 6^\circ$, the lift coefficient surpasses the baseline values by up to 15%. Furthermore, the stall angle is increased by ~2°.

In the data presented in Figure 6.18, the synthetic jets on both wingtips were activated; however, to obtain a rolling moment the synthetic jets on either the left or the right wingtip were activated separately. Figure 6.19 shows the change in the roll coefficient with respect to the corresponding values of the synthetic jets' momentum coefficient. The dashed lines correspond to the starboard synthetic jets activated, and the solid lines represent the port synthetic jets activated. At $\alpha = 0^\circ$, a very small roll moment is obtained and it is similar for all the momentum coefficients that were tested. When the angle of attack is increased to 6°, proportional control is obtained, where as C_μ increases, ΔC_r decreases. Note that

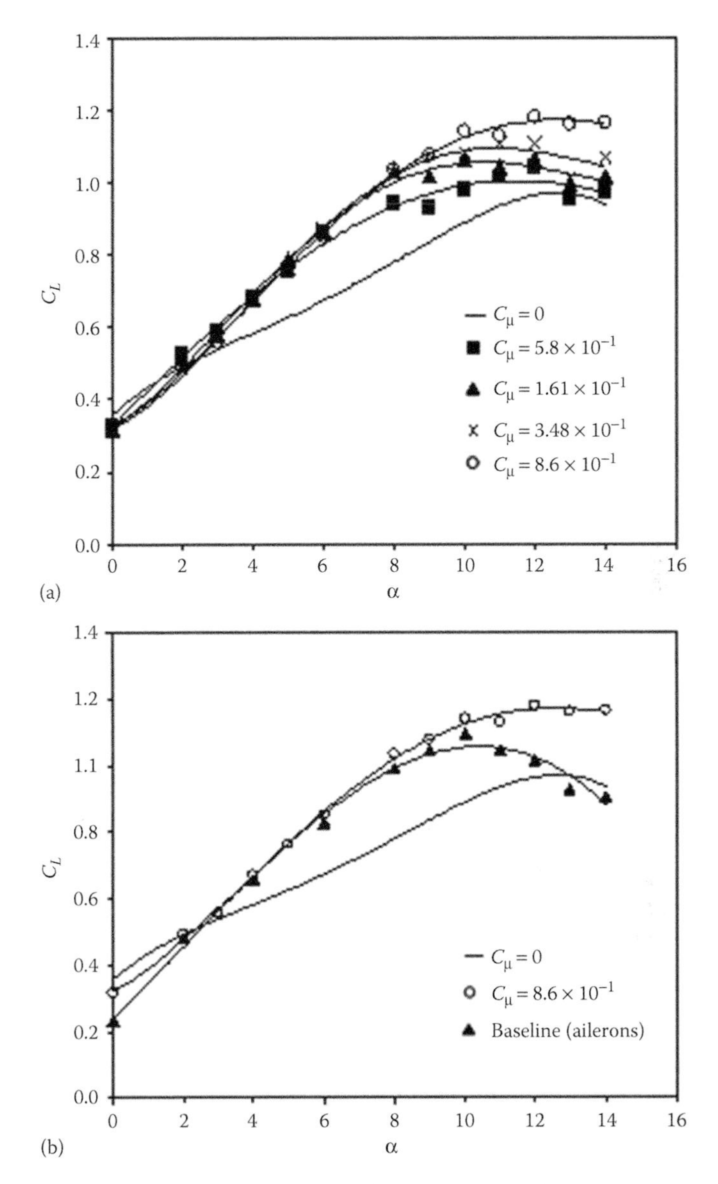

FIGURE 6.18 The lift coefficient versus angle of attack: (a) synthetic jets wingtips and (b) comparison with the baseline Cessna. $Re = 67,300$. (From Ciuryla, M. et al., *J. Aircr.*, 44, 645, 2007. With permission.)

similar trends were obtained at all angles between 5° and 9°, and the 6° angle is shown here as a representative case. At higher angles of attack, where the flow is completely separated over the wingtips (without flow control), the effect of the momentum coefficient is different. At $\alpha = 10°$, for low C_μ, increase in C_μ yields a higher rolling moment, whereas for $C_\mu > 1.73 \times 10^{-3}$, as C_μ increases ΔC_r decreases. At angles of attack past the stall angle,

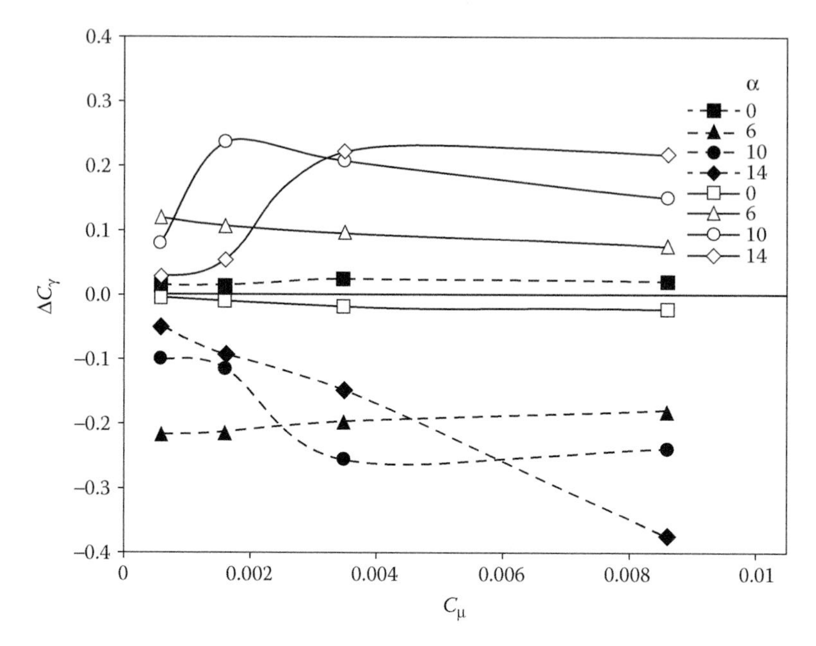

FIGURE 6.19 The change of the roll moment (with respect to the unforced case) versus the momentum coefficient at different angles of attack. The dashed lines correspond to the starboard synthetic jets activated, while the solid lines represent the port synthetic jets activated. (From Ciuryla, M. et al., *J. Aircr.*, 44, 646, 2007. With permission.)

low momentum coefficient is not sufficient to reattach the flow over the wingtip and thus low rolling moment is achieved. However, high momentum coefficient jets can reattach the flow and create a rolling moment. It is noteworthy that the magnitudes of the rolling moments obtained with synthetic jet actuators is similar to those obtained with ailerons at deflection angles of 12° and smaller.

A simple closed-loop control routine was developed to suppress separation over the wingtips using a shear stress sensor to detect the separation and the synthetic jet actuators to reattach the flow. The separation was observed qualitatively, using tuft flow visualization, and detected quantitatively, using the shear stress sensor. Figure 6.20a and b present the time trace of the shear stress sensor output voltage, and the tuft flow visualization over the wingtip, respectively, at $\alpha = 0°$. At this low angle of attack, the flow is completely attached over the wingtip as indicated by the tufts and the very low root mean square (rms) of the shear stress sensor output. Note that the shear stress sensor was not calibrated and is used here only to detect separation based on the rms of its output signal.

As the angle of attack is increased to $\alpha = 8.5°$, the flow over the wingtips is separated as indicated by the high rms values from the shear stress (Figure 6.21a) and confirmed by the tufts, which clearly show reversed flow over parts of the wingtip (Figure 6.21b). The time trace of the shear stress sensor exhibits two main frequencies of ~10 and ~100 Hz, which correspond to the model vibration and the shedding frequency of the separated flow over

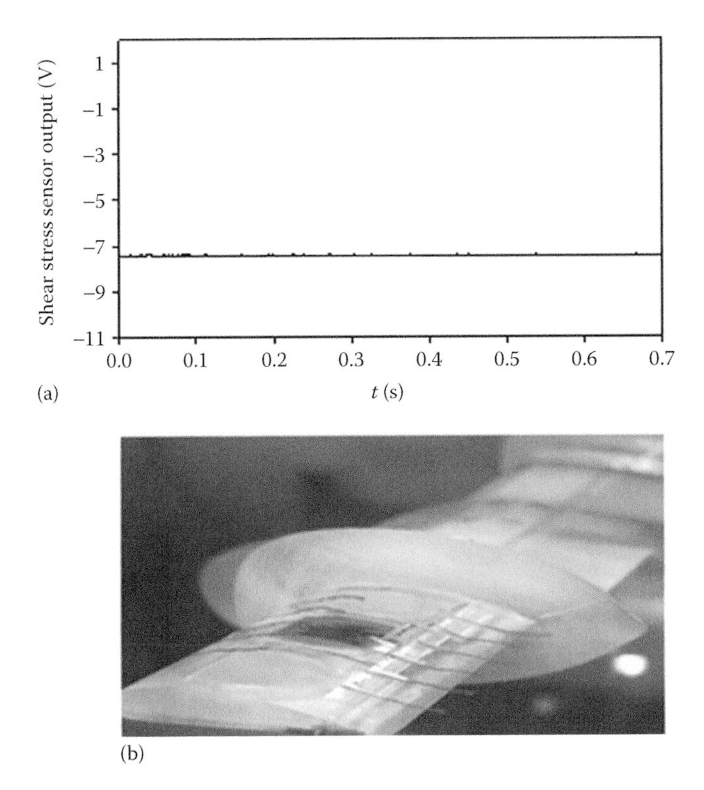

FIGURE 6.20 (a) The time trace of the shear stress sensor output and (b) tuft flow visualization over the wingtip. $\alpha = 0°$, synthetic jets off. (From Ciuryla, M. et al., *J. Aircr.*, 44, 646, 2007. With permission.)

the wingtip, respectively. Note that visual examination of the wingtips at these conditions confirmed separation and vibration through the buffeting of the wingtips.

When the synthetic jets are activated with $C_\mu = 8.6 \times 10^{-3}$, the flow is completely reattached as indicated by the low rms levels of the shear stress sensor (Figure 6.22a) and the tufts (Figure 6.22b). Note the presence of the low-amplitude periodic oscillations in the shear stress sensor signal at the synthetic jets actuation frequency, due to its proximity to the synthetic jet actuators. Moreover, the actuation of the synthetic jets eliminated completely the low frequency model vibrations (visually confirmed). In addition, in the anemometry system used in these experiments, a decrease in the output voltage of the sensor corresponds to an increase in the shear stress; thus, when the synthetic jets are activated the dc level of the shear stress sensor is much lower than that of the uncontrolled case (Figure 6.21a), which is indicative of an increase in the shear, as expected when the flow is reattached. From the shear stress data it was concluded that either the sensor's dc level or its rms (or both) can be used to detect separation. In the present work, we used the rms as the separation indicator.

From the data in Figures 6.20 through 6.22, it is clear that a stall detection system can be implemented simply by continuously monitoring the rms of the shear stress sensor

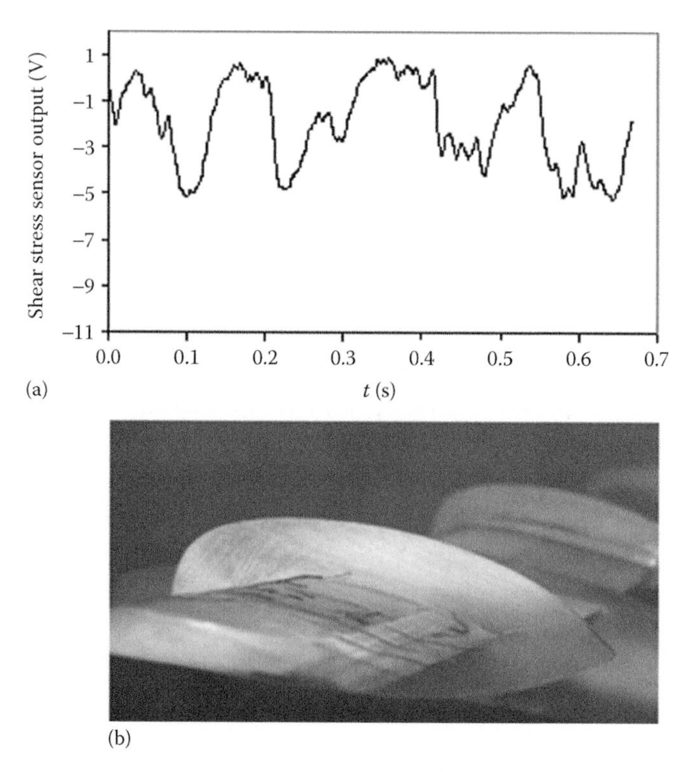

FIGURE 6.21 (a) The time trace of the shear stress sensor output and (b) tuft flow visualization over the wingtip. $\alpha = 8.5°$, synthetic jets off. (From Ciuryla, M. et al., *J. Aircr.*, 44, 647, 2007. With permission.)

reading. Thus, a simple closed-loop stall suppression system was implemented by choosing a threshold rms value which, if reached, would trigger the activation of the synthetic jets to reattach the flow. In what follows, the angle of attack was slowly increased until stall occurred. Then, based on the rms threshold criterion, the synthetic jets were automatically activated through a D/A board, using a LabVIEW code.

Figure 6.23a–c presents the time trace of the shear stress sensor output as the angle of attack increases monotonically for different rms thresholds. Note that in these figures, the time axis has been arbitrarily set to zero when the flow is on the verge of separation from the airfoil's upper surface. In the first experiment, the rms threshold was set to 0.5 V (Figure 6.23a). As time progresses (i.e., as the angle of attack increases), the shear stress output voltage increases (corresponding to a decrease in the shear stress). When the angle of attack is ~8° ($t = 0$), the flow separates, which is indicated by the increase in the rms as well as the jump in the shear stress reading (marked by the arrow). It takes about 2.5 seconds before the control is triggered and the synthetic jets are activated, resulting in a complete flow reattachment, as indicated by the significant decrease in the shear stress sensor output voltage. Figure 6.23 also includes a sketch of the roll moment (from the sting balance reading), which shows that when the flow is separated a roll moment is generated

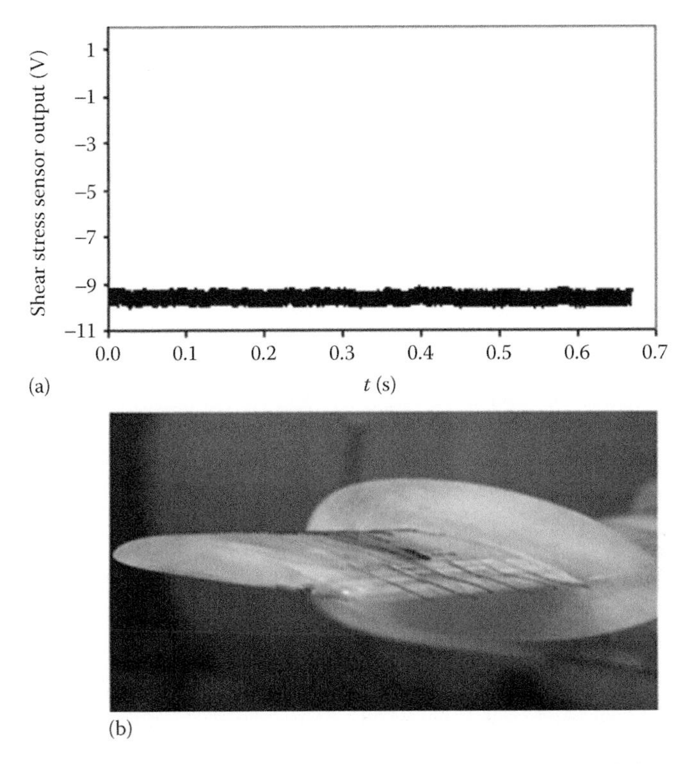

FIGURE 6.22 (a) The time trace of the shear stress sensor output and (b) tuft flow visualization over the wingtip. $\alpha = 8.5°$, synthetic jets on ($C_\mu = 8.6 \times 10^{-3}$). (From Ciuryla, M. et al., *J. Aircr.*, 44, 647, 2007. With permission.)

(due to a slight asymmetry of the model that yields an asymmetric flow separation on the wings). When the synthetic jets are activated, the roll moment diminishes due to flow reattachment on both wings, resulting in a symmetric flow over the wingtips.

Next, the rms threshold was reduced to 0.25 V (Figure 6.23b). Again, the flow separates and it takes ~0.5 seconds before the synthetic jets are activated (compared to 2.5 s for the 0.5 V rms threshold case). When the rms threshold is reduced to 0.1 V (Figure 6.23c), the threshold was reached before the wingtips ever stalled. This suggests that through the selection of an appropriate shear stress rms threshold, synthetic jets can not only recover a wing from a stall, but also avoid it altogether. Furthermore, a roll moment is not generated.

This method demonstrated what was considered a "simple closed-loop stall suppression system," because the rms value on the feedback sensor was only used to trigger the activation of the synthetic jet actuators at a constant frequency and input voltage. Additionally, once the threat of flow separation was detected and the synthetic jets were activated, they remained on and the feedback sensor was no longer monitored. Clearly, in a realistic application, a more complex feedback system would be desired, continually utilizing the information from the feedback sensor to control the amplitude and frequency of the actuation or turn off the actuation after the threat of flow separation (and wing stall)

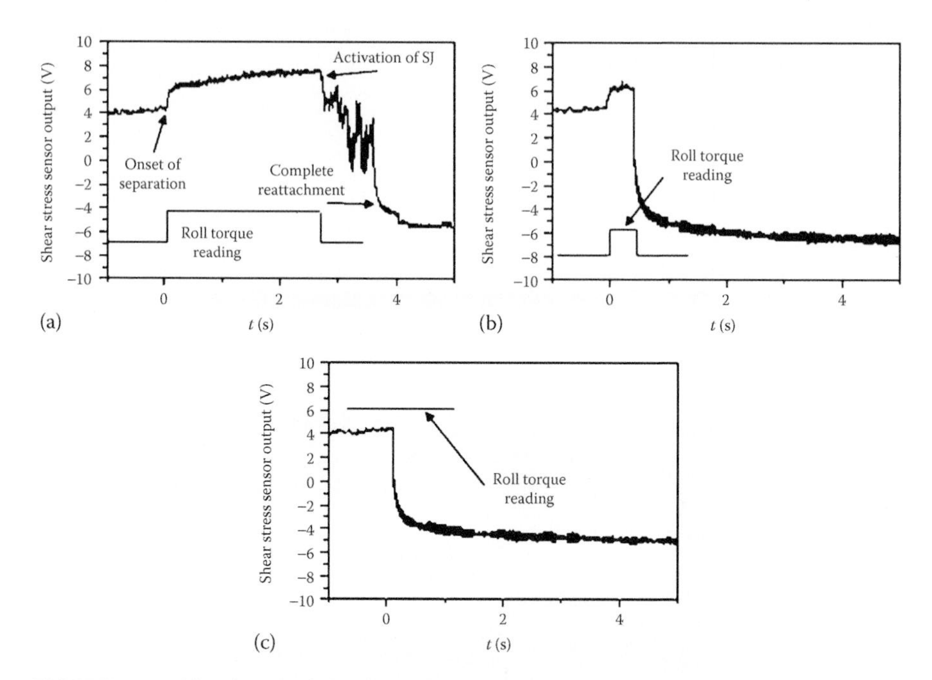

FIGURE 6.23 The change of the shear stress sensor output with time (as the angle of attack increases). (a) rms = 0.5 V, (b) 0.25 V, and (c) 0.1 V. (From Ciuryla, M. et al., *J. Aircr.*, 44, 648, 2007. With permission.)

has passed. At the time this work was conducted, it demonstrated the first implementation of closed-loop synthetic jet flow control on a 3D configuration (to the best of the authors' knowledge), and several more complex implementations of closed-loop synthetic jet-based flow control have been successfully accomplished.

6.4 Summary and Conclusions

In this chapter, a detailed discussion of synthetic jet-based separation control was presented initially outlining several key operational considerations before detailing a collection of experimental studies. In Section 6.2, three operational considerations were detailed to produce effective control of flow separation with synthetic jet actuators. Specifically, these are as follows: (1) the location of the actuation, (2) the role that the synthetic jet actuation frequency plays, and (3) the proper quantification and calibration of the synthetic jet amplitude. The importance of each of these parameters were then demonstrated in Section 6.3 through the collection of six experimental studies covering examples from bluff body separation behind a circular cylinder to separation over finite wings and vehicle stall on a scale wind tunnel model of a Cessna 182 aircraft.

The importance of properly locating the fluidic actuation was briefly mentioned in Section 6.2; however, it was better demonstrated in Section 6.3.1.2 where a single synthetic jet actuator was utilized to modify the flow separation and the wake of a finite-span circular cylinder. From the surface pressure measurements, it was demonstrated that as the synthetic jet orifice's position was rotated relative to the oncoming flow, the effect on the flow separation drastically changed. Specifically, the extent and size of flow reattachment were both significantly altered by only rotating the synthetic jet position from $\theta_{jet} \approx 50^{\circ}$ to 100°. Even with these large variations, the separation control still proved to be most effective in the local vicinity of the flow separation, that is, near $\theta_{jet} = 80^{\circ}$. This demonstrated that, as a first step of locating the fluidic control, near the point of separation is a valid choice; however, a thorough understanding of the flow field with and without control is still vital as the effectiveness can be drastically different for different applications.

The role that the synthetic jet actuation frequency plays in controlling the flow separation has been extensively studied over the past decade and was thoroughly discussed in Section 6.2.1. Clearly, the versatility of synthetic jets allows them to function effectively in both the low- and high-frequency regimes (by driving them with different pulse modulation waveforms). In this chapter, the low-frequency regime was considered where the synthetic jet actuation frequency was matched to the inherent dominate frequencies within the flow (i.e., the unstable modes of a separating shear layer). In this mode of operation, the fluidic actuation amplifies the unstable modes and causes a large time-periodic form of flow reattachment to occur; however, the effectiveness diminishes as the actuation frequency is moved away from the dominate frequency in the flow. Another mode of operation that was discussed was to actuate the synthetic jets at a high frequency, at least one order of magnitude above the most dominate frequency in the flow field. This produces a much more steady form of flow reattachment and is less dependent on exactly matching the frequency of the fluidic actuation to the dominate frequency in the flow field. The experimental example of separation control on 2D airfoil in Section 6.3.2 provided a clear demonstration and comparison of both of these modes of operation. Again, this displays that to effectively control flow separation, a thorough understanding of the unforced flow field is required a priori, including the expected dominate frequencies within the flow field.

The third operational consideration (outlined in Section 6.2.2) was the proper quantification and calibration of the synthetic jet amplitude. One of the primary benefits of active synthetic jet control is that the actuation amplitude can be adjusted, and thus, the minimum amount of energy can be added to the flow when needed. It also allows for the possibility of proportional control through fluidic actuation that was demonstrated in Section 6.3.4 on the scale wind tunnel model of the Cessna 182. However, to accurately scale these effects and compare them with other experiments and methods, a consistent and accurate normalization of the actuator amplitude is required. As a result, both the blowing coefficient (C_b) and momentum coefficient (C_{μ}) were presented to provide this comparison. However, neither one is sufficient on its own to completely quantify the actuator strength. In fact, certain flow situations have been shown to scale better with each of these parameters independent of the other. To be thoroughly accurate, researchers should compare and document the influence of both parameters.

Beyond the operational considerations that were presented in Section 6.2, the examples presented in Section 6.3 displayed several other critical features of synthetic jet-based separation control. First, as demonstrated in the separation and wake control of a finite cylinder (Section 6.3.1.2), separated flows are globally receptive to very small local perturbations. This was demonstrated by the fact that a single synthetic jet actuator of negligibly small thickness relative to the cylinder's span was able to modify the separation line across almost the entire span of the cylinder and thus significantly alter the wake distribution of the bluff body. Thus, if properly implemented, relatively weak local active fluidic actuation can induce significant global effects on a flow field.

Second, with a push toward applying flow control in real systems, where the flow fields are inherently more three dimensional, one needs to understand the efficacy of flow control in these types of complex flows. The application of a single synthetic jet on a swept-back finite wing (Section 6.3.3.1) demonstrated an application of synthetic jet-based separation control in a controlled 3D environment. However, it was observed that the three-dimensionalities of the synthetic jet actuation (i.e., the streamwise edge vortices formed from the finite span of the synthetic jet orifice) are also critically important and appear to dominate the interaction. These secondary streamwise vorticity concentrations seem to contribute to the flow reattachment, as they enhance mixing near the surface through inducing velocity into the boundary layer near the surface. Clearly, this demonstrates that even in the 2D flow scenarios, three-dimensionalities are present and should be considered.

Additionally, along with the 3D effects, unsteady flows are common in real-world applications, and synthetic jet actuators have the potential to control these time-varying situations owing to their relatively high bandwidth. In Section 6.3.3.2, synthetic jet-based separation control was implemented on a dynamically pitching finite-span model of a wind turbine blade. In this example, actuation has the capability to reduce the hysteresis loop in the forces and moment and the structural vibrations.

Finally, the example of a simple closed-loop stall suppression system on a wind tunnel model of the Cessna 182 (Section 6.3.4) demonstrated that through sensing and controlling the flow field in time, incipient separation could be identified and mitigated. Clealy, the combination of the two works (Sections 6.3.3.1 and 6.3.3.2) displays the potential of synthetic jets to control unsteady flows.

Looking toward the future, the application of synthetic jet-based separation control has a very positive outlook. Over the past two decades, the basic research has demonstrated a high level of success such that many companies are now interested in ways that separation control (and more specifically synthetic jet-based separation control) can be intelligently implemented into the design of real-world engineering systems. Synthetic jets are used for controlling the separation in compact inlet ducts so that the size and weight of the UAV's propulsion system can be significantly reduced, enhancing the effectiveness of control surfaces at high deflection angles, which could potentially improve a vehicles overall efficiency in cruise through reducing the required size constraints of control surfaces.

On the basic research side, there is still much more work to be done as well. Specifically, in the flow control community, much work in separation control is still limited to relatively 2D flow fields. Implementation and understanding of how synthetic jet-based separation control can be effectively implemented in highly 3D flow fields, where strong vortical structures dominate, is still needed. Additionally, opportunities exist to further understand the time-dependent growth and interaction of the flow structures produced by the synthetic jets and their interaction with the surrounding flow field.

With a strong societal push toward efficiency improvement in just about every aspect of our daily lives. Synthetic jet-based separation control provides the opportunity to significantly enhance many engineering systems prevalent throughout the world. From the fields of renewable energy production (i.e., wind and ocean energy) to the aircraft industry, to the design and construction of intelligent buildings, and many more, there are countless application areas that can benefit from the intelligent implementation of flow control and even more that have yet to be realized.

References

Afgan, I., C. Moulinec, R. Prosser, and D. Laurence. 2007. Large eddy simulation of turbulent flow for wall mounted cantilever cylinders of aspect ratio 6 and 10. *International Journal of Heat and Fluid Flow*, 28: 561–574.

Ahuja, K. K. and R. H. Burrin. 1984. Control of flow separation by sound. *AIAA Paper* Williamsburg, VA, October 15–17.

Amitay, M. and F. Cannelle. 2006. Evolution of finite span synthetic jets. *Physics of Fluids*, 18(5): 054101.

Amitay, M. and A. Glezer. 1999. Aerodynamic flow control of a thick airfoil using the synthetic jet actuators. *Proceedings of the 3rd ASME/JSME Joint Fluids Engineering Conference*, San Francisco, CA, July 18–23.

Amitay, M. and A. Glezer. 2002a. Role of actuation frequency in controlled flow reattachment over a stalled airfoil. *AIAA Journal*, 40: 209–216.

Amitay, M. and A. Glezer. 2002b. Controlled transients of flow reattachment over stalled airfoils. *International Journal of Heat and Fluid Flow*, 23: 690–699.

Amitay, M. and A. Glezer. 2006. Flow transients induced on a 2-D airfoil by pulse-modulated actuation. *Experiments in Fluids*, 40: 329–331.

Amitay, M., A. Honohan, M. Trautman, and A. Glezer. 1997. Modification of the aerodynamic characteristics of bluff bodies using fluidic actuators. *AIAA Paper* 97-2004, Snowmass, CO, June 29–July 2.

Amitay, M., M. Horvath, M. Michaux, and A. Glezer. 2001a. Virtual aerodynamic shape modification at low angles of attack using synthetic jet actuators. *AIAA Paper* 2001-2975, Anaheim, CA, June 11–14.

Amitay, M., V. Kibens, D. E. Parekh, and A. Glezer. 1999. Flow reattachment dynamics over a thick airfoil controlled by synthetic jet actuators. *AIAA Paper* 99-1001, Reno, NV, January 11–14.

Amitay, M., D. Pitt, and A. Glezer, A. 2002. Separation control in duct flows. *Journal of Aircraft*, 39: 616–620.

Amitay, M., D. R. Smith, V. Kibens, D. E. Parekh, and A. Glezer. 2001b. Modification of the aerodynamics characteristics of an unconventional airfoil using synthetic jet actuators. *AIAA Journal*, 39(3): 361–370.

Amitay, M., A. E. Washburn, S. G. Anders, and D. E. Parekh. 2004. Active flow control on the stingray UAV: Transient behavior. *AIAA Journal*, 42(11): 2205–2215.

Amitay, M., A. E. Washburn, S. G. Anders, D. E. Parekh, and A. Glezer. 2003. Active flow control on the stingray UAV: Transient behavior. *AIAA Paper* 2003-4001, Orlando, FL, June 23–26.

Chang, R.-C., F.-B. Hsiao, and R.-N. Shyu. 1992. Forcing level effects of internal acoustic excitation on the improvement of airfoil performance. *Journal of Aircraft*, 29(5): 823–829.

Ciuryla, M., Y. Liu, J. Farnsworth, C. Kwan, and M. Amitay. 2007. Flight control using synthetic jets on a Cessna 182 model. *Journal of Aircraft*, 44: 642–653.

Crook, A., A. M. Sadri, and N. J. Wood. 1999. The development and implementation of synthetic jets for the control of separated flow. *AIAA Paper* 99-3176, Norfolk, VA, June 28–July 1.

DeMauro, E. P., C. M. Leong, and M. Amitay. 2012. Modification of the near wake behind a finite-span cylinder by a single synthetic jet. *Experiments in Fluids*, 53(6): 1963–1978. doi:10.1007/s00348-012-1413-2.

Elimelech, Y., J. D. Vasile, and M. Amitay. 2011. Secondary flow structures due to interaction between a finite-span synthetic jet and a 3-D cross flow. *Physics of Fluids*, 23: 094104. doi:10.1063/1.3632089.

Farivar, D. J. 1981. Turbulent uniform flow around cylinders of finite length. *AIAA Journal*, 19(3): 275–281.

Glezer, A. and M. Amitay. 2002. Synthetic jets. *Annual Review of Fluid Mechanics*, 34: 503–529.

Glezer, A., M. Amitay, and A. Honohan. 2005. Aspects of low- and high-frequency actuation for aerodynamic flow control. *AIAA Journal*, 43: 1501–1511.

He, Y., A. W. Cary, and D. A. Peters. 2001. Parametric and dynamic modeling for synthetic jet control of a post-stall airfoil. *AIAA Paper* 2001-0733, Reno, NV, January 4–7.

Ho, C.-M. and P. Huerre. 1984. Perturbed free shear layers. *Annual Review of Fluid Mechanics*, 16: 365–422.

Holman, R., Y. Utturkar, R. Mittal, B. L. Smith, and L. Cattafesta. 2005. Formation criterion for synthetic jets. *AIAA Journal*, 43: 2110–2116.

Hsiao, F.-B., C.-F. Liu, and J.-Y. Shyu. 1990. Control of wall-separated flow by internal acoustic excitation. *AIAA Journal*, 28(8): 1440–1446.

Huang, L. S., L. Maestrello, and T. D. Bryant. 1987. Separation control over an airfoil at high angles of attack by sound emanating from the surface. *AIAA Paper* 87-1261, Honolulu, HI, June 8–10.

Kawamura, T., M. Hiwada, T. Hibino, I. Mabuchi, and M. Kumada. 1984. Flow around a finite circular cylinder on a flat plate. *Bulletin of JSME*, 27(232): 2142–2151.

Maldonado, V., J. Farnsworth, W. Gressick, and M. Amitay. 2009. Active control of flow separation and structural vibrations of wind turbine blades. *Wind Energy*, 13(2/3): 221–237.

Matejka, M., P. Pick, J. Nozicka, and P. Prochazka. 2009. Experimental study of influence of active methods of flow control on the flow field past a cylinder. *Journal of Flow Visualization and Image Processing*, 16: 353–366.

Norberg, C. 1993. An experimental investigation of the flow around a circular cylinder: Influence of aspect ratio. *Journal of Fluid Mechanics*, 258: 287–316.

Oster, D. and I. Wygnanski. 1982. The forced mixing layer between parallel streams. *Journal of Fluid Mechanics*, 123: 91–130.

Parekh, D. E., S. P. Williams, M. Amitay, A. Glezer, A. E. Washburn, I. M. Gregory, and R. C. Scott. 2004. Active flow control on the stingray UAV: Aerodynamic forces and moments. *AIAA Paper* 2003-4002, Orlando, FL, June 23–26.

Roberts, F. A. 1985. Effects of periodic disturbances on structure of mixing in turbulent shear layers and wakes. PhD thesis, California Institute of Technology, Pasadena, CA.

Sahni, O., J. Wood, K. E. Jansen, and M. Amitay. 2011. 3-D interactions between a finite-span synthetic jet and a cross-flow. *Journal of Fluid Mechanics*, 671: 254–287.

Seifert, A., T. Bachar, D. Koss, M. Shepshelovich, and I. Wygnanski. 1993. Oscillatory blowing: A tool to delay boundary-layer separation. *AIAA Journal*, 31: 2052–2060.

Seifert, A., A. Darabi, and I. Wygnanski. 1996. Delay of airfoil stall by periodic excitation. *Journal of Aircraft*, 33(4): 691–698.

Smith, B. L. and A. Glezer. 2002. Jet vectoring using synthetic jets. *Journal of Fluid Mechanics*, 458: 1–34.

Smith, B. L. and G. Swift. 2001. Synthetic jets at large reynolds number and comparison to continuous jets. *AIAA Paper* 2001-3030, Anaheim, CA, June 11–14.

Smith, D. R., M. Amitay, V. Kibens, D. E. Parekh, and A. Glezer. 1998. Modification of lifting body aerodynamics using synthetic jet actuators. *AIAA Paper* 98-0209, Albuquerque, NM, June 15–18.

Shuster, J. M. and D. R. Smith. 2007. Experimental study of the formation and scaling of a round synthetic jet. *Physics of Fluids*, 19: 045109.

Sumner, D., J. L. Heseltine, and O. J. P. Dansereau. 2004. Wake structure of a finite circular cylinder of small aspect ratio. *Experiments in Fluids*, 37: 720–730.

Uematsu, Y. and M. Yamada. 1993. Aerodynamic forces on circular cylinders of finite height. *Journal of Wind Engineering & Industrial Aerodynamics*, 51: 249–265.

Williams, D., M. Acharya, J. Bernhardt, and P. Yang. 1991. The mechanism of flow control on a cylinder with the unsteady bleed technique. *AIAA Paper* 91-0039, Reno, NV, January 11–12.

Wygnanski, I. 2000. Some new observations affecting the control of separation by periodic excitation. *AIAA Paper* 2000-2314, Denver, CO, June 19–22.

7

Application of Synthetic Jets to Controlling Dynamically Changing Flows

David R. Williams
Illinois Institute of Technology

7.1 Introduction

The majority of research related to the development or applications of flow control techniques has been done under steady-state conditions. Even though synthetic jet actuators are unsteady in the sense that they produce an oscillating exit velocity, the actuator frequency and external flow conditions are usually fixed, which makes it a stationary problem. However, in many practical situations, it becomes necessary to adjust the actuator to the changing flow or flight conditions, which make problems inherently unsteady. Just as conventional ailerons must be proportionally deflected on an aircraft wing during a turning maneuver, so must the modern active flow control (AFC) actuators provide variable levels of control at some rate determined by the maneuver. A key question in the application of AFC becomes how do we achieve time-varying flow control with modern AFC actuators? Following that we want to know how fast the AFC can be applied, that is, what is the achievable bandwidth with AFC? This chapter examines some of the issues related to these questions by considering the "lessons learned" from recent attempts to control lift on wings in an unsteady freestream.

For small angles of attack (α), conventional control surfaces (ailerons, elevators, and rudders) operate in attached flow conditions that require only small deflections and these can be modeled with linear transfer functions (Yechout et al. 2003). At higher angles of attack (typically $\alpha > 10^\circ$), the flow over the airfoil begins to separate, and control surfaces become less effective. When the angle of attack is increased even further ($\alpha > 15^\circ$), the separation point moves upstream until it reaches the leading edge; at this point, the flow is fully separated and conventional control surfaces often become completely ineffective.

Synthetic jet actuators enable some level of control during separated flow conditions by interacting with naturally occurring flow instabilities. Enhanced disturbance amplitudes in the separating shear layer lead to increased entrainment and reduced size of the separated flow resulting in increased lift. An example of the lift enhancement that can be achieved with steady-state AFC on a low-aspect ratio three-dimensional (3D) wing is shown in Figure 7.1 (Collins 2008). The lower curve, corresponding to the naturally occurring lift without AFC, shows that separation occurs for $\alpha > 15^\circ$ and causes stall. Separation is delayed when the flow control is activated, and a maximum in lift coefficient occurs near $\alpha > 26^\circ$. However, when the wing is pitching $[\alpha(t)]$, changing height (plunging) $[h(t)]$, or accelerating $[u(t)]$, additional forces such as added mass will influence the lift. Dynamic stall vortices may form when the pitch rate is large (Leishman 2000), and response of the lift to actuation can occur in the form of time delays, hysteresis, and amplitude changes, making it more challenging to control the lift during a maneuver.

The discussion in this chapter will focus primarily on airfoil lift enhancement by control of separated flows using synthetic jet and pulsed-blowing actuators, although the techniques discussed have broad applications to a wide range of flows, flight vehicles, and actuators. A few examples of dynamically changing flows that can benefit from time-varying actuation are as follows:

1. *Separation control/delay to reduce stall speed.* AFC is turned "on" prior to reaching the critical angle of attack. The objective is to reduce stall speed and hence reduce the landing speed, approach attitude, and landing distance (Maines et al. 2009). The deployment of conventional flaps is done gradually (in steps) as the aircraft slows down. Similarly, AFC must be coordinated with the changing flight speed.
2. *Rapid maneuvering, such as the perch-landing maneuvers for micro-air vehicles (MAVs).* The objective is to maintain a constant lift while the flight speed rapidly decreases toward zero and the vehicle angle of attack increases to 90° in order to

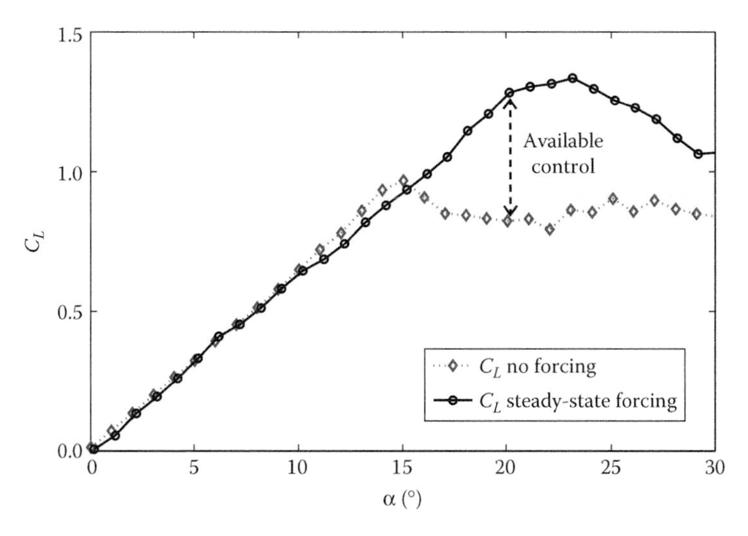

FIGURE 7.1 Typical static lift curve showing lift enhancement on a 3D wing with AFC. The vertical arrow at $\alpha = 20°$ indicates the range of "control authority" available. (From Collins, J. G., Closed loop active flow control for excitation and stabilization of leading edge vortex phenomenon, MSc thesis, Illinois Institute of Technology, July 2008. Reprinted with permission.)

land on a vertical wall. AFC will provide lift during the final stages of the maneuver when the speed is low and the angle of attack is beyond the stall angle.

3. *Gust suppression and high-cycle fatigue reduction.* The control objective is to maintain constant lift in an unsteady flow field that has significant changes in flow speed (changing dynamic pressure) and direction (changing effective angle of attack) (Williams et al. 2010; Kerstens et al. 2011). The actuation strategy invokes conventional control surface deflections for low angles of attack and small amplitude gusts and modern AFC actuators (synthetic jets) for higher angles of attack and larger amplitude gusts.
4. *Wave energy extraction.* Using variable control of the lift/drag ratio on an airfoil in proper phase and amplitude relative to the incoming gust-wave field, it is possible to extract energy from the unsteady flow field. This can be done with variable pitch control surfaces and AFC at higher angles of attack. An example of AFC for energy extraction from ocean waves is the cyclic wave energy conversion device (Siegel et al. 2011) shown schematically in Figure 7.2.
5. *Forebody vortex control on slender bodies.* The objective is to maintain lateral (yaw) control of the flight vehicle at high angles of attack ($\alpha > 50°$) during a pitch maneuver or a coning motion maneuver. This is achieved by using AFC to manipulate asymmetric vortex shedding near the tip of the body (Bernhardt and Williams 2000).

Consider an example of unidirectional, transient control, such as the perch-landing maneuver for a MAV approaching a vertical wall. The MAV must decelerate from cruise

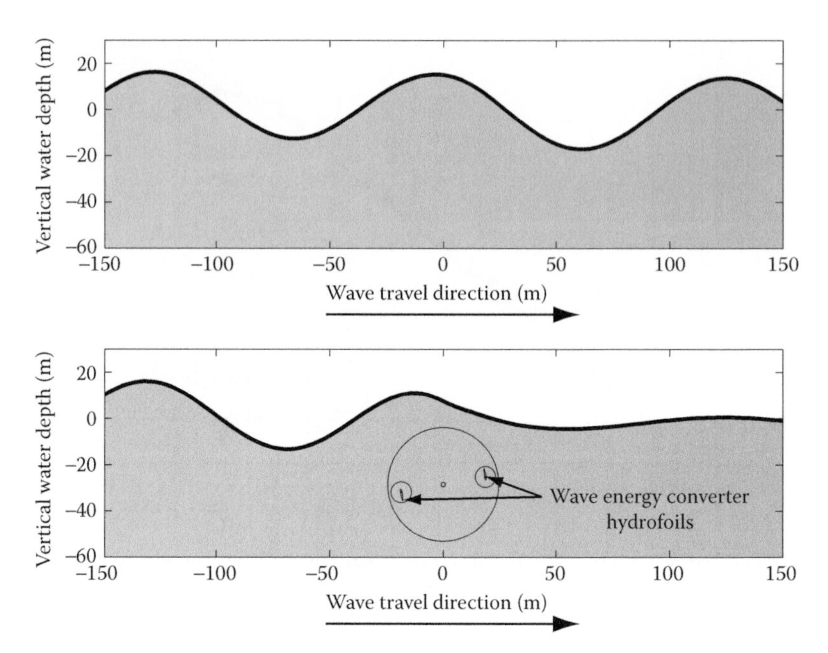

FIGURE 7.2 A schematic of unsteady flow control for wave energy extraction with a cycloidal wave energy converter. (From Siegel, S. et al., *Proceedings of the 30th International Conference on Ocean, Offshore and Arctic Engineering*, Rotterdam, the Netherlands, June 19–24, 2011. Reprinted with permission.) The surface wave amplitude is reduced downstream of the energy extraction device. Note: Wave heights enlarged ×20.

flight conditions (low angle of attack, high speed) to a landing configuration of nearly 90° and zero forward flight speed. The changes in speed and pitch will occur over a short spatial distance of only a few chord lengths. The maneuver can be initiated with an elevator, and depending on the pitch rate, the conventional stall angle will be increased to a higher angle of attack. At some angle (roughly around $\alpha = 15°–30°$), synthetic jet control of flow separation and post-stall loads acting on the flight vehicle will be necessary to maintain flight control. In this case, the actuators only act to increase the lift, and hence, it is unidirectional.

By contrast, the examples of "gust suppression" and "energy extraction" control require bidirectional actuation, which can increase or decrease the instantaneous lift. The objective of a gust suppression system is to maintain constant lift when the external flow velocity and flow incidence angle are changing. In this case, the actuators must produce both positive and negative lift increments at correct amplitude and phase to counteract the effect of a random gust field. A "lift biasing" technique is required in the situations where bidirectional control is needed.

The designer of an AFC system must be able to match the dynamic response of the actuator to the maneuver, which requires some understanding of the unsteady aerodynamic loads acting on the wing and some knowledge of the transient response of the flow to changes in actuation. Important questions for the designer to consider are as follows:

1. How fast can a maneuver be executed?
2. How fast can a separated flow be reattached?
3. How can negative lift increments be generated?

Answers to these questions can be found by applying the tools of control theory to the AFC problem.

As mentioned earlier, the majority of AFC research focuses on steady-state conditions, that is, typically the lift, drag, and pitching moment response of a wing are measured at fixed angle of attack, constant flow speed, and constant actuator operating conditions. The actuator amplitude and frequency are swept over a range of values to find optimum operating conditions, which are usually defined as the operating parameters producing largest increment in lift. The early studies with synthetic jets (Seifert et al. 1996, 1998, 1999; Amitay et al. 2001; Wu et al. 1998; Glezer and Amitay 2002) used a steady-state actuation approach to demonstrate the effectiveness of AFC. Data in Figure 7.1 show the lift enhancement achievable with AFC actuators on a semicircular wing by comparing the lift curve response with and without pulsed-blowing actuation. A similar low-aspect ratio semicircular wing with synthetic actuators is shown in Figure 7.3. The static control authority can be defined as the lift increment increase above the unforced lift. Data in Figure 7.1 show a positive lift increment above $\alpha = 16^\circ$, reaching a maximum near $\alpha = 22^\circ$, then at larger angles of attack the lift begins to decrease.

The maps of the steady-state response provide the first-order estimate of the forces that can be expected during a dynamic maneuver. In Figure 7.1, the dashed vertical line at $\alpha = 20^\circ$ shows the quasi-steady boundaries of achievable control. If the maneuver is "slow enough" (discussed in Section 7.2), then the prediction by the quasi-steady map may be sufficiently accurate for control purposes. As the pitch rate increases, the lift curves and the corresponding response to actuation change, and it becomes necessary to consider the unsteady aerodynamic effects in the application of AFC.

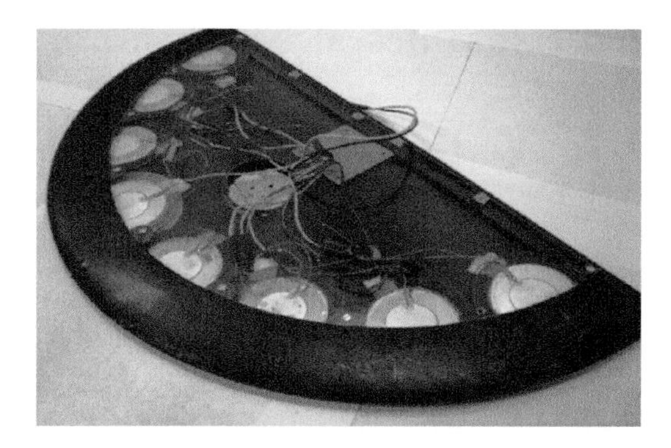

FIGURE 7.3 Photograph of the semicircular wing with eight synthetic jet actuators distributed radially along the leading edge. Each actuator has two exit ports at the leading edge. (Adapted from Quach, V. et al., Transient response of a wing to arbitrary actuator input. *The 3rd International Conference Jets Wakes and Shear Flows*, Cincinnati, OH, 2010.)

There are common features in the process of implementing AFC in dynamic situations, even though there may be significant differences in the control objectives. The application of control theory to the dynamic flow control problem results in what can be called a "procedure" for designing controllers. Researchers at the Technische Universitaet Berlin were one of the first groups to apply modern control techniques to dynamic flow control problems, and examples of their approach can be found in the work by Henning et al. (2007), Pastoor et al. (2008), and Heinz et al. (2010). The end result of the "procedure" is some type of controller (such as feedback, feedforward, or open-loop), whose performance can be quantitatively evaluated and analyzed. The tools of control theory provide powerful insight into the achievable performance and expected limitations of the overall system. Often one is able to gain insight into the reasons for the performance limitations, which can be helpful in improving the design of the flow control system.

The general steps for designing an AFC system for a dynamic maneuver are listed as follows:

1. Obtain a static map of the system output (lift, drag, yaw) relative to an input to the actuator (amplitude of input to actuator, actuator frequency, actuator duty cycle, etc.)
2. Determine the amount of "biasing" required using the static map
3. Determine a "plant model" that contains the dynamics of the system response to actuation
4. Identify a model of the system response to external disturbances
5. Build a controller based on these models that achieves the desired system performance
6. Verify controller performance with an independent set of input conditions

Before examining the steps in the design procedure in more detail, it is helpful to consider the timescales (frequencies) that arise when synthetic jet actuation is used in a dynamic maneuver.

7.2 Frequencies and Timescales of the Actuator, Flow Response, and Maneuver

Dimensionless scaling parameters for the dynamics of the maneuver determine whether a maneuver is "slow" or "fast." Three types of maneuvers considered in this chapter are pitching $[\alpha(t)]$, plunging $[h(t)]$, and longitudinal accelerations $[u(t)]$. Because a number of different length scales are defined for flow control problems, one must be careful about the definitions of various dimensionless quantities. For example, the fluid dynamics community often uses the chord length (c) as a characteristic length scale, whereas the aerodynamics community will use the half-chord ($b = c/2$). For pitch and plunge maneuvers, it is common to normalize the pitch rate [$\dot{\alpha}$ (radians/s)] and plunge rate [$\dot{h}$ (m/s)] by the freestream speed (U) and half-chord length (b) to define the dimensionless frequencies (k), where $k_{\text{pitch}} = \dot{\alpha}2\pi b/U$ and $k_{\text{plunge}} = \dot{h}2\pi/U$. Periodically oscillating flows are normalized by the frequency (f_{osc}), and the half-chord, so $k_u = 2\pi b f_{\text{osc}}/U$, but the Strouhal number is also commonly used where $St = cf_{\text{osc}}/U$.

Several timescales and frequencies must be considered when doing dynamic control of separated flow with synthetic jet or pulsed-blowing actuators. It is necessary to have some estimates of the following:

1. The actuator operating (or carrier) frequency
2. The optimum actuator frequency (or frequencies) to interact with flow instabilities
3. The time required for the maneuver

Even though these timescales arise from different flow physics, they may overlap with one another in a given application. Harmonic frequencies, frequency lock-on effects, and combination mode (sum and difference) frequencies may appear when nonlinear effects are present. A brief description of the three timescales is given in Sections 7.2.1 through 7.2.3, in the order of the fastest to the slowest process.

7.2.1 Actuator Carrier Frequency

The operating (or carrier) frequency for the synthetic jet ($f_{carrier}$) is determined by the geometry and hardware of the actuator. The amplitude of the fluctuating velocity at the exit of a synthetic jet is strongly dependent on carrier frequency, and maxima may occur at one or more resonant frequencies (Gallas et al. 2003a, 2003b; Sharma 2007; Krishnan and Mohseni 2009a, 2009b). Best performance from synthetic jets is usually obtained when they are operated close to resonant frequency to produce the largest velocity output. Methods used to design synthetic jets and to predict their resonant frequencies (transfer functions) are discussed in Chapters 1 and 2. It should be emphasized that the operating frequency is not the same as the actuator frequency that will be used to control the flow instability. Usually, the operating frequency will be an order of magnitude higher than the fluid dynamic frequencies needed for flow control. The higher frequency actuator output can be treated as carrier wave, which permits the use of the amplitude modulation technique to produce lower frequencies that are more applicable for flow control. For example, the operating frequency for the synthetic jet actuators shown in Figure 7.3 was 350 Hz, while the frequency to obtain the largest lift increment on the wing was 29 Hz; the original and modulated frequencies are illustrated in Figure 7.4. This type of amplitude modulation approach was proven to be effective in separation control and turbulent free shear flow control experiments by Wiltse and Glezer (1998), Amitay and Glezer (2001), and Margalit et al. (2005).

In contrast to synthetic jets, other types of actuators for AFC such as zero-net-mass voicecoil-driven actuators and pulsed-blowing actuators may not have a "carrier frequency"; however, it is still important to document the dependence of the actuator exit velocity on frequency. Each actuator system will have its own frequency-dependent response (transfer function) determined by the plumbing and hardware design. Resonances are the rule in pulsed-blowing and voicecoil-driven actuators, and the resonant frequencies depend on the specific design of the actuator.

7.2.2 Fluid Dynamic or Active Flow Control Frequency

In modern AFC systems, one attempts to control the forces in a flow by manipulating inherent flow instabilities. The optimum actuator frequency (f_{afc}) is the one that produces

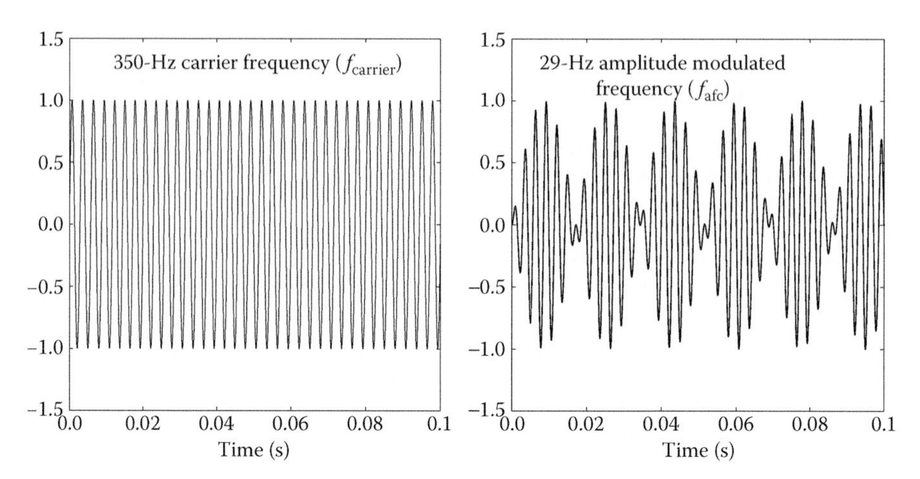

FIGURE 7.4 Illustration of a 29-Hz amplitude modulation of the 350-Hz high-frequency carrier wave to obtain a lower synthetic jet frequency that is useful for active flow control.

the largest force increment. This frequency is usually identified by conducting frequency sweeps under steady-state, open-loop forcing conditions; for example, the angle of attack and flow speed over a wing are fixed and the lift is measured with the actuator operating at a specific frequency. The frequency is then changed to different values covering a range of Strouhal numbers. In the case of a separated flow, the optimum value will be determined by some type of flow instability that is dependent on the specific angle of attack. With synthetic jet actuators, the "best" frequency for AFC (f_{afc}) will most likely be a lower magnitude than the actuator carrier frequency described earlier. In the two-dimensional (2D) separated airfoil problem, there is currently no consensus on the scaling law that should be used to determine the optimum frequency (f_{afc}) for control. The issue of appropriate scaling laws for 2D separated flows was examined in detail by Stalnov et al. (2010a, 2010b). The issue of determining the optimum frequency for flow control is even more complicated because multiple instabilities may coexist in the flow. For example, Raju et al. (2008) showed that in a 2D, steady, separated flow, there can be up to three distinct instabilities depending on the angle of attack, which are listed as follows:

1. *Shear layer* (f_{shear}). The separated shear layer supports the inviscid Kelvin–Helmholtz instability. Free shear layer frequencies scale with the momentum thickness (θ) and the average velocity across the shear layer (U_{avg}).
2. *Wake* (f_{wake}). The wake frequency is related to the vortex shedding from the body and is often the dominant unsteady force when a wing is at low angle of attack (fully attached flow) or at a high enough angle that the wing is stalled (Raju et al. 2008). In the latter case, Fage and Johansen (1927) showed a Strouhal number scaling $St = f_{wake}c \sin(\alpha)/U \sim 0.15\text{–}0.2$ applied to 2D inclined flat plates. Data for the dominant lift frequency obtained from a low-aspect ratio semicircular wing are compared in Figure 7.5 to the Fage–Johansen Strouhal number, and reasonable agreement is seen even though the wing is 3D.

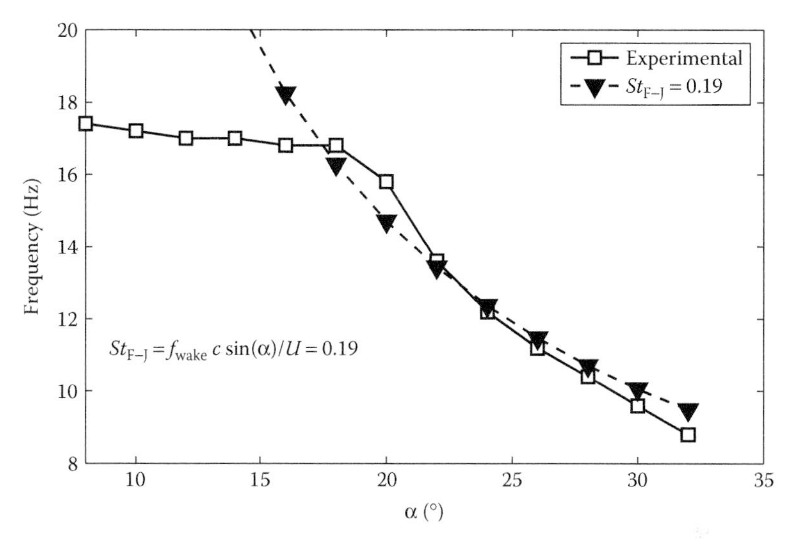

FIGURE 7.5 Dominant lift oscillation frequency of the semicircular wing and comparison with the Fage and Johansen (1927) Strouhal number. (From Collins, J. G., Closed loop active flow control for excitation and stabilization of leading edge vortex phenomenon, MSc thesis, Illinois Institute of Technology, July 2008. Reprinted with permission.)

3. *Separated-reattached flow* (f_{sep}). When the separated flow reattaches on the airfoil and forms a separation bubble, a separate instability identified by Mittal and Kotapati (2004) occurs that is associated with the shedding of large-scale vortices from the separation bubble region. Raju et al. (2008) report that the separation bubble frequency (f_{sep}) scales with U/L_{sep}, where L_{sep} is the length of the separation bubble.

The range of frequencies (bandwidth) over which the actuator must operate is determined by the maneuver; so depending on the type of maneuver, the AFC controller may have to compensate for multiple instabilities.

7.2.3 Flow or Maneuvering Timescale

The external flow or the timescale of the maneuver will be the longest time or the lowest frequency that must be considered. If the maneuver is periodic, such as in the ocean wave energy extraction seen in Figure 7.2, then a specific frequency (f_{osc}) can be defined. On the other hand, if the maneuver is simply transient, such as the perch-landing case, then the maximum pitch rate ($\dot{\alpha}$) or maximum plunge rate ($\dot{h}$) may be used to define the timescale. The flow maneuvering timescale determines whether or not the quasi-steady approximation can be used. As k-values become large, dynamic force effects in the form of added mass, wake influence, and dynamic stall vortex formation become significant. For the example of a longitudinally oscillating freestream flow, a quasi-steady model predicts that lift and velocity are in phase with each other, but measurements show that the actual phase difference between the lift and freestream flow was 20° at $k_{osc} > 0.09$. As an approximation, one should expect unsteady flow effects to be important when $k > 0.05$ (Leishman 2000).

7.3 Static Maps, Quasi-Steady Control, and Biasing

7.3.1 Static Maps

The static maps describe the steady-state lift increment dependence on the actuator input signal when using open-loop forcing. The line labeled "Available Control" in Figure 7.1 indicates the boundaries of control that can be achieved from an actuator acting on the semicircular wing at $\alpha = 20°$. The static maps show how the lift (output) will vary from the lower lift limit to the upper lift limit as the actuator input is changed. With synthetic jet actuators, the static maps can be obtained in several different ways. The input signal to the actuator can occur by amplitude modulating the carrier frequency at the optimum f_{afc}. Another approach would be to frequency modulate the carrier. A third approach to achieve control over the output lift would be a variable duty cycle signal. The semicircular planform wing is used as an example of control by varying the driving voltage to the actuator.

The static map obtained for a synthetic jet actuator control of a semicircular wing fixed at $\alpha = 16°$ is shown in Figure 7.6. Each curve corresponds to the lift response at freestream speeds ranging from 5 to 9 m/s as the voltage to the carrier signal is varied. The carrier frequency is 350 Hz, which is amplitude modulated at $f = 20$ Hz. A square-root dependence of the lift response and saturation can be seen. It is interesting, and somewhat

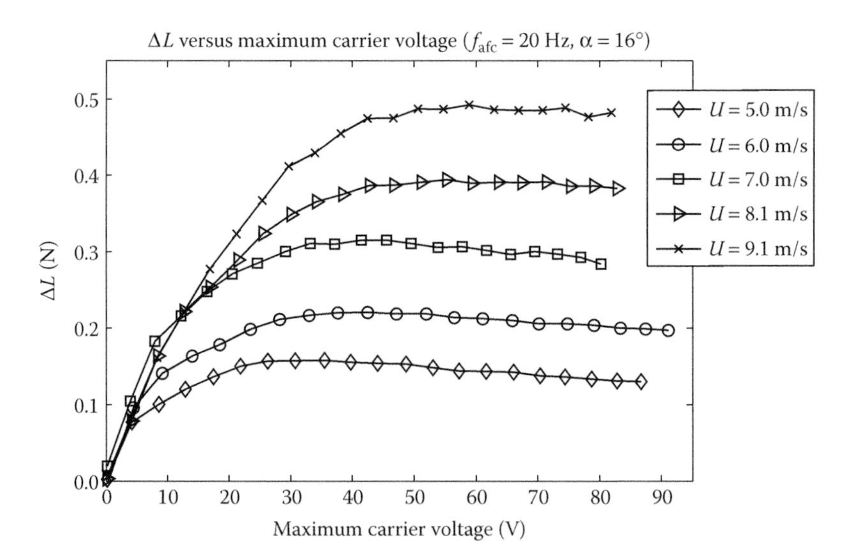

FIGURE 7.6 Static map for semicircular wing with synthetic jet actuator. $f_{carrier} = 350$ Hz, $f_{afc} = 20$ Hz, $\alpha = 16°$. (From Quach, V. et al., Transient response of a wing to arbitrary actuator input. *The 3rd International Conference Jets Wakes and Shear Flows*, Cincinnati, OH, 2010.)

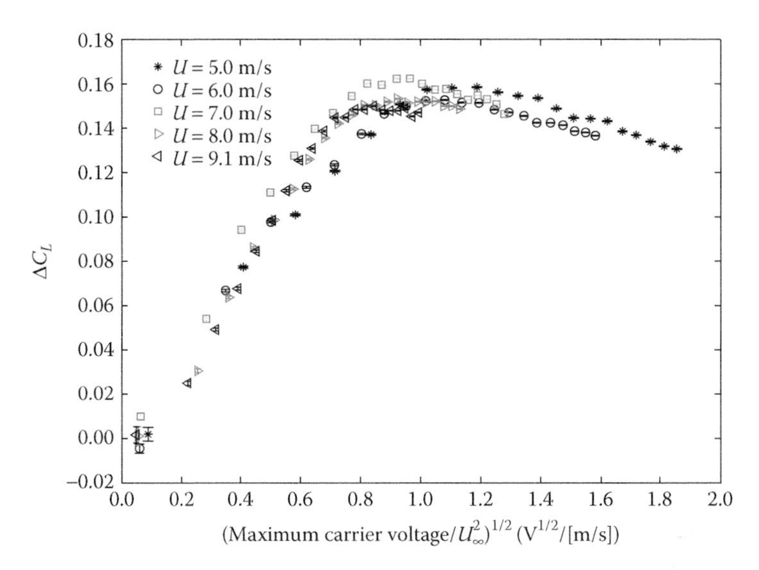

FIGURE 7.7 Static map for lift coefficient plotted against the scaled voltage from a semicircular wing with synthetic jet actuator. (From Quach, V. et al., Transient response of a wing to arbitrary actuator input. *The 3rd International Conference Jets Wakes and Shear Flows*, Cincinnati, OH, 2010.)

counterintuitive, that the lift increment increases with increasing external flow speed when the actuator amplitude is fixed. The explanation for this behavior is found by normalizing the data.

Data in Figure 7.6 were replotted in Figure 7.7 as the lift coefficient $C_L = L/0.5\rho SU^2$, where ρ is the density and S is the planform area, and the voltage to the actuator is scaled as $\sqrt{\text{Voltage}/U^2}$. The lift coefficient collapses to a single curve with this scaling, and a nearly linear response from the lift increment is obtained at low voltages prior to saturation at $\sqrt{\text{Voltage}/U^2} = 0.8\text{V}^{0.5}/(\text{m/s})$. The specific values of lift increment and saturation voltage levels will depend on the specifics of the actuator design and configuration within the wing.

The synthetic jet data shown in Figures 7.6 and 7.7 can be compared to similar data obtained with a pulsed-blowing actuator. In Figure 7.8, the lift coefficient increment obtained with a pulsed-blowing actuator used on a semicircular wing is plotted against the control variable to the actuator, which is the supply pressure. In this case, a good collapse of the data is obtained when the lift coefficient increment is plotted against the square root of the actuator supply pressure coefficient ($C_P = \Delta P/0.5\rho SU^2$), where ΔP is the supply pressure to the actuator. The square root of the actuator pressure coefficient is equivalent to the actuator jet velocity normalized by the freestream speed (u_{jet}/U).

7.3.2 Quasi-Steady Control and Biasing

If the maneuver to be controlled is slow enough that unsteady effects are not important ($k_{u,\text{pitch,plunge}} < 0.05$), then the static map is sufficient to design a controller. The static

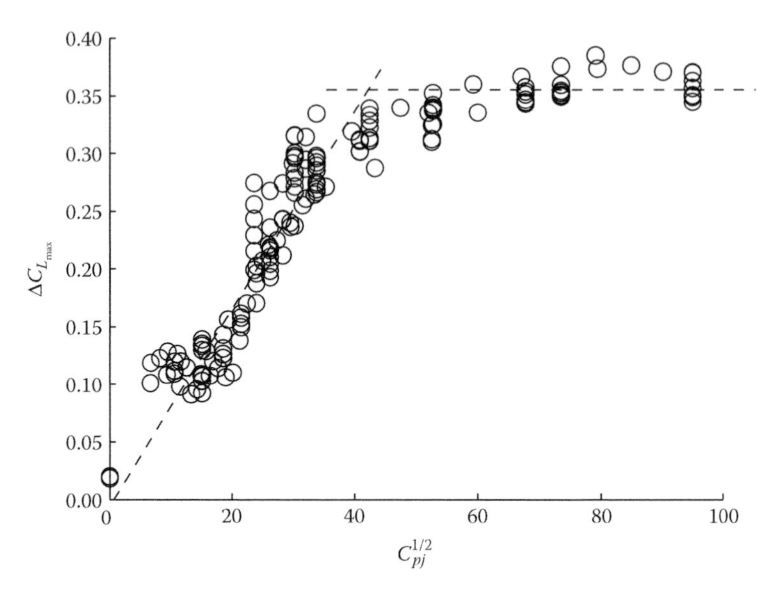

FIGURE 7.8 Static map for a pulsed-blowing actuator. The lift coefficient dependence on the square root of the pressure coefficient shows a nearly linear response prior to the saturation amplitude.

maps can be inverted to determine the required actuator input to obtain a desired lift increment. As an example, we consider the quasi-steady approach to the gust suppression problem. The objective for the controller is to maintain constant lift by compensating for the changing lift caused by a varying freestream speed or dynamic pressure. The wing is in a stalled state, so the actuator will compensate for the changing lift by adjusting the lift coefficient increment. Because the gusts have both increasing and decreasing velocity amplitudes, a bidirectional control will be needed to compensate for the increases and decreases in the lift.

Bidirectional control can be achieved by "biasing" the actuation, which refers to shifting the lift increment baseline to an intermediate level in the achievable range of control. The static map shown in Figure 7.9 illustrates biasing for a pulsed-blowing actuator. This static map was obtained by varying the duty cycle of the pulsed jet frequency while keeping the actuator supply pressure constant. In this particular case, the 0% duty cycle corresponds to no actuation, and the 100% duty cycle is continuous pulsing at 29 Hz. The lift coefficient increment ranges from $\Delta C_L = 0$ to $\Delta C_L = 0.4$ as the duty cycle is increased from 0% to 100% and the variation in ΔC_L is proportional to the square root of the duty cycle. By using a 40% duty cycle, the lift increment is centered. Positive lift increments are obtained by operating the duty cycle between 40% and 100%, while negative lift increments are obtained in the range of 0% to 40% duty cycle.

Static maps are sufficient for quasi-steady control applications, but can be inaccurate in dynamic situations. For fast, high-bandwidth control, the transient response of the actuator becomes important, and for rapid maneuvers, the unsteady aerodynamic effects can be

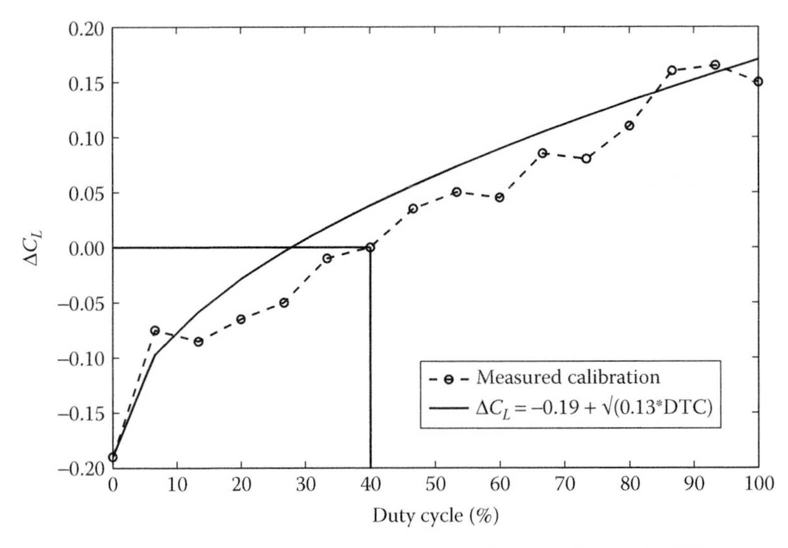

FIGURE 7.9 Static map of lift coefficient vs. duty cycle of pulsing for a pulsed-blowing actuator at a fixed pressure coefficient. (From Williams, D. et al., *The 47th AIAA Aerospace Sciences Meeting*, Orlando, FL, 2009.)

significant. Models that capture the dynamics of the lift response to actuation and the lift response to external flow disturbances are needed. Methods for obtaining such models and their implementation into a controller are discussed in Section 7.4.

7.4 Dynamic Response to Actuation—The Plant Model

The transient response of the flow to actuation generally has a response that is different from the one predicted by the static map, that is, the amplitude and phase of the lift do not follow the static map, and the so-called plant dynamics must be modeled. The model of the system dynamics including the actuator is referred to as the "plant model." In principle, the Navier–Stokes equations can be used to obtain solutions for the lift response to actuator input, but for practical applications low-dimensional models are required. Ideally, the flow physics of the system response to actuation would be known, so that a low-dimensional *physics-based model* can be derived. If the response to input is linear, then a transfer function (plant model) can be used. Modern system identification techniques can be used (e.g., Ljung 2009) to determine linear models. If nonlinear response occurs, then look-up tables, genetic algorithms, or neural-net models may be useful; for instance, a neural-net model was used by Bernhardt and Williams (2000) to model the nonlinear side force on a slender forebody as it pitched up to high angles of attack.

One method of characterizing the dynamics of a system is to examine the system's transient response to short-duration pulses or to step inputs from the actuator. The transient lift response to short-duration pulses will give an approximate "impulse response" model. There have been a few experimental investigations into the transient lift response to transient actuator inputs on 2D wings (Amitay and Glezer 2006; Woo et al. 2008) and on 2D flaps (Darabi and Wygnanski 2004a, 2004b). Their investigations concentrated on identifying the fundamental flow physics and improving the range of achievable control, not on identifying models for flow control. Darabi and Wygnanski (2004a, 2004b) studied the effect of step changes in zero-net-mass actuation (voicecoil actuator) on the transient separation and reattachment process on separating flow over a flap. Amitay and Glezer (2006) and Woo et al. (2008) studied the transient effect of single and multiple pulses from synthetic jet actuators. In both cases, an initial lift reversal occurred within two convective time units ($t^+ = tU/c$) after the actuator was first turned on. The lift reversal was followed by a rapid increase to its maximum value. For a step input from the actuator, the lift establishes its steady-state value within 20 convective times. Using a train of pulses as input, Woo et al. (2008) showed that significant increases in lift could be achieved if the correct pulse timing was used. These results emphasize the need for a dynamic plant model if the actuation levels will need to change within 20 convective times.

Linear, time invariant systems can be described by their impulse response functions. Williams et al. (2009b) showed that a short duration ($\Delta t^+ < 0.5$) single-pulse input from the actuator approximated an impulse to the system, and the resulting transient lift output could be used as a filter kernel to approximate the lift response to arbitrary actuator inputs. The lift coefficient increment response to a single-pulse input is shown in Figure 7.10a for flow over a semicircular wing with pulsed-blowing actuators. The abscissa is the time normalized by the convective time. Each curve in Figure 7.10a corresponds to a different freestream flow speed, where the pulse duration time was adjusted to maintain a constant dimensionless pulse time $\Delta t^+ = 0.49$. The supply pressure to the actuator was fixed, so the pressure coefficient decreases with increasing flow speed. The initial lift reversal seen by previous investigators also occurs in this experiment at $t^+ = 1.5$. The rapid increase to a maximum lift occurs between $2 < t^+ < 6$, and this is followed by a slower decay back to the original value. The peak of the transient lift is approximately twice that obtained with continuous pulsed-blowing (steady state) actuation at 29 Hz. The peak lift increment amplitude has the same square-root dependence on the actuator pressure coefficient as observed with the continuous pulsing case.

It was also shown by Williams et al. (2009b) that a linear model (convolution integral) could predict the transient lift response following an arbitrary actuator input. An example of the lift response to a 1.4-Hz square-wave input from the actuator is shown in Figure 7.10b. The predicted lift is shown by the solid line and the measured lift is shown by the dashed line. Without actuation, the mean lift coefficient is $C_L = 0.8$ as shown by the dash-dot line. With AFC actuation, the lift oscillations increase, and the mean lift increases to $C_L = 1.2$. Both the time delays and changes in lift amplitude are reasonably well predicted using the linear model.

From the control theory perspective, the initial lift reversal that occurs with single-pulse input shown in Figure 7.10a corresponds to a time delay in the control system. This is known as nonminimum-phase behavior, which is indicative of a right-half plane (rhp) zero in the complex plane of the system transfer function. The time delay limits the bandwidth that is achievable from the controller, and this places an upper limit on how fast a maneuver

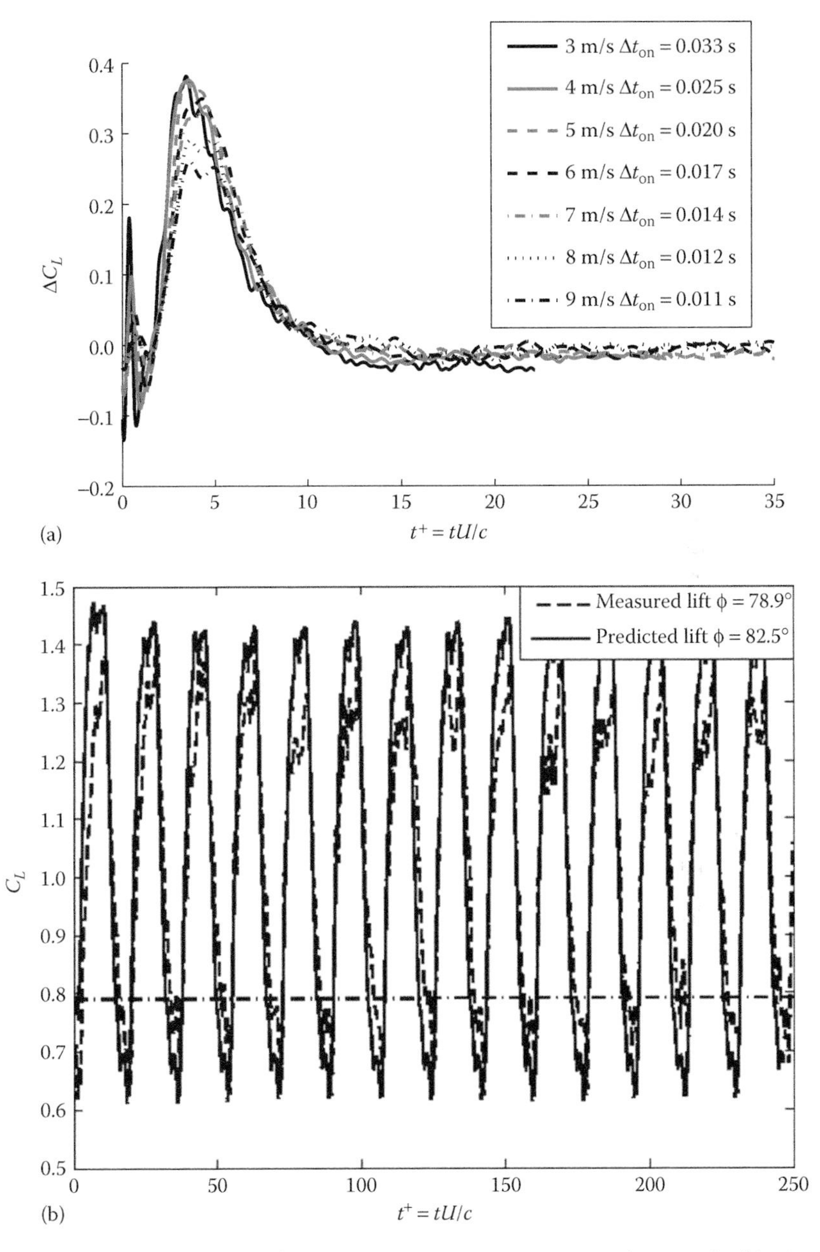

FIGURE 7.10 Transient lift coefficient increment response to an input from a pulse-blowing jet actuator. (Data from Williams, D. et al., *AIAA J.*, 47, 3031–3037, 2009b.) (a) Lift response to a single-pulse input; (b) comparison of measured lift (dashed line) and predicted lift (solid line) is shown for a square-wave forcing frequency of 1.4 Hz. Note that the baseline lift coefficient without actuation is 0.8 and that continuous actuator input increases the mean to $C_L = 1.2$.

can be controlled with an AFC system. The bandwidth limitations resulting for the gust suppression application are discussed in Section 7.6.

The convolution integral approach is often not the most useful form of a system model when it comes to control design. State-space or Laplace transforms are simpler to implement when designing a controller. Ljung (2009) describes a variety of modern system identification techniques that can be used to obtain state-space, transfer function, and frequency response models for the flow system. The models obtained are generally referred to as "black box" models, because one simply correlates the output with the input. If some understanding of the flow physics can be incorporated into the plant models, then it is known as a "gray box" model.

Better system identification models (compared to pulse input) can be obtained if a pseudorandom binary signal (PRBS) is used as the transient to the actuator. In this approach, step inputs to the actuator are given with a randomized delay times occurring between the steps. A variety of system identification techniques are described by Ljung (2009) that can be used to obtain plant models for dynamic AFC. The "prediction error method" and other black-box approaches have been used successfully on separated flow control problems to fit models to the experimental data (Henning et al. 2007; Heinz et al. 2010). The control designer must first choose the form and order of the model before the fit is made, so the process involves some trial-and-error. After a plant model is obtained from a specific set of input data, it is necessary to verify that its performance is acceptable with an independent data set.

Depending on the specific control objective, the plant model alone may be sufficient to complete the controller design without the need for a model of the external unsteady aerodynamic effects. For example, Williams et al. (2010) designed a controller to suppress lift fluctuations on a wing in a randomly gusting flow based on the plant model without including a specific model for the unsteady aerodynamics. The changes in lift resulting from gusts were treated as external disturbance errors that were compensated for by the closed-loop controller. This approach produced acceptable performance, but even better gust suppression was achieved by Kerstens et al. (2011) when a specific model for the unsteady aerodynamics was included. Section 7.5 describes models for external disturbances.

7.5 Response to External Disturbances

When the external flow is unsteady and/or the wing is maneuvering, then the wing's response to the unsteady flow is also dynamic. If there is a priori knowledge of the incoming disturbance or the imminent maneuver, then the required actuator response can be predicted in advance by using a feedforward controller. To do this requires a model of the system's response to external disturbances. Classical unsteady aerodynamic theories (Johnson 1980; Leishman 2000) can be used to predict the lift response when the flow is attached to the wing and the wake is thin. Recently, for pitching and plunging maneuvers, McGowan et al. (2011) showed that Theodorsen's (1935) corrections to the quasi-steady effective angle of attack could be used even at high dimensionless frequencies provided the noncirculatory contribution to lift dominates over the circulatory viscous effects. If the flow is highly separated or highly 3D, which is often the case with low-aspect ratio wings, then it is unlikely that the classical methods will be sufficiently accurate to obtain the control objectives. In this case, we return to system identification techniques to obtain black-box models to represent the unsteady aerodynamic response to external disturbances.

Returning to the gust suppression example, an unsteady aerodynamics model for the low-aspect ratio wing was obtained by experimental measurements in a wind tunnel (Kerstens et al. 2011). The time-varying lift response was measured over a range of freestream oscillating frequencies. The ratio of the lift force to the velocity amplitude and the phase between the lift and velocity are plotted against frequency in Figure 7.11. The solid line is the black-box model for the unsteady aerodynamics (fourth-order transfer function) obtained using a frequency response system identification method. The model is used as the "external disturbance" model in the controller described in Section 7.6.

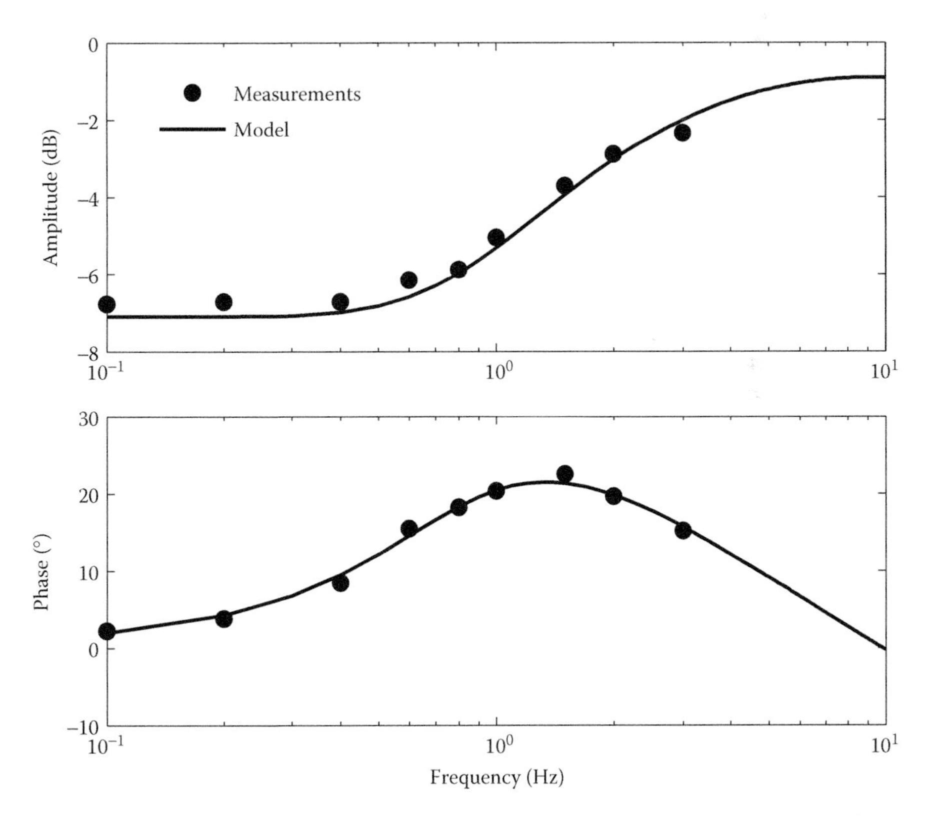

FIGURE 7.11 Frequency response model for lift force variation with external velocity fluctuations—a comparison of experimental measurements (dots) with the model (solid line). (From Kerstens, W. et al., *AIAA J.*, 49, 1721–1728, 2011. Reprinted with permission.)

7.6 Examples of Controllers Used for Dynamic Active Flow Control

The design of the controller and choice of the control architecture are strongly dependent on the specific problem and control objective. There is no universal control architecture that will apply to all unsteady AFC scenarios, so the process of selecting a controller requires some experience and trial-and-error. A few examples of different control architectures used for AFC are discussed in Sections 7.6.1 and 7.6.2.

7.6.1 Forebody Vortex Control Example

Strong asymmetric side forces and yaw moments are produced on slender bodies at high angles of attack ($\alpha > 35^{\circ}$); this is known as the "forebody problem" (Lamont 1982). Fundamental investigations into the origins of the side forces showed that small disturbances in the tip region of the slender body were being amplified in the wake behind the body. Open-loop forcing experiments showed that the side forces and yawing moments could be controlled with very small actuator inputs under static or quasi-steady conditions (Roos and Magness 1993; Bridges and Hornung 1994). Bernhardt and Williams (2000) explored the use of closed-loop control for two different objectives involving dynamic maneuvers; the first control objective was to eliminate the side force during pitching maneuvers, and the second was to enhance flight vehicle maneuverability by modulating the side force to control the yaw angle during a "coning motion" maneuver. The first case required the actuators to maintain a symmetric pressure distribution as the model pitch angle increased. Static maps of the differential pressure dependence on actuator input, pitch angle, and pitch rate were highly nonlinear. A neural network model was an effective way to approximate the strongly nonlinear, experimentally measured data. Several different control architectures were investigated, and it was determined that the neural net model of the plant combined with a proportional-derivative compensator was able to maintain the vortex symmetry during the desired pitch maneuver. By contrast, even though the same model and actuators were used, a different control architecture was needed when the control objective was to execute a yaw maneuver to a desired yaw angle. In this case, a proportional-integral controller with a velocity feedback compensator produced the best results.

One benefit from the analysis of the closed-loop control system is that often one can gain insight into important flow physics and limits of achievable performance. In both cases described earlier, time delays related to the finite convection speed of disturbances from the actuator through the vortex system behind the model were identified and were determined to be partially responsible for limiting the rate at which control could be maintained. Such time delays set upper limits on how fast a maneuver can be executed.

7.6.2 Gust Suppression Example

Three different control architectures were examined in the gust suppression experiments. In each case, the wing was a low-aspect ratio, semicircular planform that was at a fixed angle of attack in a fully stalled condition. Either pulsed-blowing jets or synthetic jets were used as the actuators, which were positioned radially along the leading edge. The first approach was based on a quasi-steady lift control (Williams et al. 2008) and used a feedforward algorithm obtained by inverting the nonlinear static map. The inverted static map provided a nonlinear control algorithm that was capable of suppressing both the fundamental and, to a lesser extent, the first harmonic of the freestream oscillation. The lift oscillation amplitude was reduced by 10 dB at the fundamental frequency. But the quasi-steady approach was only effective at low frequencies ($k_{OSC} < 0.03$). Time and phase delays between the lift response and the freestream oscillation (an aerodynamic property) and time delays related to the response to actuation became significant at $k_{OSC} > 0.03$, which made the quasi-steady controller ineffective. It became clear that the plant dynamics and the unsteady aerodynamics would need to be incorporated into the controller for AFC to be effective.

An *ad hoc* modification to the quasi-steady feedforward controller was designed, which accounted for both the aerodynamic and the plant time delays at a single frequency (Williams et al. 2009a). By compensating for the two time delays, it was possible to get very substantial reduction (17 dB) in fluctuating lift at a design frequency five times larger ($k_{OSC} = 0.15$). But the bandwidth was extremely narrow, and control was only effective at the specific design frequency, that is, the gust suppression only worked for the case of a harmonic oscillation. The control effect rapidly diminished when the freestream frequency was changed, and if a random transient disturbance was applied, the control did not work totally. Models for the complete system dynamics were needed that would provide the phase and amplitude information over the entire range of operating frequencies.

The procedure for designing the broadband controller came from the Measurement and Control Group at the Technische Universitaet Berlin. The prediction error method of system identification was used by Williams et al. (2010) to obtain black-box models for the separated flow response to actuation. Instead of feedforward control, a robust H_∞ controller was designed based on the feedback of the lift coefficient increment. No explicit unsteady aerodynamic model was included in the initial architecture, because the amplitude changes and delays related to unsteady aerodynamics would be treated as external disturbances to be compensated by the robust controller. The performance of the controller was tested by generating randomized changes in the freestream speed. The controller showed good suppression and was able to maintain a constant lift force with a 50% reduction in the root mean square amplitude of lift fluctuation. However, the bandwidth of the controlled frequencies was limited to $k < 0.04$, which was not much of an improvement over the quasi-steady controller.

The next improvement came when Kerstens et al. (2011) designed a new controller that explicitly accounted for the unsteady aerodynamics in the control architecture. The unsteady aerodynamics model described in Section 7.5 and shown in Figure 7.11 was added as a feedforward loop. The control architecture is shown in Figure 7.12. A robust H_∞ controller was designed which used the lift increment as the feedback signal instead of the lift coefficient. In Figure 7.12, the plant model is shown by $G_P(s)$ and the feedback controller is shown by $K(s)$. The output lift from the plant is y_P, so the actual lift force is $y = y_P + y_d$. In the experiment, the measured lift is compared to the reference lift value (r). The "biasing," described in Section 7.3.2, that is needed to get negative lift increments is obtained by setting the reference lift (r) slightly higher than the largest uncontrolled lift value.

The effectiveness of this controller is shown by the lift time series in Figure 7.13. The randomized velocity changes from 6.25 to 7.25 m/s, which produces lift variations from

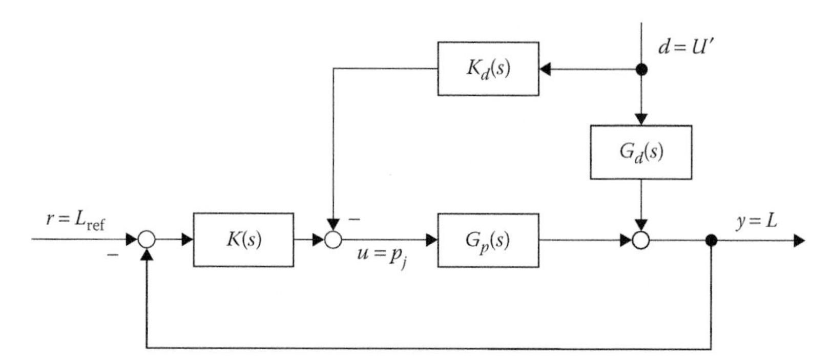

FIGURE 7.12 Controller architecture used for gust suppression control on a semicircular wing in a randomly fluctuating freestream. (From Kerstens, W. et al., *AIAA J.*, 49, 1721–1728, 2011. Reprinted with permission.)

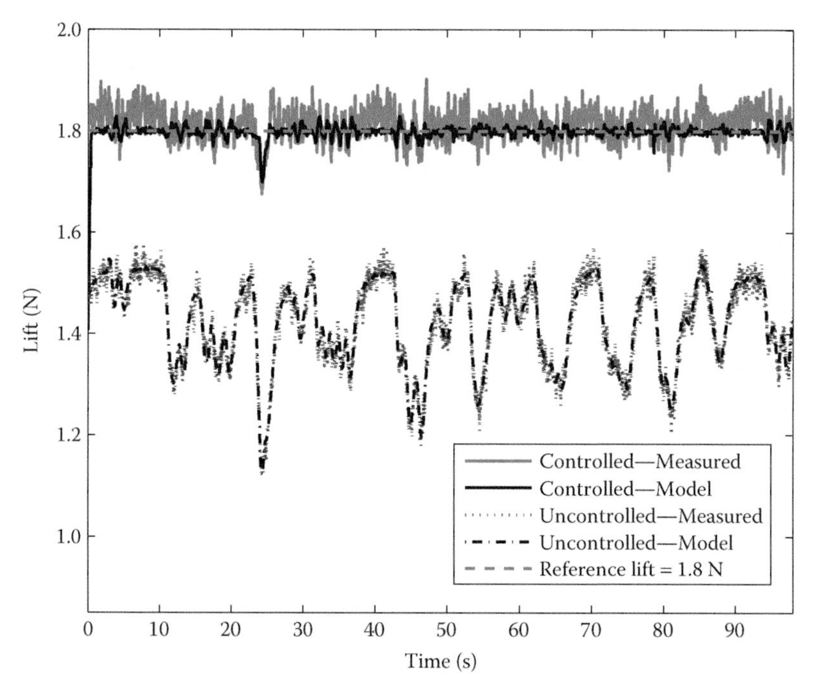

FIGURE 7.13 Lift time series in randomized longitudinal gusting flow with and without closed-loop flow control. (From Kerstens, W. et al., *AIAA J.*, 49, 1721–1728, 2011. Reprinted with permission.)

1.1 to 1.55 N. The controller is turned on with a commanded reference lift value of 1.8 N, and the measured values range from 1.7 to 1.88 N. The bandwidth of the controller was determined to be $k < 0.09$, which is a significant improvement over the previous lift coefficient-based controller.

Analysis of the controller by Kerstens et al. (2011) showed that time delays associated with the convection of the actuator disturbances in the separated shear layer were responsible for the bandwidth limitation. The high bandwidth suppression by the *ad hoc* controller relies on the harmonic nature of the flow, that is, that type of controller can only work at its design frequency in a harmonic flow. If the control objective is to suppress disturbances that are random or transient, then a more general control approach is needed as illustrated in the latter two controllers.

7.7 Summary

This chapter examined the application of pulsed-blowing and synthetic jets actuators to dynamic situations that occur when the external flow is changing or a flight vehicle is maneuvering. A number of issues arise that typically are not considered in steady-state applications of AFC, such as the need to bias the control to obtain negative changes in lift and the need to model the transient response of the control to actuator input. The design of the controller and the performance of the system will be different depending on the specific control objectives, but there are common features in the design of an AFC system for dynamically changing situations. These similarities in the design of controllers that use synthetic jet or pulsed-blowing actuators were explored. The steps needed to design a controller for dynamic flow control are summarized as follows:

1. Estimate timescales for the maneuver, plant, and actuator
    a. k values determine if it is quasi-steady problem or not.
2. Obtain static map of the system's response to steady-state actuation
    a. Determine the range of useful control
    b. If desired, set biasing needed for "negative" lift control.
    c. Identify any significant nonlinearity in the system
3. Obtain plant model (in the absence of external disturbances)
    a. The output of the system in response to transient inputs to the actuator is modeled.
        i. If the system is linear, then transfer functions or state-space models can be used.
        ii. If the response is nonlinear, then neural networks, look-up tables, or genetic algorithm may be useful.
4. Obtain external disturbance model if needed
    a. Response of the system to external gusts in the absence of AFC
    b. Response of the system to a maneuver in the absence of AFC
5. Design the control architecture
    a. Combine plant and external model
6. Verify the overall system performance with independent set of input conditions
    a. Identify performance limitations that may be related to delays in fluid dynamic system
        i. Forebody flow control limited by speed of traveling wave
        ii. Separated flow over wing response limited by vortex convection

Not all of these steps need to be implemented in every controller, because each control application will have its own specific requirements. It is hoped that the generalized procedure provides a useful template for the steps that should be considered when applying AFC to dynamically varying flows.

Acknowledgments

Many of the concepts presented in this chapter are the result of research supported by Air Force Office of Scientific Research Grants FA9550-05-1-0369 and FA9550-09-1-0189. Additional support from Alexander von Humboldt Foundation and the Illinois NASA Space Grant Consortium is gratefully acknowledged. I wish to thank Professor Rudibert King and his students Jens Pfeiffer and Notger Heinz at the Technische Universitaet Berlin for their numerous contributions to closed-loop flow control. The long-term collaboration with Professor Tim Colonius, Professor Clancy Rowely, and Professor Gilead Tadmor, who have played key roles in developing many of the control strategies and concepts presented, is gratefully acknowledged.

References

Amitay, M. and A. Glezer. 2001. Role of actuation frequency in controlled flow reattachment over a stalled airfoil. *AIAA Journal*, 40: 209–216.

Amitay, M. and A. Glezer. 2006. Flow transients induced on a 2D airfoil by pulse-modulated actuation. *Experiments in Fluids*, 40: 329–331.

Amitay, M., D. Smith, V. Kibens, D. Parekh, and A. Glezer. 2001. Aerodynamics flow control over an unconventional airfoils using synthetic jet actuators. *AIAA Journal*, 39: 361–370.

Bernhardt, J. and D. Williams. 2000. Closed-loop control of forebody flow asymmetry. *Journal of Aircraft*, 37: 491–498.

Bridges, D. H. and H. G. Hornung. 1994. Elliptic tip effects on the vortex wake of an axisymmetric body at incidence. *AIAA Journal*, 32(7): 1437–1445.

Collins, J. G. 2008. Closed loop active flow control for excitation and stabilization of leading edge vortex phenomenon. MSc thesis, Illinois Institute of Technology, Chicago, IL.

Darabi, A. and I. Wygnanski. 2004a. Active management of naturally separated flow over a solid surface. Part 1. The forced reattachment process. *Journal of Fluid Mechanics*, 510: 105–129.

Darabi, A. and I. Wygnanski. 2004b. Active management of naturally separated flow over a solid surface. Part 2. The separation process. *Journal of Fluid Mechanics*, 510: 131–144.

Fage, A. and F. C. Johansen. 1927. On the flow of air behind an inclined flat plate of infinite span. *Proceedings of the Royal Society of London A*, 116(773): 170–197.

Gallas, Q., R. Holman, T. Nishida, B. Carrol, M. Sheplak, and L. Cattafesta. 2003a. Lumped element modeling of piezoelectric-driven synthetic jet actuators. *AIAA Journal*, 41: 240–247.

Gallas, Q., G. Wang, M. Papila, M. Sheplak, and L. Cattafesta. 2003b. Optimization of synthetic jet actuators. *The 41st AIAA Aerospace Sciences Meeting and Exhibit, AIAA Paper* 2003-0635, Reno, NV, January 6–9.

Glezer, A. and M. Amitay, M. 2002. Synthetic Jets. *Annual Review of Fluid Mechanics*, 34: 503–529.

Henning, L., M. Pastoor, R. King, B. Noack, and G. Tadmor. 2007. Feedback control applied to the bluff body wake. In *Active Flow Control*. ed. King, R. Berlin, Germany: Springer-Verlag, pp. 369–390.

Heinz, N., R. King, and B. Gölling. 2010. Robust closed-loop lift control on an industry-related civil aircraft half model. In *Active Flow Control*. ed. King, R. Berlin, Germany: Springer-Verlag, pp. 125–139.

Johnson, W. 1980. *Helicopter Theory*. New York: Dover Publications.

Kerstens, W., J. Pfeiffer, D. Williams, R. King, and T. Colonius. 2011. Closed-loop control of lift for longitudinal gust suppression at low Reynolds numbers. *AIAA Journal*, 49: 1721–1728.

Kerstens, W., D. Williams, J. Pfeiffer, and R. King. 2010. Closed loop control of a wing's lift for "gust" suppression. *The 40th AIAA Fluid Dynamics Conference and Exhibit, AIAA Paper* 2010-4969, Chicago, IL, June 28–July 1.

Krishnan, G. and K. Mohseni. 2009a. An experimental and analytical investigation of rectangular synthetic jets. *ASME Journal of Fluid Engineering*, 131(12): 121101.

Krishnan, G. and K. Mohseni. 2009b. Axisymmetric synthetic jets: An experimental and theoretical examination. *AIAA Journal*, 47(10): 2273–2283.

Lamont, P. 1982. Pressures around an inclined give cylinder with laminar, transitional, or turbulent separation. *AIAA Journal*, 20: 1492–1499.

Leishman, J. 2000. *Principles of Helicopter Aerodynamics*. Cambridge, NY: Cambridge University Press.

Ljung, L. 2009. *System Identification, Theory for the User*. 2nd Edition. Upper Saddle River, NJ: Prentice Hall.

Maines, B. H., B. R. Smith, D. Merrill, S. Saddoughi, and H. Gonzalez. 2009. Synthetic jet flow separation control for thin wing fighter aircraft. *The 47th AIAA Aerospace Sciences Meeting*. Orlando, FL, January 5–8.

Margalit, S., D. Greenblatt, A. Seifert, and I. Wygnanski. 2005. Delta wing stall and roll control using segmented piezoelectric fluidic actuators. *Journal of Aircraft*, 42: 698–709.

McGowan, G., K. Granlund, M. Ol, A. Gopalarathnam, and J. Edwards. 2011. Investigations of lift-based pitch-plunge equivalence for airfoils at low Reynolds numbers. *AIAA Journal*, 49: 1511–1524.

Mittal, R. and R. Kotapati. 2004. Resonant mode interaction in a canonical separated flow. *Proceedings of the IUTAM Symposium on Laminar-Turbulent Transition*, Bangalore, India, December.

Pastoor, M., L. Henning, B. Noack, R. King, and G. Tadmor. 2008. Feedback shear layer control for bluff body drag reduction. *Journal of Fluid Mechanics*, 608: 161–196.

Quach, V., W. Kerstens, D. Williams, G. Tadmor, and T. Colonius, T. 2010. Transient response of a wing to arbitrary actuator input. *The 3rd International Conference Jets Wakes and Shear Flows*. Cincinnati, OH, September 27–30.

Raju, R., R. Mittal, and L. Cattafesta. 2008. Dynamics of airfoil separation control using zero net mass forcing. *AIAA Journal*, 46: 3103–3115.

Roos, F. and C. Magness. 1993. Bluntness and blowing for flowfield asymmetry control on slender forebodies. *The 11th AIAA Applied Aerodynamics Conference, AIAA Paper* 93-3409, August 9–11.

Siegel, S., M. Romer, J. Imamura, C. Fagley, and T. McLaughlin. 2011. Experimental wave generation and cancellation with a cycloidal wave energy converter. *Proceedings of 30th International Conference on Ocean, Offshore and Arctic Engineering*. Rotterdam, the Netherlands, June 19–24.

Seifert, A., A. Darabi, and I. Wygnanski. 1996. Delay of airfoil stall by periodic excitation. *Journal of Aircraft*, 33: 691–698.

Seifert, A., S. Eliahu, D. Greenblatt, and I. Wygnanski. 1998. Use of piezoelectric actuators for airfoil separation control, *AIAA Journal*, 36: 1535–1537.

Seifert, A. and L. Pack. 1999. Oscillatory control of separation at high Reynolds numbers. *AIAA Journal*, 37: 1062–1071.

Sharma, R. 2007. Fluid-dynamics-based analytical model for synthetic jet actuation. *AIAA Journal*, 45: 1841–1847.

Stalnov, O., A. Kribus, and A. Seifert. 2010a. Evaluation of active flow control applied to wind turbine blade section. *Journal Renewable and Sustainable Energy*, 2: 1–24.

Stalnov, O. and A. Seifert. 2010b. On amplitude scaling of active separation control. In *Active Flow Control II*. ed. King, R. Berlin, Germany: Springer-Verlag, pp. 63–80.

Theodorsen, T. 1935. General theory of aerodynamic instability and the mechanism of flutter. *National Advisory Committee for Aeronautics Report* NACA-TR-496, NACA, Jamaica Plain, MA.

Williams, D., J. Collins, G. Tadmor, and T. Colonius. 2008. Control of a semi-circular planform wing in a "gusting" unsteady freestream flow: I—Experimental issues. *The 4th AIAA Flow Control Conference*, Seattle, WA, June 23–26.

Williams, D., W. Kerstens, J. Pfeiffer, R. King, and Colonius, T. 2010. Unsteady lift suppression with a robust closed loop controller. In *Active Flow Control II*. ed. King, R. Berlin, Germany: Springer-Verlag, pp. 19–30.

Williams, D., V. Quach, W. Kerstens, S. Buntain G. Tadmor, C. Rowley, and T. Colonius. 2009a. Low-Reynolds number wing response to an oscillating freestream with and without feed forward control. *The 47th AIAA Aerospace Sciences Meeting*, Orlando, FL, January 5–8.

Williams, D., G. Tadmor, T. Colonius, W. Kerstens, V. Quach, and S. Buntain. 2009b. The lift response of a stalled wing to pulsatile disturbances. *AIAA Journal*, 47: 3031–3037.

Wiltse, J. M. and A. Glezer. 1998. Direct excitation of small-scale motions in free shear flows. *Physics of Fluids*, 10: 2026–2036.

Woo, T., T. Crittenden, and A. Glezer. 2008. Transitory control of a pitching airfoil using pulse combustion actuation. *The 4th AIAA Flow Control Conference*, Seattle, WA, June 23–26.

Wu, J.-Z., X. Lu, A. Denny, M. Fan, and J.-M. Wu. 1998. Post-stall flow control on an airfoil by local unsteady forcing. *Journal of Fluid Mechanics*, 371: 21–58.

Yechout, T., S. Morris, and D. Bossert. 2003. *Introduction to Aircraft Flight Mechanics: Performance, Static Stability, Dynamic Stability and Classical Feedback Control*. Reston, VA: AIAA Education Series.

8

Synthetic Jets in Boundary Layers

Shan Zhong
University of Manchester

8.1 Introduction

The concept of using an array of vortex-generating jets to prevent or delay boundary layer separation is well established and understood to rest on the resulting increase in momentum mixing, which "energizes" the near-wall layer, thus enhancing the ability of the boundary layer to resist the influence of an adverse pressure gradient. As a special form of vortex-generating jets, synthetic jets have received a great deal of research attention in the last two decades, due to their potential promises of delivering flow separation control for aircraft applications with zero net mass flux and the relative ease in being integrated in microelectromechanical systems (MEMS) through micro-fabrication (Gad-El-Hak 2000; Glezer and Amitay 2002; Tang et al. 2007; Zhong et al. 2007; Kotapati et al. 2010).

8.1.1 Review of Existing Research on Synthetic Jet Interaction with Crossflow

A large volume of work has been undertaken to characterize the flow structures and investigate the interaction mechanism of synthetic jets with a crossflow, either experimental or numerical or combined. So far, the research on synthetic jets reported in the literature, especially numerical simulations, has been dominated by two-dimensional (2D) or slot jets with a very large aspect ratio (greater than 75). This is because that firstly a synthetic jet issuing from a thin slot is capable of ejecting greater momentum and vorticity flux across its entire spanwise extent than those issuing from an array of circular orifices with a diameter equal to the slot width, and secondly, its nominal 2D nature simplifies both measurements and numerical simulations (Leschziner and Lardeau 2010).

Many of the early studies of 2D slot synthetic jets were conducted to investigate the effect of actuation frequency, location, and injection strength on a separated flow. For example, Amitay et al. (1997) modified the global aerodynamic forces on a 2D circular cylinder using synthetic jets at a Reynolds number of 1.31×10^5. They found that the synthetic jets induce a local "transpirating" recirculation bubble, which displaces the local streamlines well outside the undisturbed boundary layer, causing the lift coefficient of the cylinder to increase and the drag coefficient to decrease. In a subsequent experiment by Amitay et al. (2001), the above cylinder model is modified to form a 24% thick aerofoil section with a cylindrical leading edge. They showed that in the absence of control, the aerofoil stalls at an angle attack greater than 5°; however, with control, fully attached flow can be achieved at 17.5°. They also found that although relatively high levels of power are required to affect the flow when the actuator is located at a more upstream position, the interaction of the jets with a crossflow is more robust and can yield a higher level of lift-to-drag ratio that may not be achieved when the jets are closer to the separation point. McCormick (2000) also applied a slot synthetic jet near the leading edge of an aerofoil section at a Reynolds number of 5×10^5 and managed to extend the stall angle by 5° and increase the maximum lift by 25%. Using the laser sheet smoke visualization technique, he illustrated that an optimal forcing level exists at which the flow is completely attached with no large-scale vortical structures occurring above the aerofoil surface.

A series of efforts also have been made to simulate the synthetic jet actuation. The National Aeronautics and Space Administration (NASA) test case provided by Greenblatt et al. (2006) has been used widely for validating computational methods. In their experiment, a slot synthetic jet is injected upstream of a dune-shaped hump at a Strouhal number of 0.77 based on the length of the separated shear layer in the uncontrolled flow. It is shown by Rumsey et al. (2006) that the Reynolds-averaged Navier–Stokes equations (RANS) methods tend to overpredict the recirculation-zone length considerably due to the inherent incapability in the current turbulence models in modeling separated turbulent flows. On the contrary, the scale-resolving approaches, such as large eddy simulation (LES) and direct numerical simulations (DNS), can yield reasonable agreements with the measurements, albeit at a high computational cost (Avdis et al. 2009).

The discussion on the role of resonance between the frequency of actuation and the natural instability modes in the uncontrolled flow pertains virtually exclusively to 2D slot jets because the resonance relies on the actuated disturbances being able to interact with 2D instability modes (Leschziner and Lardeau 2010). A significant enhancement in flow separation delay is achieved by tuning the actuation frequency toward the 2D instability modes as shown in the experiment by Greenblatt et al. (2006) and the numerical simulations by Dandois et al. (2007) and Kotapati et al. (2010). The optimal frequency for flow control is found to scale with the advection time, leading to a Strouhal number based on the length of the aerofoil chord or the circulation zone in the separated flow between 0.5 and 1, in accordance with the shedding mode of the uncontrolled flow (Dandois et al. 2007; Leschziner and Lardeau 2010). It is shown that actuating a 2D slot synthetic jet in this frequency range encourages the formation of large-scale spanwise roller vortices, which entrain high-momentum fluids to the near-wall region, resulting in a strong delay of flow separation (Greenblatt et al. 2006).

A different approach, whereby the actuators are driven at much higher frequencies than the characteristic flow frequencies, has also been investigated. At such a high frequency, the actuation is spectrally decoupled from any macroscales in the controlled flow such that the flow control effect becomes invariant to changes in frequency (Amitay et al. 2001). It is shown by Amitay et al. (1997, 2001), Glezer and Amitay (2002), and Mittal and Rampunggoon (2002) that the process by which control is effected is via the modification of the apparent aerodynamic shape of the lifting surfaces, that is, virtual shaping. Virtual shaping has been utilized not only for separation delay, but also for induction of controlled changes in global aerodynamic forces and moments on aerofoils at low incidences without the need of moving surfaces (Glezer 2010).

A synthetic jet, issuing into a boundary layer through an orifice of circular, elliptical, or rectangular shape with an aspect ratio less than 5, results in the formation of complex three-dimensional (3D) structures. Here we will concentrate on the synthetic jets issuing from circular orifices. Detailed studies of the interaction of vortex rings produced by circular synthetic jets with a laminar crossflow were conducted by Crook and Wood (2001), Sauerwein and Vakili (1999), Jabbal and Zhong (2008, 2010), Sau and Mahesh (2008), and Zhou and Zhong (2008, 2010), with the former three being experimental and the other two computational. In these studies, the shape and evolution of vortical structures produced by synthetic jets are examined. Measurements, which quantify the effect of synthetic jets on the boundary layer, were also performed by a number of researchers (Cater and Soria 2002; Zaman and Milanovic 2003; Garcillan et al. 2005). Their results show deflected mean streamlines and a presence of counter-rotating vortex pairs produced by the synthetic jet injection. These features are also reproduced by the LES simulations by Wu and Leschziner (2009), who simulated the experimental cases of Garcillan et al. (2005). The experiment on circular synthetic jets in a turbulent flow of Schaeffler and Jenkins (2006) has provided the validation data set for many 3D simulations (Qin and Xia 2008; Dandois et al. 2006; Rumsey 2009). Circular orifices of different orientations also have received some research attention. Pitched synthetic jets issuing through a circular orifice duct, which are inclined downstream, are found to produce similar flow structures as normal circular jets. However, they produce a region of high momentum near the wall and a smaller jet penetration in the boundary layer (Zaman and Milanovic 2003; Zhong et al. 2005b). In their dye visualization experiment, Zhong et al. (2005a) showed that synthetic jets from an inclined circular orifice duct, which is skewed by 90° to the oncoming freestream flow, produce an asymmetric counter-rotating vortex pair with one branch being stronger than the other.

However, there are only a few experimental studies of circular synthetic jets issuing into a separated flow, for example, Zhang and Zhong (2010a), Crook et al. (1999), and Zhang and Zhong (2010b), with the latter two undertaking in a separated turbulent flow. According to the oil visualization experiments undertaken on a 2D circular cylinder model by Crook et al. (1999), the delay of flow separation using an array of circular synthetic jets is associated with the formation of counter-rotating vortex pairs within the boundary layer, not dissimilar to those formed by the conventional vane devices (Ashill et al. 2002). Among the only two computational studies, that is, Ozawa et al. (2010) and Leschziner and Lardeau (2010), the latter was carried out at the corresponding experimental condition of Zhang and Zhong (2010b), with both the near-field behavior of the jet and profiles of the streamwise turbulent stress showing a good agreement with the experiment.

It is worth mentioning that a limited amount of research has also been carried out on finite-span synthetic jets with an aspect ratio between 5 and 75. For example, Gilarranz et al. (2005) illustrated a significant increase in lift on a NACA 0015 wing at high incidence angles using finite-span synthetic jets with an aspect ratio of 22. Smith (2002) performed hotwire measurements on an array of finite-span jets with an aspect ratio of 45, aligned in the spanwise and streamwise direction, respectively, in a turbulent boundary layer. The orifices aligned in streamwise direction exhibit longitudinal vortices embedded in the boundary layer, whereas those aligned in the spanwise direction create a wave-like feature downstream. More recently, Sahni et al. (2011) explored the detailed flow structures in the vicinity of finite-span synthetic jets with an aspect ratio of 21 over a NACA 4421 aerofoil using a combined experimental and computational approach. They found that at a low blowing rate, the 2D spanwise vortex roller formed near the slit is perturbed by the crossflow, leading to the formation of small and organized streak-like secondary flow structures further downstream. As the blowing rate continues to increase, three-dimensionalities are formed closer to the slit, eventually resulting in a train of counter-rotating coherent vortices that lift off the aerofoil surface.

8.1.2 Motivation for Further Research

So far, within the large volume of research on synthetic jets, a majority of the studies have targeted at investigating various global characteristics and manifestations of synthetic-jet injection, and only few have been devoted to elucidating the detailed flow-physical processes involved. Furthermore, despite the many successes in demonstrating the control effect on aerodynamic bodies of various shapes in laboratory experiments, most flow-control experiments reported so far have been undertaken by setting the operating conditions of the actuators by trial-and-error, subject to the crude criterion that the ratio of the jet peak velocity to the freestream velocity should not be less than unity. As a result, the level of injection effectiveness has been observed to vary in an unpredictable manner with varying geometric and injection parameters (Crook et al. 1999; Gilarranz et al. 2005; Dandois et al. 2007). Therefore, to design effective flow control devices that are capable of delivering a required flow control effect at minimum energy expenditure, a sound understanding of the fluid mechanics of the interaction between synthetic jets and a boundary layer is required.

In this chapter, the key results from a series of carefully designed experimental studies undertaken by the author's research group at the University of Manchester between 2006 and 2011, aiming at achieving a better understanding of the flow physics of synthetic jets interacting with a grazing boundary layer, are presented. Such an improved understanding is acquired via gaining an insight into the nature of the vortical structures produced and their impact on the boundary layer using various experimental techniques supplemented with results from numerical simulations. The experiments are carried out in a logical order starting from a circular synthetic jet issuing perpendicularly into a flat-plate laminar boundary layer, then to an array of synthetic jets issuing upstream of a separated laminar boundary layer, and finally to an array of synthetic jets issuing upstream of a separated turbulent boundary layer.

The layout of this chapter is as follows: Firstly, to assist the understanding of the results presented here, the important dimensionless parameters, which determine the behavior of synthetic jets issuing into a boundary layer, will be discussed. The experimental setup and results from each experiment will be then presented in order. Finally, a summary of the key findings will be given.

8.2 Dimensionless Parameters

For a circular synthetic jet issuing perpendicularly into a crossflow boundary layer, the physical parameters, which the behavior of the synthetic jet will depend on, include the jet velocity ($\overline{U}_O$), the actuation frequency (f), the jet orifice diameter (D_O), the fluid density (ρ), the fluid viscosity (μ), the freestream velocity (U_∞), and the boundary layer momentum thickness (θ). Based on the dimensional analysis, these physical parameters can be combined to yield four dimensionless parameters, that is, the dimensionless stroke length (L), the jet Reynolds number (Re_j), the jet-to-freestream velocity ratio (VR), and the ratio between the boundary layer momentum thickness and the orifice diameter (ε).

The dimensionless stroke length is defined as

$$L = \frac{L_O}{D_O} = \frac{\overline{U}_O T}{D_O} \tag{8.1}$$

where:

L_O is the stroke length, which, according to the slug model of Smith and Glezer (1998), represents the length of the fluid column expelled during the blowing stroke

$\overline{U}_O$ is the time-averaged blowing velocity over the entire actuation cycle defined as

$$\overline{U}_O = \frac{1}{T}\int_0^{T/2} \tilde{u}_O(t)\mathrm{d}t \tag{8.2}$$

where:

$\tilde{u}_O(t)$ is the instantaneous space-averaged velocity at the orifice exit

T is the diaphragm oscillation cycle, $T = 1/f$

The definition of $\overline{U}_O$ was originally proposed by Smith and Glezer (1998) to facilitate a comparison between synthetic jets and steady jets. Since L can be interpreted as the ratio between the inertia force and the unsteady force (Zhou et al. 2009), its magnitude determines the extent of impact from the suction stroke of the actuation cycle, with a greater influence of suction being asserted on synthetic jets that are formed at low values of L, in which case the vortex ring has not moved sufficiently far away from the orifice at the onset of the suction stroke. Holman et al. (2005) and Milanovic and Zaman (2005) have shown that, in the absence of a crossflow, *a round synthetic jet cannot form* when L is approximately less than 0.5. On the contrary, although the synthetic jet can form when $3 > L > 0.5$, the vortex ring will be weakened by the onset of suction cycle (Guo and Zhong 2007). For a synthetic jet issuing into a boundary layer, this effect of suction is expected to be confined to the upstream branch of the vortex, hence resulting in an asymmetric structure. Furthermore, the amount of circulation that can be contained in the primary vortex ring structure ceases to increase as L increases above 4. The extra vorticity ejected during the blowing cycle will then be shed in the form of the secondary vortices trailing behind the primary vortex ring (Gharib et al. 1998; Crook and Wood 2001; Jabbal et al. 2006). For a synthetic jet issuing into a boundary layer, the shedding of secondary vortices increases the complexity of the vortical structures formed as a result of the interaction between the synthetic jet and the boundary layer (Zhou and Zhong 2010).

The jet Reynolds number is defined based on the stroke length as

$$Re_L = \frac{\overline{U}_O L_O}{\nu} \tag{8.3}$$

Studies of synthetic jets at quiescent conditions undertaken by several researchers (Glezer 1988; Jabbal et al. 2006; Tang et al. 2007) show that this Reynolds number is proportional to the synthetic jet vortex circulation (Γ/ν) and as such it is a good indicator of the vortex strength.

The jet-to-freestream velocity ratio is defined as

$$\mathrm{VR} = \frac{\overline{U}_O}{U_\infty} \tag{8.4}$$

VR indicates the relative strength between the jet and freestream velocity, and hence, it determines the trajectory of the vortical structures as they propagate downstream (Milanovic and Zaman 2005). As such, at a low VR the vortical structures will reside in the boundary layer for a longer period of time and the shear in the boundary layer will have a greater influence on them, as opposed to those formed at a high VR (Zhong et al. 2005b).

The ratio between the boundary layer momentum thickness and the orifice diameter (ε), characterizes the scaling of the jet orifice to the local boundary layer. For flow separation control applications to minimize its disturbance to the flow, as a rule of thumb the jet orifice diameter is not expected to be greater than, say, 20% of the local boundary layer thickness.

On the contrary, to ensure a sufficient impact of the synthetic jets on the boundary layer, the orifice diameter is not expected to be much less than, say, 5% of the local boundary layer thickness. Hence, the orifice diameter would be roughly of the same order of the momentum thickness of the boundary layer to be controlled. For convenience, ε may be replaced by $d = \delta/D_O$. It is shown by Tang et al. (2007) that the Strouhal number defined based on the freestream velocity and the boundary layer thickness (δ)

$$Str_\delta = \frac{f\delta}{U_\infty} \tag{8.5}$$

is a useful dimensionless frequency, where Str_δ = VR.d/L. This is because that $1/Str_\delta$ gives an approximate measure of the distance between the consecutive structures produced by the synthetic jets as a fraction of the local boundary layer thickness. Hence, Str_δ can be used to replace d as one of the dimensionless parameters, which characterizes the impact of a normal circular synthetic jet on a boundary layer.

Assuming incompressible flow and a sinusoidal diaphragm motion, for a simple synthetic jet actuator as shown in Figure 8.1, which has a cavity diameter of D_C and an orifice diameter D_O, the mean time-averaged blowing velocity at the orifice exit during the entire cycle can be expressed as

$$\overline{U}_O = \alpha f \Delta \left(\frac{D_C}{D_O}\right)^2 \tag{8.6}$$

The coefficient α depends on the shape of the diaphragm with $\alpha = 1$ for a piston diaphragm and 1/3 for a diaphragm that exhibits a cone shape. The jet peak velocity is π times of $\overline{U}_O$. From this equation, it can be seen that for a given actuator, $\overline{U}_O$ is proportional to the diaphragm oscillation frequency and displacement. Therefore, the dimensionless stroke

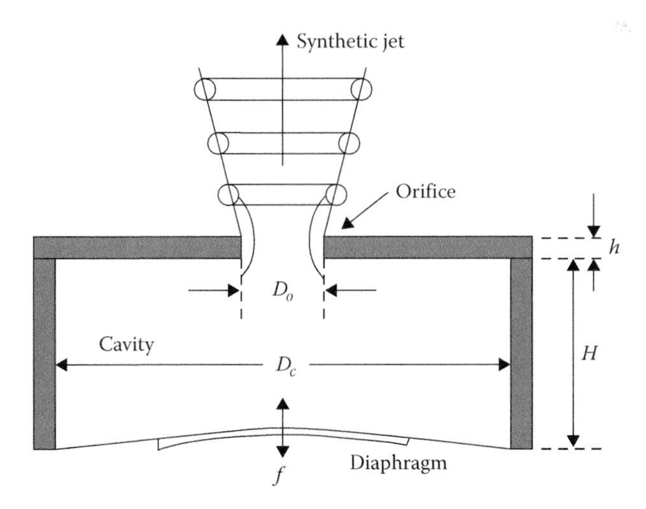

FIGURE 8.1 Schematic of a synthetic jet actuator. (From Zhou, J. and S. Zhong, *Comput. Fluids*, 39, 1296–1313, 2010. Reprinted with permission.)

length is independent of the diaphragm frequency and proportional to the diaphragm displacement. Since the predicted $\overline{U}_O$ given by Equation 8.6 is found to agree well with the measured values for our actuators (Jabbal and Zhong 2010), it is used to calculate the dimensionless parameters presented here.

8.3 Synthetic Jets in a Zero-Pressure-Gradient Laminar Boundary Layer

To understand the nature of the vortical structures formed as the result of the interaction between normal circular synthetic jets and a boundary layer, a flow visualization experiment is conducted in a tilting water flume. The results from numerical simulations are used to provide further insight into these vortical structures and their impact on the near-wall flow.

8.3.1 The Experimental Setup

In this experiment, the test plate, along which a zero-pressure-gradient laminar boundary layer develops, is placed horizontally across the whole width of the test section of the water flume with its test surface facing downward. The synthetic jet actuator is mounted to the back of the test plate and the center of its orifice is located at 0.7 m downstream of the leading edge of the plate (Figure 8.2). The cavity of the synthetic jet actuator has a diameter of D_C = 45 mm and a height of H = 20 mm. The orifice diameter D_O is 5 mm and its height

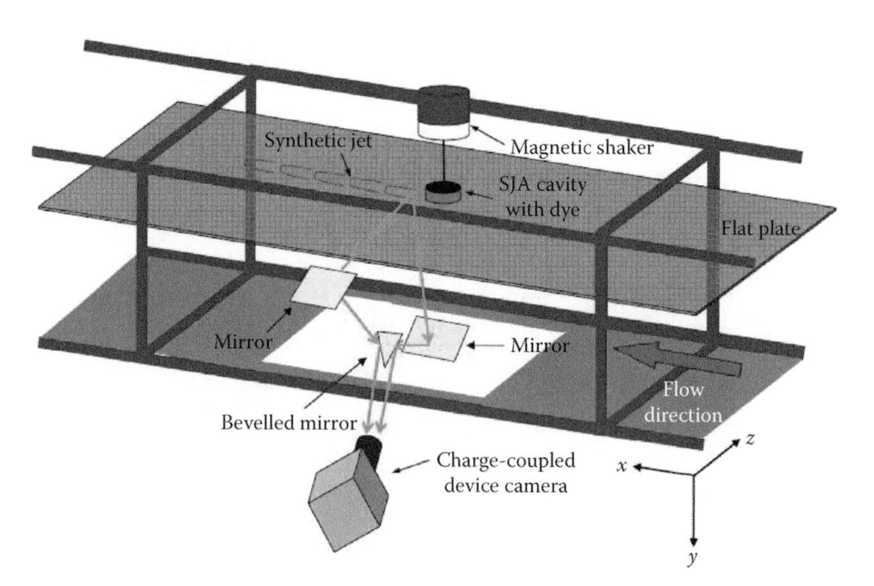

FIGURE 8.2 A schematic layout of the setup for the dye flow visualization experiment. (From Zhou, J. and S. Zhong, *Comput. Fluids*, 38, 393–405, 2008. Reprinted with permission.) CMOS, complementary metal-oxide-semiconductor; SJA, synthetic jet actuator.

h is also 5 mm (Figure 8.1). The piston diaphragm is made to oscillate in a sinusoidal manner at predetermined oscillation displacements and frequencies. In the experiment, the freestream velocity is fixed at 0.05 m/s. The boundary layer approaching the orifice of the synthetic jet is laminar and its thickness is about four times of the orifice diameter.

In the dye flow visualization experiment, dye is introduced into the actuator cavity by gravity. A video camera is used to capture the dye pattern from two orthogonal views simultaneously from side and below. The thermal footprints of the flow structures are visualized using a layer of thermochromic liquid crystals which is coated on the heated test surface. The liquid crystals change color in response to the passage of the flow structure, which cause a localized variation in convective heat transfer. Detailed information of the experimental setup can be found in the work by Jabbal and Zhong (2008).

8.3.2 The Typical Vortical Structures

Figure 8.3 shows the dye visualization images of the vortical structures produced by synthetic jets operated at the same L but different VR due to the use of two different actuation frequencies. In the lower VR case (VR = 0.27, L = 2.7, Re_L = 182), the vortical structures produced by the synthetic jets appear as highly stretched hairpin-like vortices, which are located in the boundary layer (Figure 8.3a and b). In contrast, in the higher VR case (VR = 0.54, L = 2.7, Re_L = 364), the structures appear as titled vortex rings, which

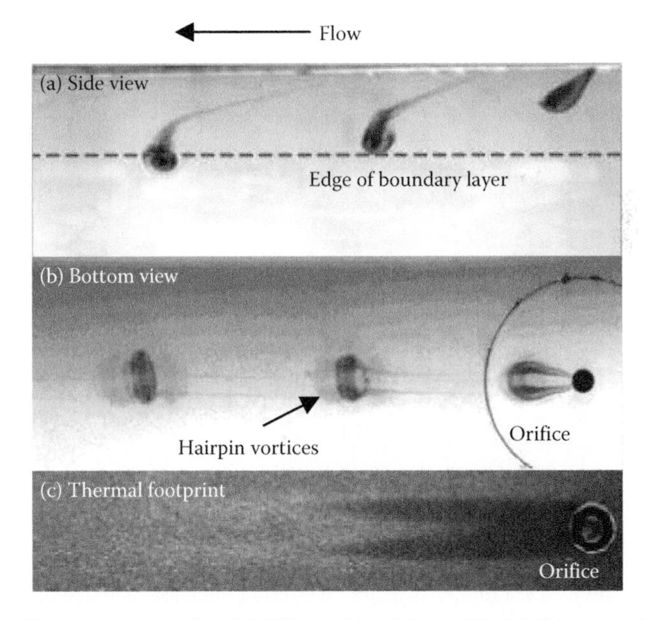

FIGURE 8.3 Hairpin vortices at L = 2.7, VR = 0.27, and Re_L = 182. (a) Side view of dye visualization image; (b) bottom view of dye visualization image; and (c) thermal footprints revealed by the liquid crystal coating. (From Jabbal, M. and S. Zhong, *Phys. Fluids*, 22, 063603, 2010. Reprinted with permission.)

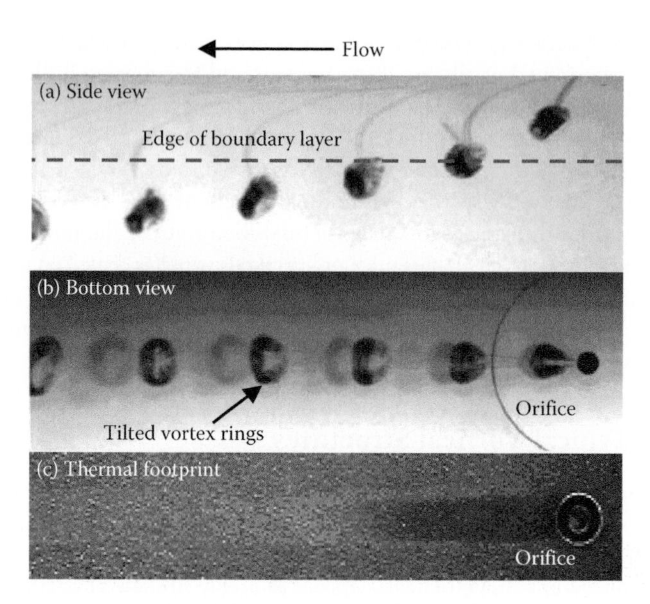

FIGURE 8.4 Hairpin vortices at $L = 2.7$, VR = 0.54, and $Re_L = 364$. (a) Side view of dye visualization image; (b) bottom view of dye visualization image; and (c) thermal footprints revealed by the liquid crystal coating. (From Jabbal, M. and S. Zhong, *Phys. Fluids*, 22, 063603, 2010. Reprinted with permission.)

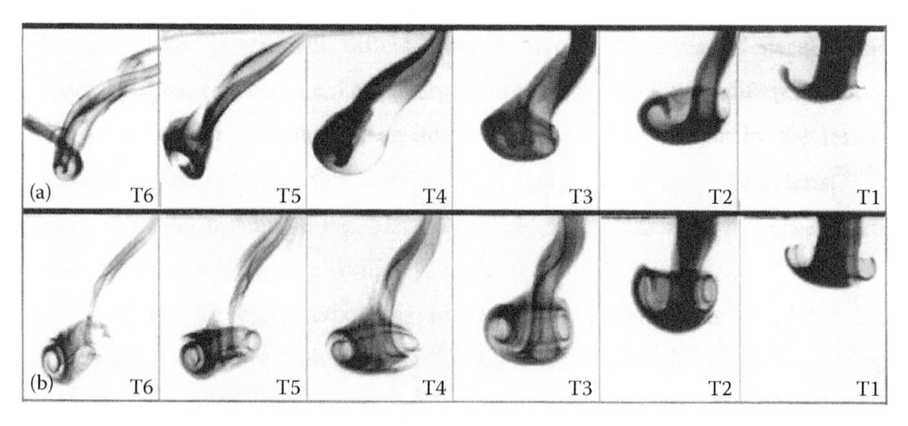

FIGURE 8.5 Time sequence of close-up development of vortical structures near the orifice exit: (a) $L = 2.7$, VR = 0.27, and $Re_L = 182$; (b) $L = 2.7$, VR = 0.54, and $Re_L = 364$.

are attached to the wall via highly stretched legs and penetrate the edge of the boundary layer shortly downstream (Figure 8.4a and b).

A close-up view of the development of the structure near the orifice in the above two cases is provided in Figure 8.5. The sequence of images for each case was not taken at matching phases during the diaphragm oscillatory cycle, but at arbitrary instants to illustrate the rollup process. In the case of the hairpin vortex (Figure 8.5a), a rollup on the upstream side of the vortex is visible at instant T1 and T2, which is gradually weakened by the vorticity in the boundary layer. Conversely, the strength of the downstream branch becomes intensified due to stretching. Anticlockwise tilting is also evident by the motion of the vortex head as the structure convects downstream. Based on the spinning solid cylinder concept suggested by Chang and Vakili (1995), Zhong et al. (2005b) attribute this behavior to the moment of the magnus forces acting on the vortex rollers. By the time of T4, a hairpin vortex has formed with its head inclined downstream and its tail attached to the orifice. In the case of the tilted vortex ring (Figure 8.5b), it is seen that the initial vortex experiences a symmetrical rollup and the ring shape is retained. This is because the strength of the jet is increased relative to the freestream, with the ensuing vortical structure penetrating the boundary layer. Consequently, the boundary layer shear has little influence on this structure.

8.3.3 The Near-Wall Effect

To obtain an insight into the complete structure of the aforementioned vortical features and their impact in the near-wall region, numerical simulations are performed using FLUENT 6.2 with the incompressible laminar flow model at identical flow conditions. The geometry of the computational flow domain and boundary conditions used in the simulation are shown in Figure 8.6. Since the flow field is expected to be symmetric relative to the central plane of the orifice, only a half of flow field is modeled to save the computational time. The inlet flow condition is specified using the Blasius velocity profile. Both the orifice duct

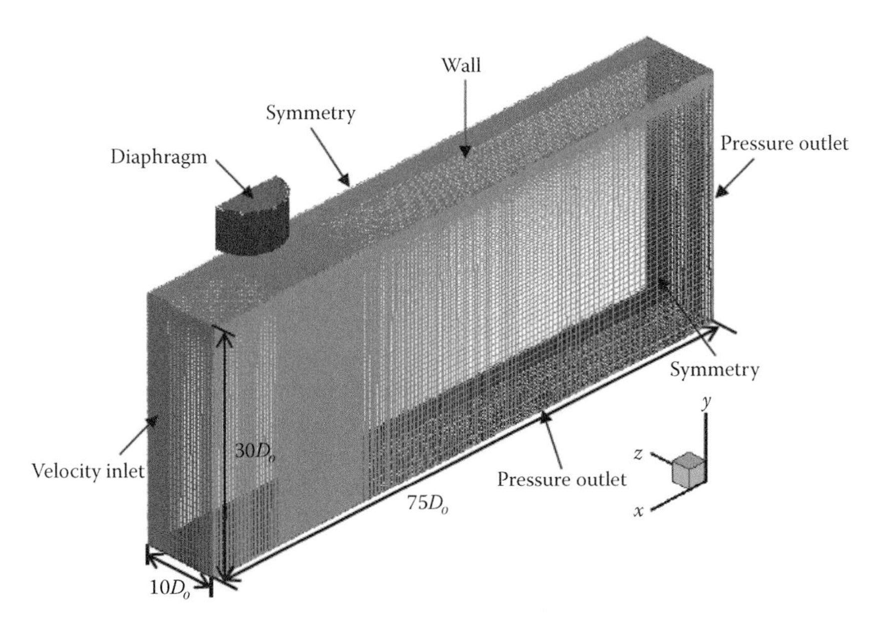

FIGURE 8.6 Computational domain and boundary conditions for numerical simulations of a single normal circular synthetic jet issuing into a laminar boundary layer. (From Zhou, J. and S. Zhong, *Comput. Fluids*, 38, 393–405, 2008. Reprinted with permission.)

and actuator cavity are included in the simulation, and a velocity boundary condition is applied at the neutral position of the diaphragm. Details of the computational setting and the mesh and time step independence studies can be found in the work by Zhou and Zhong (2008).

The Q-criterion is employed to reveal the coherent structures in the flow (Hunt et al. 1988). A positive Q isosurface represents a region where the strength of rotation overcomes that of the strain, allowing the coherent structures to be extracted from a shear layer. Interestingly, the numerical simulations not only reproduce the primary structures observed in the experiment but also reveal a hierarchy of coherent vortices consisting of primary, secondary, and tertiary structures (Figure 8.7), thereby providing further insight into the formation of the two different patterns of surface shear stress observed by Jabbal and Zhong (2008).

As shown in Figure 8.7a, the legs of the hairpin vortex trailing along the wall also induce a pair of streamwise vortices (secondary vortices) of the opposite sign outboard. Each leg of hairpin vortex and its induced streamwise vortex create a downwash between them, which brings high-momentum fluid from the outer part of the boundary layer to the wall. Consequently, each hairpin vortex produces a pair of streamwise streaks with high local shear stress in the near-wall region. By the Reynolds analogy, the streaks of high wall shear stress are associated with regions of high wall heat transfer rate, and hence, they are seen as two low-temperature streaks as revealed by the temperature-sensitive liquid crystal surface coating (Figure 8.3c).

In the case of tilted vortex rings, the rings are connected to the wall via a pair of counter-rotating legs (secondary vortices), which further induce a pair of tertiary streamwise vortices of an opposite sense directly underneath as shown in Figure 8.7b. It is the tertiary streamwise vortices that produce a downwash that brings high-momentum fluids to the near-wall region on the central plane of the orifice. Consequently, each tilted vortex ring structure produces a single streak with high wall shear stress or high heat transfer rate in the near-wall region as shown in Figure 8.4c. Schematic drawings depicting the hierarchy of these structures and their near-wall impact are given in Figures 8.8 and 8.9.

8.3.4 The Parameter Map

Since the nature of the vortical structures produced by the interaction between synthetic jets and a boundary layer changes as the actuator operating condition varies, it will be beneficial to produce a parameter map from which the condition of formation for different structures can be established. Therefore, numerical simulations of a total of 34 cases covering a reasonably wide range of dimensionless parameters were conducted (Zhou and Zhong 2010). All the computed results are then incorporated into a parameter space defined by L and Str_δ shown in Figure 8.10.

In Figure 8.10, the cases with two-streak and one-streak wall shear stress patterns as well as the transitional cases where the pattern changes from one-streak to two-streak or vice versa are shown using different symbols. The instantaneous coherent structures of some cases are inserted and lines of constant VR are added to assist the interpretation of data. It can be seen that the boundary between hairpin vortices, which produce the two-streak pattern, and tilted vortex rings, which produce the one-streak pattern, is correlated with

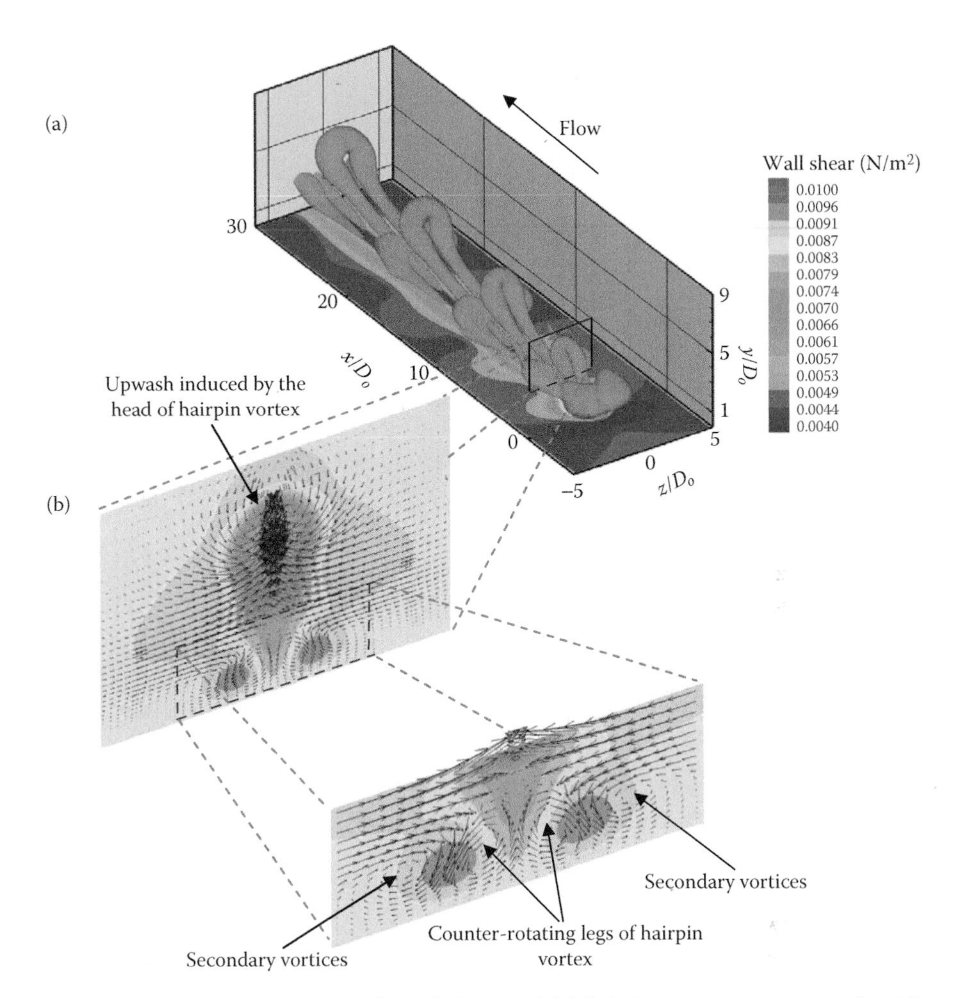

FIGURE 8.7 Instantaneous isosurface of $Q = 0.1$ of (a) hairpin vortex structures at $L = 2.7$, VR = 0.27, and $Re_L = 182$; and (b) tilted vortex ring structures at $L = 2.7$, VR = 0.54, and $\mathrm{Re}_L = 364$. The inserts show the velocity vectors and streamwise velocity contours at the crossplane indicated. (From Zhou, J. and S. Zhong, *Comput. Fluids*, 38, 393–405, 2008. Reprinted with permission.)

VR = 0.4 closely for $Str_\delta < 1$. For $Str_\delta > 1$, however, the boundary deviates from VR = 0.4, approaching a horizontal line of $L = 1.6$. The boundary between hairpin and tilted vortex ring type of vortical structures around VR = 0.4 is consistent to the finding from the experiment by Jabbal and Zhong (2008).

The correlation of the boundary with VR is not surprising because VR determines the trajectory of the vortical structure in the boundary layer. At relatively low VR, because of its proximity to the wall, the upstream branch of the initial vortex ring ejected from the orifice will be weakened, firstly by the impact of the actuator suction cycle and then by the vorticity in the boundary layer, both of which encourage the formation of hairpin vortices.

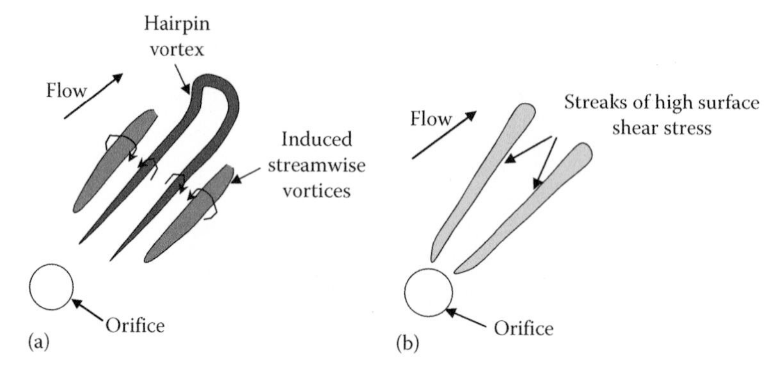

FIGURE 8.8 Schematics of (a) a hairpin vortex and its induced streamwise vortices and (b) the resultant surface shear stress pattern streaks produced by a single synthetic jet at relatively low L and VR. (From Zhang, S. and S. Zhong, *AIAA J*., 48, 611–623, 2010. Reprinted with permission.)

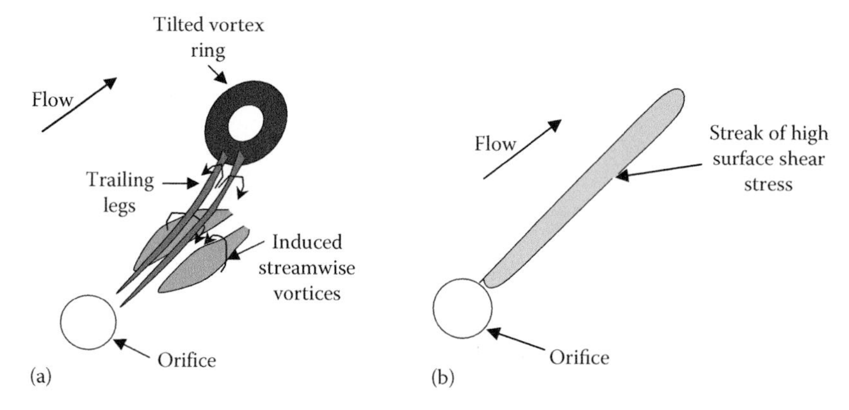

FIGURE 8.9 Schematics of (a) a tilted vortex ring and its secondary and tertiary structures and (b) the resultant surface shear stress pattern streaks produced by a single synthetic jet at relatively high L and VR. (From Zhang, S. and S. Zhong, *AIAA J*., 48, 611–623, 2010. Reprinted with permission.)

The deviation of this trend at higher Str_{δ} is caused by the strong interaction between adjacent hairpin vortices due to their close proximity to each other, which keeps them closer to the wall, such that hairpin vortices are formed at higher VRs.

8.4 Interaction of Synthetic Jets with a Separated Laminar Boundary Layer

On the basis of understanding the nature of the vortical structures produced by the interaction of a single synthetic jet with a zero-pressure-gradient boundary layer, an array of three synthetic jets is applied upstream of a separated laminar boundary layer over an inclined plate in a water flow experiment. Particle image velocimetry (PIV) measurements were carried out to provide quantitative information about the control effect of synthetic

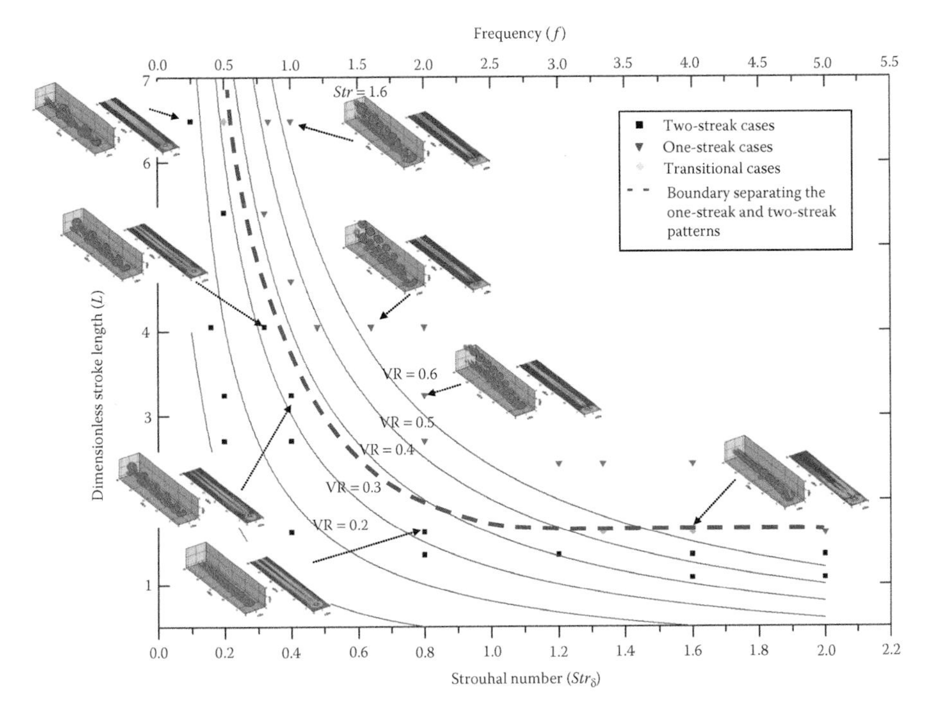

FIGURE 8.10 Parameter map of different vortical structures and their corresponding time-averaged surface shear stress patterns.

jets operating at different actuation conditions. The purpose of this work is to identify the synthetic jet operating conditions and the type of vortical structures, which could yield the best flow control effect with the minimum power consumption.

8.4.1 The Experimental Setup

The experiments are conducted in the same tilting flume used for the flat plate experiment described in Section 8.3.1. Figure 8.11a shows the sketch of the test plate and its coordinate systems. The test plate consists of three flat plates connected with each other using hinges, each having the same width as the test section. The test surface faces downward to allow the synthetic jets actuator to be mounted above the free surface of water. The boundary layer developing along the horizontal plate separates at some point along the second plate that is inclined 5° upward, creating a separated boundary layer to which flow control with synthetic jets can be applied.

The array of synthetic jets is issued through three circular orifices aligned normally to the oncoming flow direction. The orifices have a diameter of $D_O = 2$ mm and a depth of 3 mm (Figure 8.11b). The three jets are produced by an oscillating piston diaphragm attached to a single cylindrical cavity with a diameter of 45 mm and a height of 12 mm. The array is located at 450 mm downstream of the leading edge of the horizontal plate, leaving a distance of 80 mm ($40D_O$) to the start of the inclined plate and a distance of

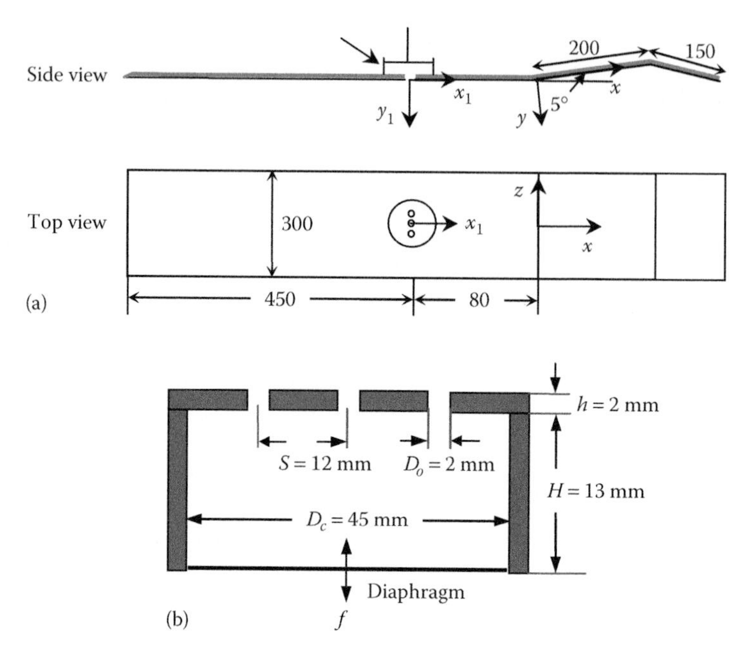

FIGURE 8.11 Schematics of (a) the test plate and coordinate system (units in mm) and (b) the synthetic jet actuator. (From Zhang, S. and S. Zhong, *AIAA J.*, 48, 611–623, 2010. Reprinted with permission.)

about 120 mm ($60D_O$) to the baseline separation line. The spacing between synthetic jets is fixed at 12 mm ($6D_O$).

In this study, to provide quantitative information about the time-averaged control effect of synthetic jets on the separated boundary layer, PIV measurements are undertaken on a spanwise plane parallel to the inclined plate at a distance of about 1.5 mm below. All the tests in this study are performed at a freestream velocity of 0.1 m/s. The local boundary layer thickness at the location of the synthetic jet array is about 10 mm. To investigate the flow control effectiveness of the synthetic jets at a range of actuator operating conditions, the diaphragm oscillating frequency is varied between 1 and 16 Hz. At a given frequency, the diagram peak-to-peak displacement is varied to yield a range of dimensionless stroke length from 1 to 16. The actuation frequencies are higher than the highest T–S wave frequency (0.67 Hz) and the shear layer frequency (0.4 Hz). Hence, a coupling between the actuation frequencies and these instability frequencies is not expected to occur. Detailed descriptions of this experimental setup can be found in the work by Zhang and Zhong (2010a).

8.4.2 Footprints of Synthetic Jets in the Separated Flow

To assess the flow control effect of synthetic jets operated at different conditions, the contours of time-averaged streamwise velocity superimposed with the streamlines on the

PIV measurement plane parallel to the inclined plate are examined. Due to the limited space in this paper, only the results at $f = 4$ Hz are shown here. As shown in Figure 8.12a, a visible separation delay is not seen at $L = 2$ and VR = 0.16. At $L = 3$ and VR = 0.24, the delayed separation is associated with the appearance of two high-speed streaks located between two adjacent synthetic jets (Figure 8.12b). The separation is delayed further as L increases to 6 and VR increases to 0.48 (Figure 8.12c). Interestingly, the level of separation delay reduces at $L = 8$ and VR = 0.64 with the emergence of a three-streak pattern, which is aligned with the centers of the three synthetic jets (Figure 8.12d). The level of separation delay peaks up again at $L = 11$ and VR = 0.88 (Figure 8.12e), and it then drops slightly as L and VR increase further to $L = 15$ and VR = 1.2 (Figure 8.12f).

Figure 8.13 shows the mapping of the flow patterns observed in the separated flow for all test conditions. Different flow patterns are indicated using different symbols, and four constant VR lines are added to assist the interpretation of results. It is seen that at a velocity ratio of less than about 0.3, the control effect is not appreciable. A VR between 0.3 and 0.7 approximately results in a separation delay associated with the presence of the two-streak flow pattern, whereas a VR between 0.7 and 1.5 approximately results in a separation delay largely associated with the three-streak flow pattern. A velocity ratio greater than 1.5 appears to provoke a global alternation of the separated flow. According to Crook (2002), the synthetic jets become turbulent when the jet peak velocity to the freestream velocity ratio exceeds 4 (equivalent to VR = 1.27). Hence, the global alternation of the separated flow observed at high VRs in the present experiment is likely to be associated with the breakdown of the laminar boundary layer into turbulent caused by the strong turbulent synthetic jets, and these cases are not of interest to the present study.

8.4.3 The Vortical Structures Responsible for Flow Separation Delay

The streaky patterns observed in the separated flow are no doubt caused by the vortical structures produced by the synthetic jets. The laser-induced florescence images taken at $L = 2$ and 4 at $f = 8$ Hz confirm that the vortical structures, which produce the two-streak and three-streak velocity patterns, are indeed the hairpin and tilted vortex ring type structures, respectively (Figure 8.14).

Based on the finding from single synthetic jets, a tilted vortex ring type structure will result in a high-speed streak directly downstream of each orifice. Hence, the connection between the three-streak flow pattern and the tilted vortex ring structures is expected. However, the link between the two-streak flow pattern and the hairpin-like structures is less obvious because each hairpin-like structure is expected to produce a pair of high-speed streaks. In fact, the two-streak flow pattern observed here is due to the merger of the high-speed regions produced by the adjacent jets, which results in a stronger single high-speed region midway between two adjacent orifices and the weakening of the high-speed regions outboard of the two side orifices by the end effect of the array. This is confirmed by the numerical simulation results of the corresponding cases by Zhou and Zhong (2010) (Figure 8.15). As shown in Figure 8.15, the merger of neighboring streaks results in the formation of two profound high-speed streaks in the case of hairpin vortices, whereas the three streaks remain individually distinct in the case of tilted vortex rings.

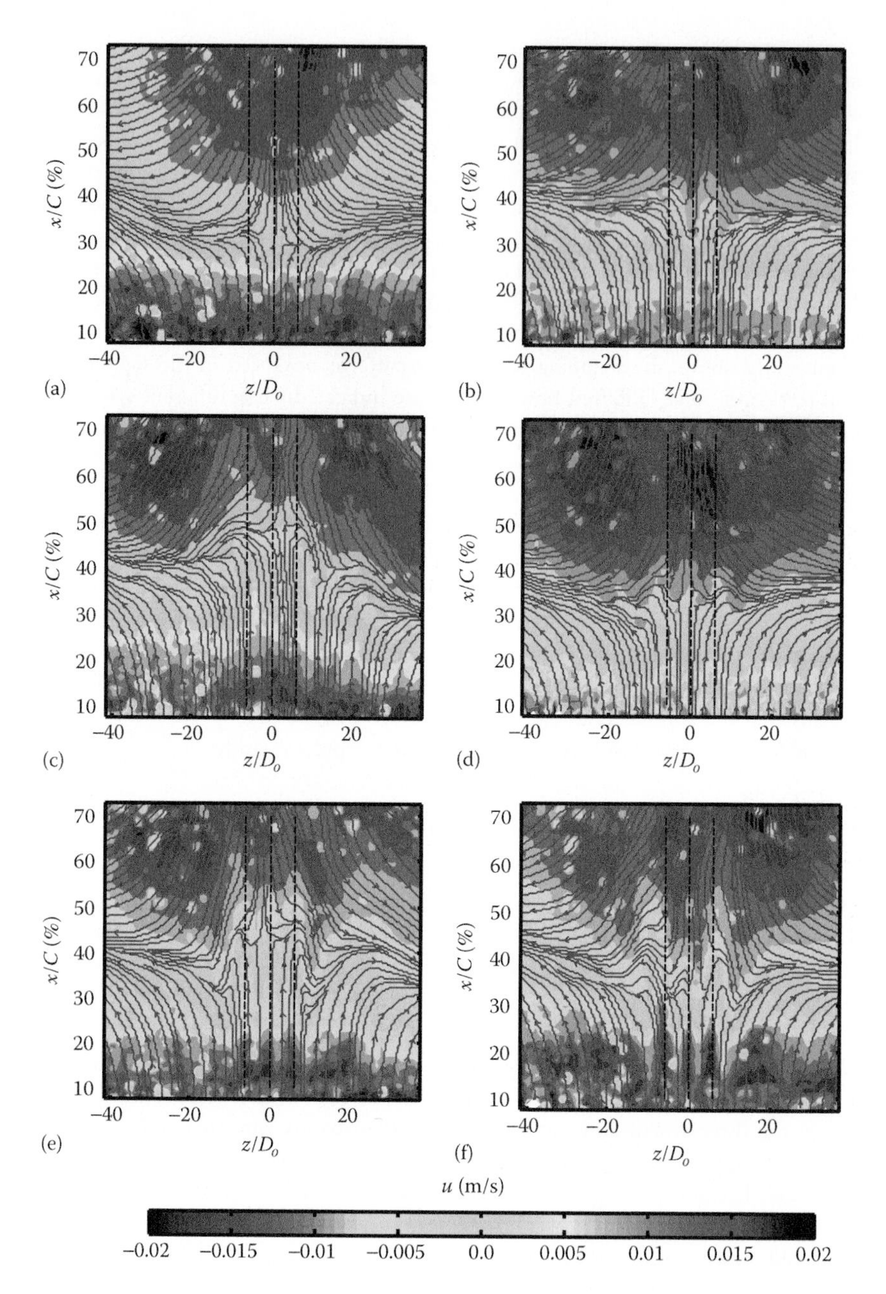

FIGURE 8.12 Contours and streamlines of averaged velocity on a plane parallel to the inclined plate showing the footprints induced by synthetic jets operated at f = 4 Hz (Str_δ = 0.8). (a) L = 2, VR = 0.16; (b) L = 3, VR = 0.24; (c) L = 6, VR = 0.48; (d) L = 8, VR = 0.64; (e) L = 11, VR = 0.88; and (f) L = 15, VR = 1.2. Dashed lines mark the centerlines of three synthetic jets; flow is from bottom to top. (From Zhang, S. and S. Zhong, *AIAA J.*, 48, 611–623, 2010. Reprinted with permission.)

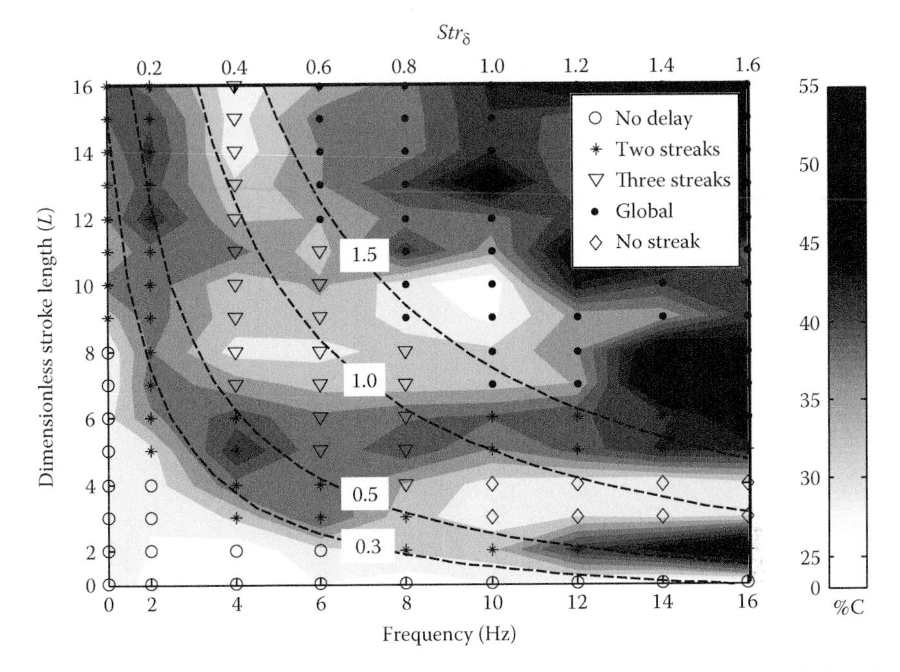

FIGURE 8.13 Contours of separation control effect at various frequencies and stroke lengths (dashed lines: VR = constant). (From Zhang, S. and S. Zhong, *AIAA J.*, 48, 611–623, 2010. Reprinted with permission.)

8.4.4 Evaluation of Flow Control Effectiveness

To evaluate the flow control effectiveness of the synthetic jet array quantitatively, the distance between the time-averaged mean separation line spanning a distance of $6D_O$ across the middle jet and the start of the inclined plate is deduced from the PIV data, and it is then normalized by the length of the inclined plate. The flow control effectiveness calculated by this way is presented as a contour in the space of dimensionless stroke length and diaphragm frequency in Figure 8.13.

As shown in Figure 8.13, on the one hand, the flow control effectiveness is not appreciable at VR less than 0.3, and on the other hand, the strong effect of separation delay obtained at VR > 1.5 is not of interest in the present study. At $Str_\delta < 1$, a stronger flow separation delay occurs at VR around 0.5. Nevertheless, a region with a profound flow control effect also occurs at high frequencies near the bottom right-hand corner of Figure 8.13. In both regions, the hairpin structures are observed.

In the present experiment, actuation frequencies up to 40 times of the characteristic frequency of the separated laminar flow are used. At the frequencies tested here, the separation delay is always associated with the presence of distinct streamwise high-speed streaks in the separated flow. It is, therefore, believed that a key mechanism of separation delay using synthetic jets is to increase mixing by introducing coherent structures into the boundary layer, which entrain high-speed fluid from the outer part of the boundary

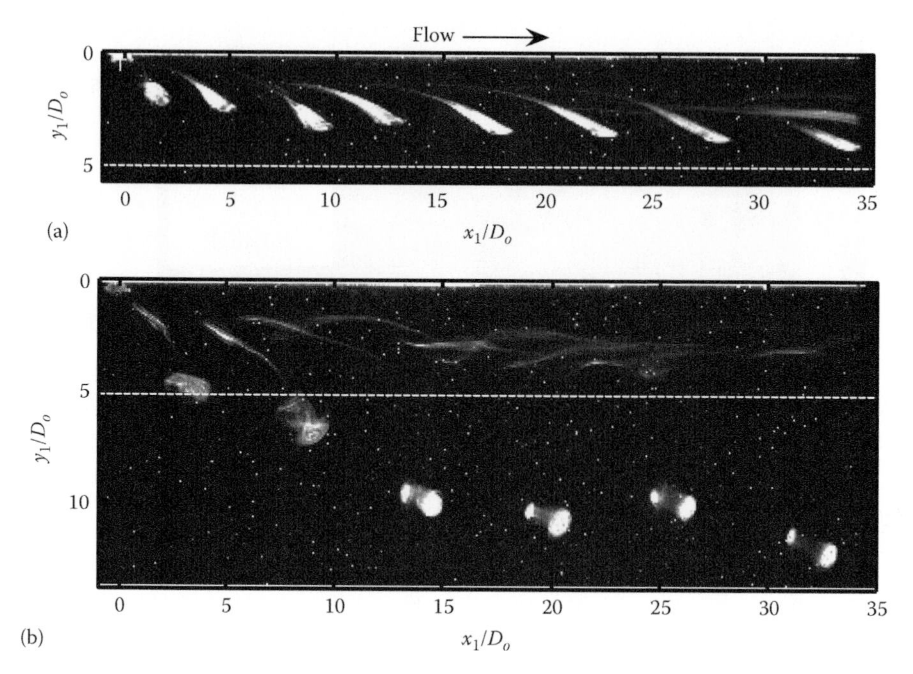

FIGURE 8.14 LIF visualization on the central plane of the middle synthetic jet operated at $f = 8$ Hz ($Str_\delta = 0.8$). (a) $L = 2$, VR = 0.32; (b) $L = 4$, VR = 0.64 (dashed line: edge of undisturbed boundary layer). (From Zhang, S. and S. Zhong, *AIAA J.*, 48, 611–623, 2010. Reprinted with permission.)

layer to the near-wall region. This seemingly contradiction to the finding of Amitay et al. (1997, 2001), who attribute the delay of separation to the virtual shaping of the local surface contour by synthetic jets, is believed to be caused by the use of circular orifices here, which encourages the formation of 3D vortical structures. In the work of Amitay et al. (1997, 2001), the synthetic jets are issued through a 2D slot.

Overall, the hairpin vortical structures appear to be capable of producing a stronger effect of separation delay than the tilted vortex ring-like structures. The hairpin structures, which are produced at relatively low VR, are capable of remaining in the boundary layer and persisting further downstream. Hence, they are likely to be more effective in enhancing mixing, provided that they have sufficient vorticity strength.

8.4.5 The Optimal Actuator Operating Condition

Based on the results from the present study, a flow control effect can be achieved with VR > 0.3. For practical applications, it would be beneficial to identify an optimal actuator operating condition, which delivers the maximum level of separation delay with minimum energy expenditure. The time-averaged power consumption of the present synthetic jet actuator is hence deduced from the temporal variations of current and voltage supplied

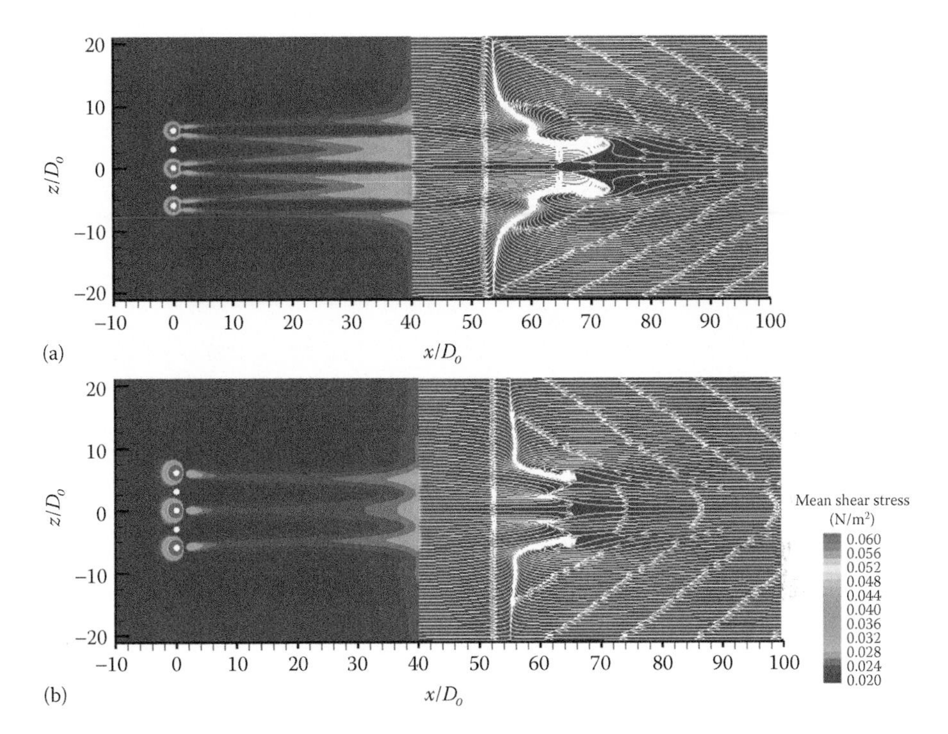

FIGURE 8.15 Contours of surface shear stress upstream to the start of the inclined plate located at $x/D_O = 40$ and the surface friction lines over the inclined plate obtained from computational fluid dynamics (CFD) simulations at $f = 8$ Hz ($Str_\delta = 0.8$). (a) $L = 2$, VR = 0.32; (b) $L = 4$, VR = 0.64. (From Zhang, S. and S. Zhong, *AIAA J.*, 48, 611–623, 2010. Reprinted with permission.)

to the magnetic shaker, which is used to drive the diagram. Since both the diaphragm resonant frequency and Helmholtz resonant frequency of the present actuator are significantly higher than the actuation frequencies, their effect is negligible. As it is shown in Figure 8.16, the power consumption increases sharply with an increasing diaphragm displacement, whereas it increases much slowly with an increasing frequency. As such, at the same VR, the power consumption is lower when the actuator is operated at a higher frequency.

In the present study, the best flow separation effect is obtained at the highest frequency (f = 16 Hz) tested and L = 2 at which the power consumption is also maintained at a very low level. Therefore, among the actuator operating conditions tested in this study, the optimal condition is identified as L = 2, VR = 0.65, and Str_δ = 1.6. When taking into account the actual convection velocity ($0.7U_\infty$) of the vortical structures produced by the synthetic jets deduced from the laser-induced fluorescence (LIF) images, the spacing between the consecutive hairpin-like structures is found to be about 50% of the local boundary layer thickness at the orifices of the synthetic jets. Nevertheless, as the level of interaction

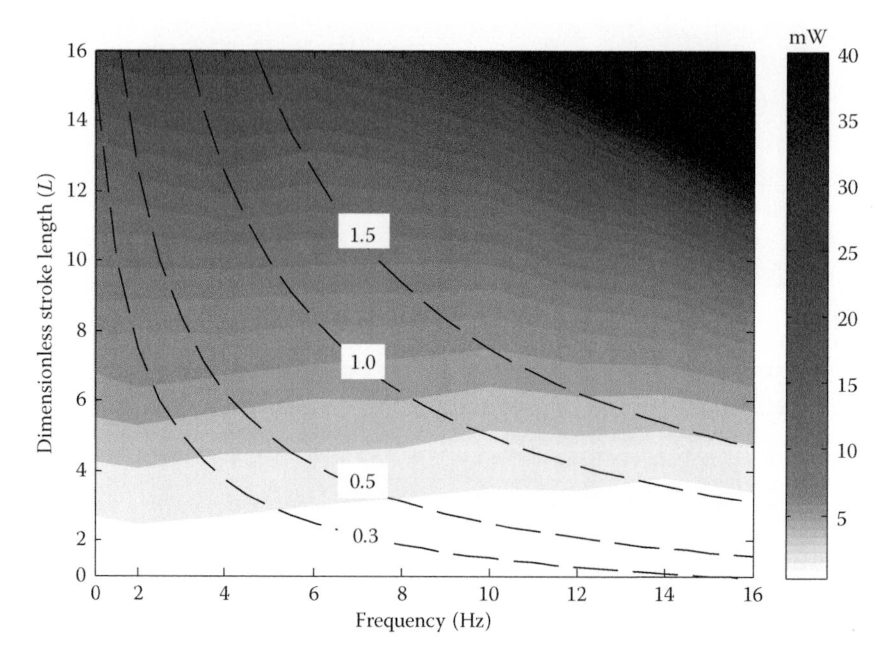

FIGURE 8.16 Contour of power consumption of the synthetic jet actuator operated at different conditions (dashed lines: VR = constant). (From Zhang, S. and S. Zhong, *AIAA J.*, 48, 611–623, 2010. Reprinted with permission.)

between neighboring synthetic jets and their flow control effectiveness is expected to alter upon changes in the jet spacing and the distance between the orifices and the baseline separation line, the generality of the above finding remains to be established in future work.

8.5 Control of Turbulent Flow Separation Over a 2D Ramp-Down Section

Through the studies presented in Sections 8.3 and 8.4, a good understanding of the flow physics involved in the interaction of circular synthetic jets with both attached and separated laminar boundary layers is acquired. In view of the importance of controlling turbulent separated flows in practical settings, flow control of the separated turbulent flow over a 2D ramp using an array of three circular synthetic jets is also undertaken. Both 2D PIV and stereo-PIV techniques are employed to provide detailed information about the flow field in the interaction zone between the separated boundary layer and synthetic jets of different intensities at a given actuation frequency. This allows the response of the separated boundary layer to the passage of synthetic jets to be examined in both time- and phase-averaged sense and the vortical structures responsible for the delay of the incipiently separated turbulent boundary layer flow to be identified.

8.5.1 The Experimental Setup

The experiments are conducted in a boundary layer tunnel with a 1.2-m-wide and 0.3-m-high test section. A 2D ramp with a height of h = 31.5 mm is mounted on the ceiling of the test section as shown in Figure 8.17. A separated flow region is created over the ramp-down section to which flow control using synthetic jets is applied.

The array of synthetic jets is issued through three circular orifices aligned normally to the oncoming flow direction (Figure 8.17b). The orifices are D_O = 5 mm in diameter and spaced 50 mm ($10D_O$) apart. Each synthetic jet is generated using a separate cylindrical cavity, which has a diameter of 100 mm and a height of 10 mm. The piston diaphragms of the actuators are made to oscillate in phase in a sinusoidal manner at predetermined oscillation displacements and frequencies.

The experiments are performed at a freestream velocity of 6.5 m/s and a freestream turbulent intensity of about 0.3%. The locations of the separation and reattachment point in the baseline case without synthetic jet actuation are located at x = 34.4 mm (x/h = 1.1) and x = 132.7 mm (x/h = 4.2), respectively. The shear layer frequency of this flow is about 40 Hz, corresponding to a Strouhal number based on the ramp height of Str_h = 0.2 (Hasan and Khan 1992). The jet array is located at 20 mm ($4D_O$) upstream of the beginning of the ramp-down section or 54.4 mm ($10.88D_O$) upstream of the location of separation line in the baseline case. The boundary layer is turbulent at this location with a thickness of 26.8 mm

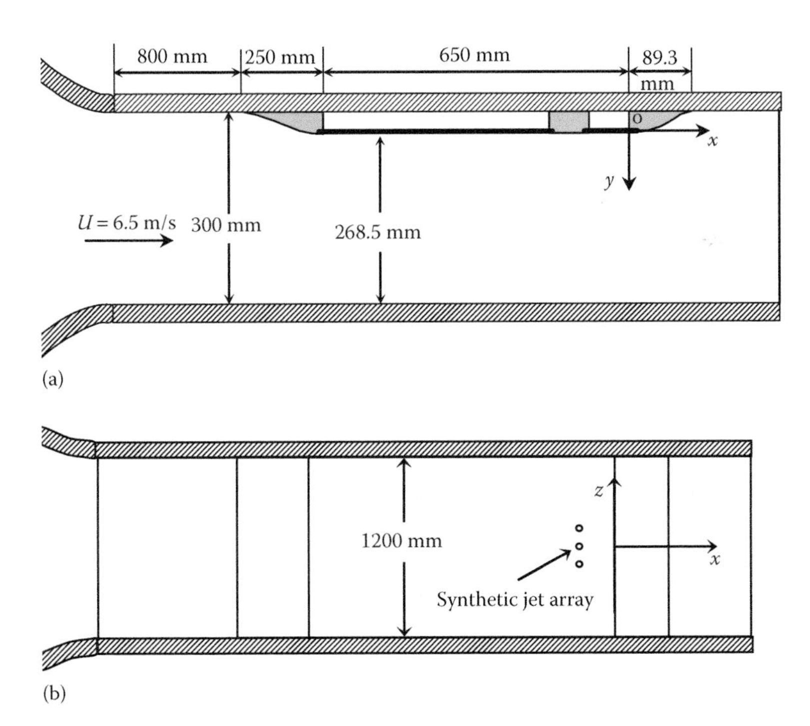

FIGURE 8.17 The geometry of the tunnel test section and the 2D ramp. (a) Side view; (b) top view. (From Zhang, S. and S. Zhong, *AIAA J.*, 49, 2637–2649, 2010b. Reprinted with permission.)

and a Reynolds number based on the local momentum thickness of 984. The synthetic jets are actuated at a frequency of 120 Hz, corresponding to a Strouhal number based on the height of the 2D ramp of $Str_h = 0.6$. Laser-Doppler anemometry (LDA), 2D PIV, and stereo-PIV techniques are employed to provide detailed information about the flow field in the interaction zone between the separated boundary layer and synthetic jets. More detailed information about the model can be found in the work by Zhang and Zhong (2010b).

8.5.2 Time-Averaged Flow Control Effects

The flow control effect obtained using 2D PIV on the central plane of the middle jet at various velocity ratios at $Str_h = 0.6$ is shown in Figure 8.18. The vertical extent of the reverse flow extracted from the time-averaged streamwise velocity distributions is used to evaluate the control effect. The top boundary of the reverse flow is taken as where the time-averaged streamwise velocity becomes zero. It appears that the height of reverse flow decreases steadily as VR increases, leading to a 50% reduction at VR = 0.5 compared to that in the baseline case. Figure 8.19 shows the time-averaged boundary of reverse flow across the span obtained using stereo-PIV at $x/h = 2.5$. It can be seen that the jet appears to affect a spanwise region within $\pm 3D_0$ from the jet centerline with an increasing VR resulting in a further reduction in the height of reverse flow region.

8.5.3 Temporal Variations in the Flow Control Effects

Figure 8.20 shows the contours of phase-averaged normalized streamwise velocity on the streamwise central plane of the middle jet at $Str_h = 0.6$ and VR = 0.5 at four different phases in a synthetic jet actuation cycle. The edge of reverse flow in the baseline case is indicated by the white line, whereas in the control cases, it is indicated by the top boundary of the darkest contour. The flow structures produced by the synthetic jet are seen as lumps of low-speed fluid, which propagate downstream firstly in the boundary layer and then over the separated shear layer. The contours of spanwise vorticity shown in Figure 8.21

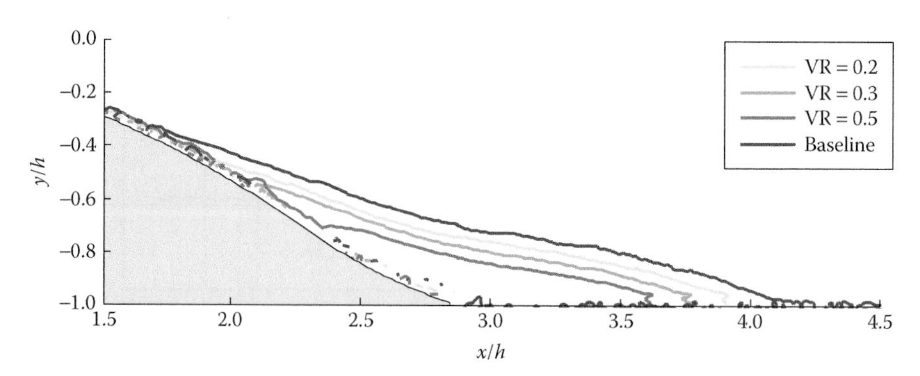

FIGURE 8.18 Control effect on the reverse flow on the central plane at different VR at $Str_h = 0.6$. (From Zhang, S. and S. Zhong, *AIAA J.*, 49, 2637–2649, 2010b. Reprinted with permission.)

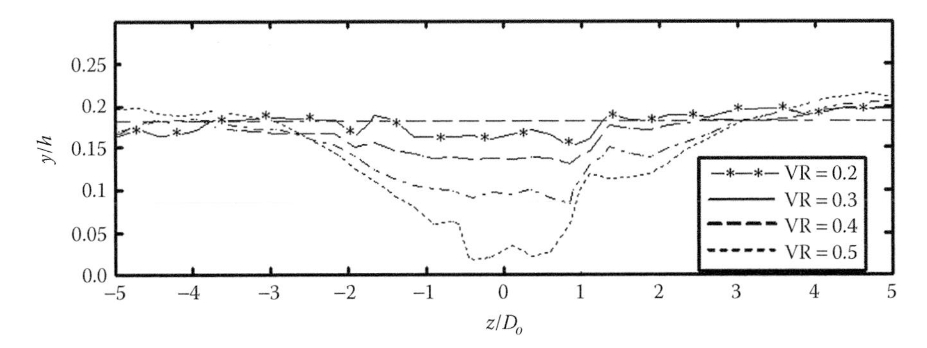

FIGURE 8.19 Control effect on the reverse flow across the span downstream of the middle jet at different VR at $Str_h = 0.6$ at $x/h = 2.5$. The dashed line indicates the time-averaged location of the reverse flow in the baseline flow.

reveal that the downstream side of the heads of these structures is associated with positive spanwise vorticity, whereas the upstream side is associated with negative vorticity. The upstream side, which has the opposite sign of vorticity in the boundary layer, becomes weaker more quickly and disappears eventually. The size of reverse flow remains to be suppressed throughout the entire actuation cycle (Figure 8.20) without an appreciable flapping of the shear layer (Figure 8.21).

In Figure 8.22a, the phase-averaged streamwise vorticity contours as the vortical structures pass through the measurement plane at $x/h = 2$ during an actuation cycle are displayed for $Str_h = 0.6$ and VR = 0.5. It is seen that a pair of counter-rotating vortices located in the outer part of the separated boundary layer comes to sight just before $t/T = 0.4$. It appears to be slightly inclined downstream and persists till $t/T = 0.85$ when both its strength and vertical location drop abruptly. A streamwise vortex pair, which has the opposite sign of rotation, is located underneath the aforementioned structure, and it persists during the entire cycle. In Figure 8.22a, the region of reversed flow is indicated by the isosurface of zero streamwise velocity in dark colour, and the attached flow appears as the light region in the middle. It is seen that the spanwise width of the attached flow remains almost constant, and this is consistent with the observation in Figure 8.20 that the flow separation is constantly suppressed during the entire cycle. By correlating the streamwise vorticity and streamwise velocity contours, that is, Figure 8.22a and b, one can see that the upper vortex pair produces an upwash ejecting low-momentum fluid away from the wall, whereas the lower vortex pair produces a downwash between its limbs bringing high-momentum fluid toward the wall.

8.5.4 Vortical Structures Responsible for Delaying Flow Separation

The physical appearance of the vortical structures observed in Figure 8.22 can be reconstructed using the Q-criterion along with the Taylor's frozen structure hypothesis (Taylor 1938). Here the convection velocity of the coherent structure is found to be 5.3 m/s from the PIV measurement on the streamwise central plane shown in Figure 8.20.

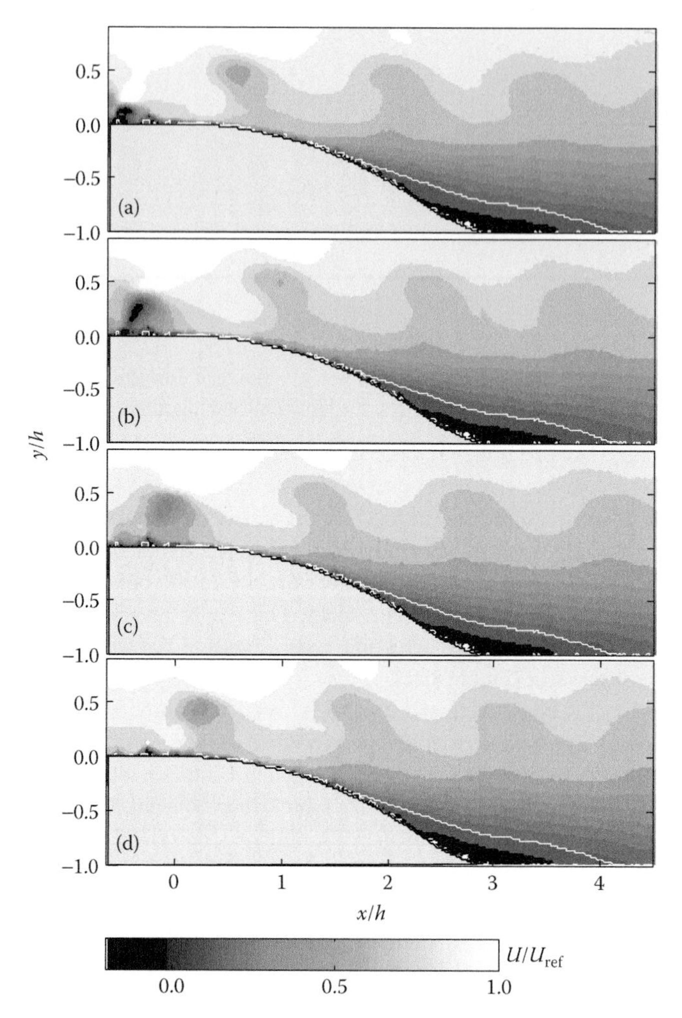

FIGURE 8.20 Contours of phase-averaged streamwise velocity at $Str_h = 0.6$ and VR = 0.5 at phase (a) 0°, (b) 90°, (c) 180°, and (d) 270° of the actuation cycle. The white line indicates the edge of the reverse flow in the baseline flow. (From Zhang, S. and S. Zhong, *AIAA J.*, 49, 2637–2649, 2010b. Reprinted with permission.)

In Figure 8.23, the isosurfaces of $Q = 5$ color coded with the streamwise vorticity ω_x using stereo-PIV data measured at three different streamwise locations ($x/h = 1, 2,$ and 2.5) along the ramp-down section are shown for VR = 0.5. In this way, an impression of how the structure develops as it propagates downstream can also be obtained. As seen in this figure, the primary structure appears as a tilted vortex ring, which has a very weak upstream side and is attached to a pair of trailing legs. The structure is inclined downstream within the local boundary layer. As it propagates downstream, it moves away from the curved surface and its trailing legs become weaker and therefore less visible. In this case, the

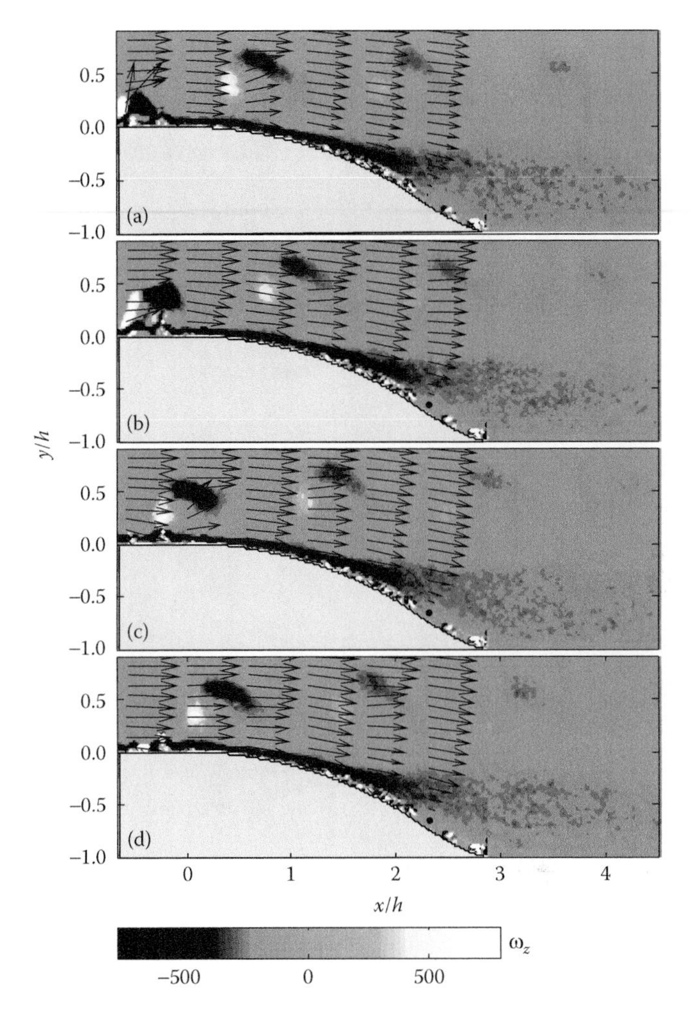

FIGURE 8.21 Contours of spanwise vorticity (ω_z) with superimposed velocity vectors at $Str_h = 0.6$ and VR = 0.5 at phase (a) 0°, (b) 90°, (c) 180°, and (d) 270° of the actuation cycle. (From Zhang, S. and S. Zhong, *AIAA J.*, 49, 2637–2649, 2010b. Reprinted with permission.)

streamwise vortex pair of the opposite sense of rotation is not obvious since it is weaker and embedded in the separated flow.

Based on the above results, a conceptual model of the structures, which are responsible for delaying flow separation, is deduced as shown in Figure 8.24. It is believed that at the synthetic jet operating conditions investigated in this study, the interaction of the synthetic jets and the turbulent boundary layer eventually results in the formation of a tilted vortex ring type structure in the layer. The legs of this structure induce a further pair of streamwise vortices of the opposite sign of rotation underneath. Both structures propagate downstream over the separated flow. It is the induced streamwise vortices that produce a downwash, which brings high-speed fluid toward the wall hence reducing the

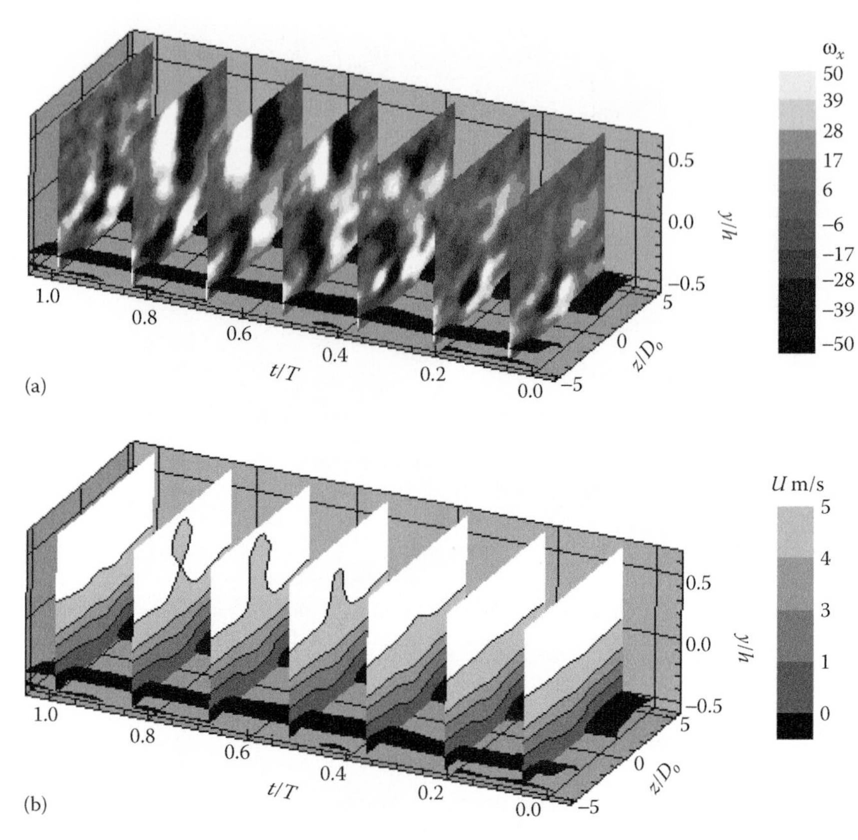

FIGURE 8.22 Contours of (a) phase-averaged streamwise vorticity and (b) phase-averaged streamwise velocity during a synthetic jet actuation cycle at $x/h = 2$ for $Str_h = 0.6$ at VR = 0.5 (isosurface indicating the reverse flow). (From Zhong, S. and S. Zhang, *Flow Turbul. Combust.*, 91, 177–208, 2013. Reprinted under the Creative Commons Attribution License.)

flow separation. This model is consistent with that obtained from our laminar flow studies shown in Figure 8.9. Therefore, one can see that despite the differences in the nature between laminar and turbulent flows, the vortical structures formed as the result of the interaction of synthetic jets with a laminar and a turbulent boundary layer at a similar velocity ratio remain similar.

8.6 Conclusion

In this chapter, the key results from a series of carefully designed experiments dedicated to study the behaviors of normal circular synthetic jets issuing into a grazing boundary layer are presented. From these studies, an improved understanding of the flow physics of the interaction between synthetic jets and a grazing boundary layer has been obtained.

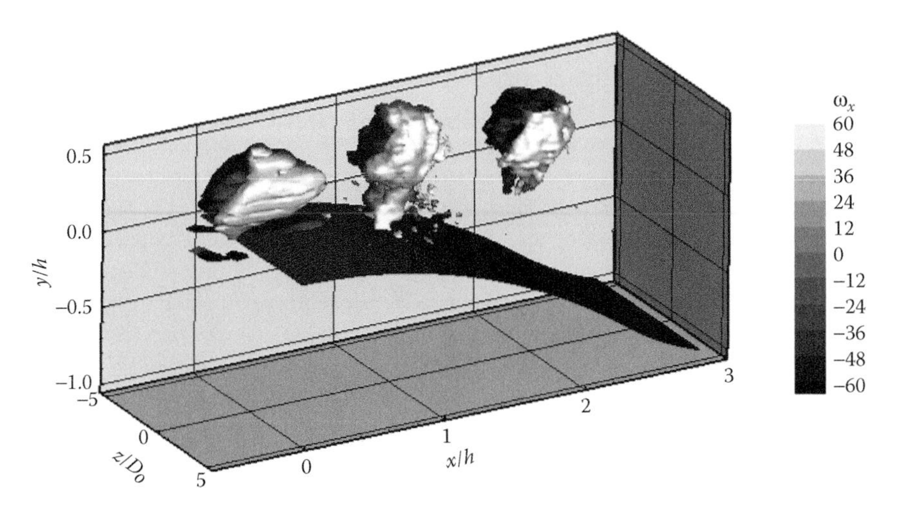

FIGURE 8.23 Isosurfaces of Q-criteria ($Q = 5$) grayscale coded with streamwise vorticity ω_x at $Str_h = 0.6$ and VR = 0.5. (From Zhong, S. and S. Zhang, *Flow Turbul. Combust.*, 91, 177–208, 2013. Reprinted under the Creative Commons Attribution License.)

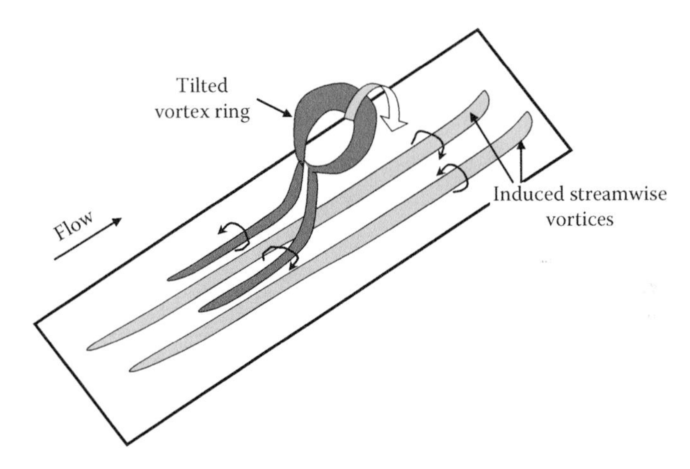

FIGURE 8.24 A conceptual model of the vortical structures, which produce the delay of flow separation.

It is found that as the actuator operating condition varies, the interaction between a single synthetic jet and a flat plate laminar boundary layer results in the formation of two typical types of vortical structures, that is, the hairpin vortex type and the tilted vortex ring type structures, each producing a distinctly different pattern of surface shear stress. The formation mechanism and the near-wall effect of these two different types of structures have been investigated and understood. A parameter map is also produced from which the parameter boundary between these typical structures is clearly established. Upon applying an array of synthetic jets upstream of a separated laminar flow over a range of actuator operating conditions, it is observed that the patterns of flow separation delay are directly related to the shear stress footprints of the vortical structures produced by the synthetic jets. In general, the hairpin vortex type structures appear to be capable of delivering a stronger flow control effect due to their closer proximity to the wall. The optimal actuator operating condition, which delivers the best flow control effect with the least energy consumption, is also identified, and it is found to produce hairpin vortical structures. Finally, based on the PIV study of synthetic jets issuing upstream of a separating turbulent boundary layer, the vortical structures, which are responsible for delaying flow separation, are identified, and the associated mechanism of flow separation delay is established. It is found that despite the differences in the nature between laminar and turbulent flows, the vortical structures formed as the result of the interaction of synthetic jets with a laminar and a turbulent boundary layer at a similar velocity ratio remain similar. The knowledge gained from these studies would be useful for designing and selecting the operating conditions for synthetic jets for effective flow separation control in practical settings.

Acknowledgment

The work presented in this paper has received funding from the Engineering and Physical Sciences Research Council (EPSRC) in the United Kingdom, Airbus UK, the Universities UK, and Manchester University.

References

Avdis, A., S. Lardeau, and M. Leschziner. 2009. Large-eddy simulation of separated flow over a two-dimensional hump with and without control by means of a synthetic slot-jet. *Flow, Turbulence and Combustion*, 83: 343–370.

Amitay, M., A. Honohan, M. Trautman, and A. Glezer. 1997. Modification of the aerodynamic characteristics of bluff bodies using fluidics actuators. *The 28th AIAA Fluid Dynamics Conference, AIAA Paper* 97-2004, Snowmass Village, CO, June 29–July 2.

Amitay, M., D. Smith, V. Kikens, D. Parekh, and A. Glezer. 2001. Aerodynamic flow control over an unconventional airfoil using synthetic jet actuators. *AIAA Journal*, 39: 361–370.

Ashill, P. R., J. L. Fulker, and K. C. Hackett. 2002. Studies of flows induced by sub boundary layer vortex generators (SBVGs). *The 40th AIAA Aerospace Sciences Meeting and Exhibit*, Reno, NV, January 14–17.

Cater, J. E. and J. Soria. 2002. The evolution of round zero-net-mass-flux jets. *Journal of Fluid Mechanics*, 472: 167–200.

Chang, Y. K. and A. D. Vakili. 1995. Dynamics of vortex rings in crossflow. *Physics of Fluids*, 7: 1583–1597.

Crook, A. 2002. The control of turbulent flows using synthetic jets. PhD thesis, University of Manchester, Manchester.

Crook, A., A. M. Sadri, and N. J. Wood. 1999. The development and implementation of synthetic jets for the control of separated flow. *The 17th AIAA Applied Aerodynamics Conference*, Norfolk, VA, June 28–July 1.

Crook, A. and N. J. Wood. 2001. Measurements and visualisations of synthetic jets. *The 39th Aerospace Sciences Meeting and Exhibit*, Reno, NV, January 8–11.

Dandois, J., E. Garnier, and P. Sagaut. 2006. Unsteady simulation of a synthetic jet in a crossflow. *AIAA Journal*, 44: 225.

Dandois, J., E. Garnier, and P. Sagaut. 2007. Numerical simulation of active separation control by a synthetic jet. *Journal of Fluid Mechanics*, 574: 25–58.

Gad-El-Hak, M. 2000. *Flow Control: Passive, Active and Reactive Flow Management.* 1st Edition, Cambridge: Cambridge University Press.

Garcillan, L., S. Liddle, S. Zhong, and N. J. Wood. 2005. Time evolution of the interaction of synthetic jets with a turbulent boundary layer. *Proceedings of CEAS/KATnet Conference on Key Aerodynamic Technologies*, Bremen, Germany, June 20–22.

Gharib, M., E. Rambod, and K. Shariff. 1998. A universal time scale for vortex ring formation. *Journal of Fluid Mechanics*, 360: 121–140.

Gilarranz, J. L., L. W. Traub, and O. K. Rediniotis. 2005. A new class of synthetic jet actuators. Part II. Application to flow separation control. *Journal of Fluids Engineering*, 127: 377.

Glezer, A. 1988. The formation of vortex rings. *Physics of Fluids*, 31(12): 3532–3542.

Glezer, A. 2010. Some aspects of aerodynamic flow control using synthetic-jet actuation. *Philosophical Transactions of the Royal Society A: Mathematical, Physical and Engineering Sciences*, 369(1940): 1476–1494.

Glezer, A. and M. Amitay. 2002. Synthetic jets. *Annual Review of Fluid Mechanics*, 34: 503–529.

Greenblatt, D., K. B. Paschal, C.-S. Yao, and J. Harris. 2006. Experimental investigation of separation control, Part 2: Zero mass-flux oscillatory blowing. *AIAA Journal*, 44(12): 2831–2845.

Guo, F. and S. Zhong. 2007. A PIV investigation of the characteristics of micro-scale synthetic jets. *Aeronautical Journal*, 111: 509–518.

Hasan, M. and A. Khan. 1992. On the instability characteristics of a reattaching shear layer with nonlaminar separation. *International Journal of Heat and Fluid Flow*, 13: 224–231.

Holman, R., Y. Utturkar, R. Mittal, B. L. Smith, and L. Cattafesta. 2005. Formation criterion for synthetic jets. *AIAA Journal*, 43(10): 2110–2116.

Hunt, J. C. R., A. A. Wray, and P. Moin. 1988. Eddies, stream, and convergence zones in turbulent flows. Center for Turbulence Research Report CTR-S88, Stanford University, Stanford, CA.

Jabbal, M., J. Wu, and S. Zhong. 2006. The performance of round synthetic jets in quiescent flow. *Aeronautical Journal*, 110: 385–393.

Jabbal, M. and S. Zhong. 2008. The near-wall effect of synthetic jets in a boundary layer. *International Journal of Heat and Fluid Flow*, 29(1): 119–130.

Jabbal, M. and S. Zhong. 2010. PIV measurements of the interaction of synthetic jets with a zero pressure gradient laminar boundary layer. *Physics of Fluids*, 22(6): 063603.

Kotapati, R. B., R. Mittal, O. Marxen, F. Ham, D. You, and L. N. Cattafesta. 2010. Nonlinear dynamics and synthetic-jet-based control of a canonical separated flow. *Journal of Fluid Mechanics*, 654: 65–97.

Leschziner, M. A. and S. Lardeau. 2010. Simulation of slot and round synthetic jets in the context of boundary-layer-separation control. *Philosophical Transactions of the Royal Society A: Mathematical, Physical and Engineering Sciences*, 370(1): 1.

McCormick, D. C. 2000. Boundary layer separation control with directed synthetic jets. *The 38th Aerospace Sciences Meeting and Exhibit*, Reno, NV, January 10–13.

Milanovic, I. M. and K. B. M. Q. Zaman. 2005. Synthetic jets in cross flow. *AIAA Journal*, 43(5): 929–940.

Mittal, R. J. and P. Rampunggoon. 2002. On the virtual aeroshaping effect of synthetic jets. *Physics of Fluids*, 14: 1533.

Ozawa, T., S. Lesbros, and G. Hong. 2010. LES of synthetic jet in boundary layer with laminar separation caused by adverse pressure gradient. *Computers & Fluids*, 39: 845–858.

Qin, N. and H. Xia. 2008. Detached eddy simulation of a synthetic jet for flow control. *Proceedings of the Institution of Mechanical Engineers, Part I: Journal of System Control Engineering*, 222(15): 373–380.

Rumsey, C. L. 2009. Successes and challenges for flow control simulations. *International Journal of Flow Control*, 1: 1–27.

Rumsey, C., T. Gatski, W. Sellers, V. Vatsa, and S. Vilken. 2006. Summary of the 2004 computational fluids dynamics validation workshop on synthetic jets. *AIAA Journal*, 44: 194.

Sahni, O., J. Wood, K. Jansen, and A. Amitay. 2011. Three-dimensional interactions between a finite-span synthetic jet and a cross-flow. *Journal of Fluid Mechanics*, 671: 254–287.

Sau, R. and K. Mahesh. 2008. Dynamics and mixing of vortex rings in crossflow. *Journal of Fluid Mechanics*, 604: 389–409.

Sauerwein, S. C. and A. D. Vakili. 1999. An experimental study of zero-mass jets in crossflow. *The 37th AIAA Aerospace Sciences Meeting and Exhibit, AIAA Paper* 99-0668, Reno, NV, January 11–14.

Schaeffler, N. W. and L. N. Jenkins. 2006. Isolated synthetic jet in crossflow: Experimental protocols for a validation dataset. *AIAA Journal*, 44(12): 2835.

Smith, D. R. 2002. Interaction a synthetic jet with a crossflow boundary layer. *AIAA Journal*, 40: 2277–2288.

Smith, B. L. and A. Glezer. 1998. The formation and evolution of synthetic jets. *Physics of Fluids*, 10: 2281–2297.

Tang, H., S. Zhong, M. Jabbal, L. Garcillan, F. Guo, N. J. Wood, and C. Warsop. 2007. Towards the design of synthetic-jet actuators for full-scale flight conditions. Part 2: Low-dimensional actuator prediction models and actuator design methods. *Flow, Turbulence and Combustion*, 78(3): 309–329.

Taylor, G. I. 1938. The spectrum of turbulence. *Proceeding of Royal Society of London A*, 164: 476.

Wu, D. K. L. and M. A. Leschziner. 2009. Large-eddy simulation of circular synthetic jets in quiescent surroundings and in turbulent cross-flow. *Computers & Fluids*, 38: 394–405.

Zaman, K. B. M. Q. and I. M. Milanovic. 2003. Synthetics in crossflow. Part 1. Round jet. *The 33rd AIAA Fluid Dynamics Conference and Exhibit*, Orlando, FL, June 23–26.

Zhang, S. and S. Zhong. 2010a. An experimental investigation of laminar flow separation control using an array of synthetic jets. *AIAA Journal*, 48(3): 611–623.

Zhang, S. and S. Zhong. 2010b. An experimental investigation of turbulent flow separation control by an array of synthetic jets. *The 5th AIAA Shear Flow Control Conference*, Chicago, IL, June 28–July 1.

Zhong, S., L. Garcillan, and N. J. Wood. 2005a. Dye visualisation of inclined and skewed synthetic jets in a cross-flow. *Aeronautical Journal*, 109: 147–155.

Zhong, S., M. Jabbal, H. Tang, L. Garcillan, F. Guo, N. J. Wood, and C. Warsop. 2007. Towards the design of synthetic-jet actuators for full-scale flight conditions. Part 1: The fluid mechanics of synthetic-jet actuators. *Flow, Turbulence and Combustion*, 78(3): 283–307.

Zhong, S., F. Millet, and N. J. Wood. 2005b. The behaviour of circular synthetic jets in a laminar boundary layer. *Aeronautical Journal*, 110: 461–470.

Zhong, S. and S. Zhang. 2013. Further examination of the mechanisms of round synthetic jets in delaying turbulent flow separation. *Flow, Turbulence and Combustion*, 91(1): 177–208.

Zhou, J. and S. Zhong. 2008. Numerical simulations of the interaction of circular synthetic jets with a laminar boundary layer. *Computers & Fluids*, 38: 393–405.

Zhou, J. and S. Zhong. 2010. Coherent structures produced by the interaction between synthetic jets and a laminar boundary layer and their surface shear stress patterns. *Computers & Fluids*, 39: 1296–1313.

Zhou J., H. Tang, and S. Zhong. 2009. Vortex roll-up criterion for synthetic jets. *AIAA Journal*, 47(5): 1252–1262.

9

Synthetic Jets for Heat Transfer Augmentation

Mehmet Arik
Özyeğin University

Yogen V. Utturkar
GE Global Research

Synthetic jets are meso- or microscale fluidic devices, which operate on the "zero-net-mass-flux" principle. However, they impart a positive net momentum flux to the external environment and are able to produce the cooling effect of a fan sans its ducting, reliability issues, and oversized dimensions. The rate of heat removal from the thermal source is expected to depend on the location, orientation, strength, and shape of the jet. In this chapter, we would like to provide an overview of the synthetic technology, ranging from its basic operating principle and mechanisms of heat transfer enhancement to discussing some possible applications.*

* Disclaimer: This chapter was prepared as an account of work sponsored by an agency of the US government. Neither the US government nor any agency thereof, nor any of their employees, makes any warranty, express or implied, or assumes any legal liability or responsibility for the accuracy, completeness, or usefulness of any information, apparatus, product, or process disclosed, or represents that its use would not infringe privately owned rights. Reference herein to any specific commercial product, process, or service by trade name, trademark, manufacturer, or otherwise does not necessarily constitute or imply its endorsement, recommendation, or favoring by the US government or any agency thereof. The views and opinions of authors expressed herein do not necessarily state or reflect those of the US government or any agency thereof.

9.1 Introduction

Synthetic jets have been routinely investigated from the standpoint of flow control (Smith et al. 1998; Crook et al. 1999; Mittal et al. 2001; Mittal and Rampunggoon 2002), thrust vectoring of jets (Smith and Glezer 2002), triggering turbulence in boundary layers (Rathnasingham and Breur 1997; Lee and Glodstein 2001), and heat transfer applications (Mahalingam et al. 2004; Holman et al. 2005; Mahalingam and Glezer 2005). In case of heat transfer applications, the cooling process can be facilitated either by direct impingement of vortex dipoles on heated surfaces (Erbas et al. 2005; Li 2005) or by employing the jets to enhance the performance of existing cooling circuits (Timchenko et al. 2004). In view of these applications, the jet performance can be assessed with various methodologies.

Integral analyses of the time- and spatially dependent jet exit velocity profiles serve to quantify the jet performance in terms of the net momentum/energy flux effectively (Mittal et al. 2001; Utturkar et al. 2003). Gallas et al. (2004) formulated accurate, low-dimensional models of the synthetic jet flow field to predict the impact of design changes on the losses in jet orifice. Similarly, Raju et al. (2005) developed scaling relationship for the vorticity flux across the jet orifice. In addition, a semianalytical model for estimating the pressure loss across the orifice for oscillatory flows was also proposed. Holman et al. (2005) derived a formation criterion for synthetic jets, which essentially stipulated a condition in terms of nondimensional parameters to prevent reentrainment of vortex dipoles into the jet cavity during the ingestion stroke. This is particularly crucial for heat transfer applications since they heavily rely on the vortex dipole dynamics for their cooling performance.

To the best of our knowledge, the heat transfer behavior of synthetic jets was first observed by Yassour et al. (1986). Their experimental setup comprised an enclosed cavity with an orifice and a loudspeaker acting as flexible membrane. A heater was aligned in front of the orifice, and the heat transfer rates on its surface were recorded. The acoustic "puffs" emanating from the orifice produced a 4× enhancement in the heat transfer. Coe et al. (1994) generalized the above-mentioned cooling concept by making use of actuation technologies (piezoelectric actuation) other than loudspeaker to create the synthetic jet action. Since then active cooling with synthetic jet has been an active topic of research (Mahalingam et al. 2004; Erbas et al. 2005; Li 2005; Mahalingam and Glezer 2005). Garg et al. (2005) performed an experimental study to characterize the cooling power of the jet, by contrasting its heat transfer capability under a given temperature difference, with that of natural convection, with the aid of an enhancement factor (EF). Very recently, Utturkar et al. (2006, 2008) investigated the sensitivity of synthetic cooling performance to its operating conditions and alignment by experimental and computational strategies.

Numerical simulations of synthetic jets have been greatly instrumental in distilling the flow field physics from the standpoint of flow control (Mittal et al. 2001; Mittal and Rampunggoon 2002) and heat transfer applications (Holman et al. 2005). Numerical simulations are able to provide finer temporal and spatial resolution of the flow domain, which could be a challenging proposition for experimental studies because of $O(10^{-3}–10^{-6})$ length scales; especially in the case of heat transfer applications, the cooling process is driven by impingement of vortex dipoles on the heated surface (Erbas et al. 2005). Consequently, a sound understanding of the vortex rollup and dipole-forming mechanism and the interaction of vortex dipole with wall are imperative to enhance the jet design for heat transfer applications. This is possible via the approach of numerical simulations, which yield a detailed description of the flow field in space and time.

Recently, synthetic jet heat transfer research has attracted more scientists in academia and industry. The effects of a small-scale rectangular synthetic air jet on the local convective heat transfer from a flat heated surface were reported in Gillespie et al. (2006). They conclude that synthetic jets can lead to substantial enhancement of the local heat transfer from heated surfaces by strong mixing that disrupts the surface thermal boundary layer. The dependence of the local heat transfer coefficient on the primary parameters of jet motion is characterized over a range of operating conditions. Average Nusselt numbers were maximized when the dimensionless plate spacing was between 14 and 18. Furthermore, heat transfer rates were maximized when the jet frequency was close to the resonance frequency of the driver cavity. The efficiency and mechanisms of cooling a constant heat flux surface by impinging synthetic jets were investigated experimentally and compared to cooling with continuous jets (Pavlova and Amitay 2006). Effects of jet formation frequency and Reynolds number (*Re*) at different nozzle-to-surface distances were investigated. High formation frequency (1200 Hz) synthetic jets were found to remove heat better than low formation frequency (420 Hz). Moreover, synthetic jets are about three times more effective in cooling than continuous jets at the same *Re*. Using particle image velocimetry, it was shown that the higher formation frequency jets are associated with breakdown and merging of vortices before they impinge on the surface. Heat transfer and acoustic aspects of the small-scale synthetic jets are presented by Arik (2007). The designed and developed synthetic jets provided peak air velocities of 90 m/s from a 1-mm hydraulic diameter rectangular orifice. The jets are driven by a sine wave with an operating frequency between 3 and 4.5 kHz, providing the highest thermal performance for the current jets. It is found that the enhancement can be between 4 and 10 times, which depends on the heater size showing that smaller sizes provide the best jet effectiveness. It is also noted that jet noise can be as large as 73 dB, but possible abatement techniques can decrease this noise level to as low as 30 dB.

9.2 Fundamental Heat Transfer Studies

9.2.1 Experimental Studies

9.2.1.1 Heat Transfer Experiments

A number of experiments have been proposed in the literature (Garg et al. 2005). First, experimental studies have been on simple thin foil heaters to measure localized surface temperatures and the heat transfer to the ambient via synthetic jet cooling (Figure 9.1). A large Plexiglas enclosure was employed to minimize ambient air drifts to the experimental rig. Thin foil heaters sandwiched between thin aluminum backing plates (with 125 μm thickness) were used in the experimental study. Mapping the temperature profile of the localized points at the heater was critical, and thus, the entire surface was visible to an infrared (IR) camera. While the jet was impinging on one of the heater surfaces, the temperature measurements were performed from the back surface using the IR camera. It was ensured that a large fraction of the electrical power input to the heater was dissipated from the surface facing the jet, and the back surface heat transfer and electrical leads were

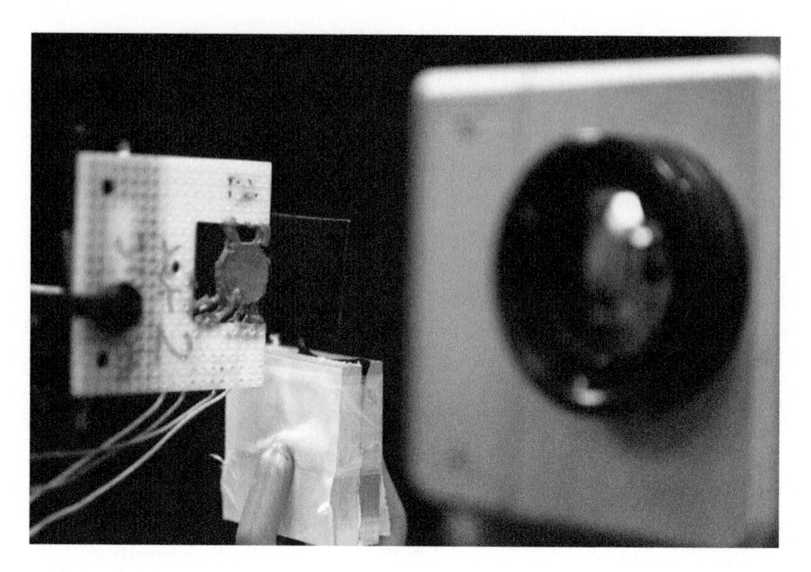

FIGURE 9.1 Experimental setup. (From Utturkar, Y. et al., *J. Heat Transfer*, 130, 062201, 2008. Reprinted with permission.)

assumed to contribute to the total heat losses. In earlier experiments by Garg et al. (2005), a slightly different approach by attaching the heater to a Plexiglas frame is utilized.

The heater temperature profile was first obtained under natural convection conditions. The front and back side of the heater was tested at the same heater power condition. In this new procedure, a few critical experimental settings were ensured before the synthetic jet experiment was started. Firstly, the average temperature between the front and back heater surfaces was ascertained to be varying less than 1°C. Then, the natural convection heat loss curve based on the difference between the average surface temperature and ambient temperature was obtained. This curve was vital to the accuracy of the heat transfer enhancement calculations, as explained in Section 9.2.1.2.

For accurate IR thermal imaging, the emissivity of the surface should be known with high accuracy. Heater surfaces were coated with known emissivity paint, and the emissivity (equal to 0.97) was measured by using the same methodology used by Garg et al. (2002). Steady-state temperature profiles were captured by using an IR camera. A minimum pixel size of 300 μm was achieved via appropriate lenses. A typical temperature profile under natural conditions has been illustrated in Figure 9.2a. Similarly, Figure 9.2b shows the influence of the jet on the heater temperature. It is seen that the maximum temperature in Figure 9.2a is 71.2°C, while the maximum temperature difference between the hottest spot and the coolest spot (at the side) is 10°C. This difference is expected because of the heat dissipation from the heater sides.

While IR measurements provided local temperature gradients and heat transfer information, a more robust and automated experimental setup has been necessary to accelerate the technology development. A fully automated experimental test rig (Mittal and Rampunggoon 2002) was designed and built to measure the synthetic jet heat transfer performance. Synthetic jet operating conditions (voltage and frequency), heater surface

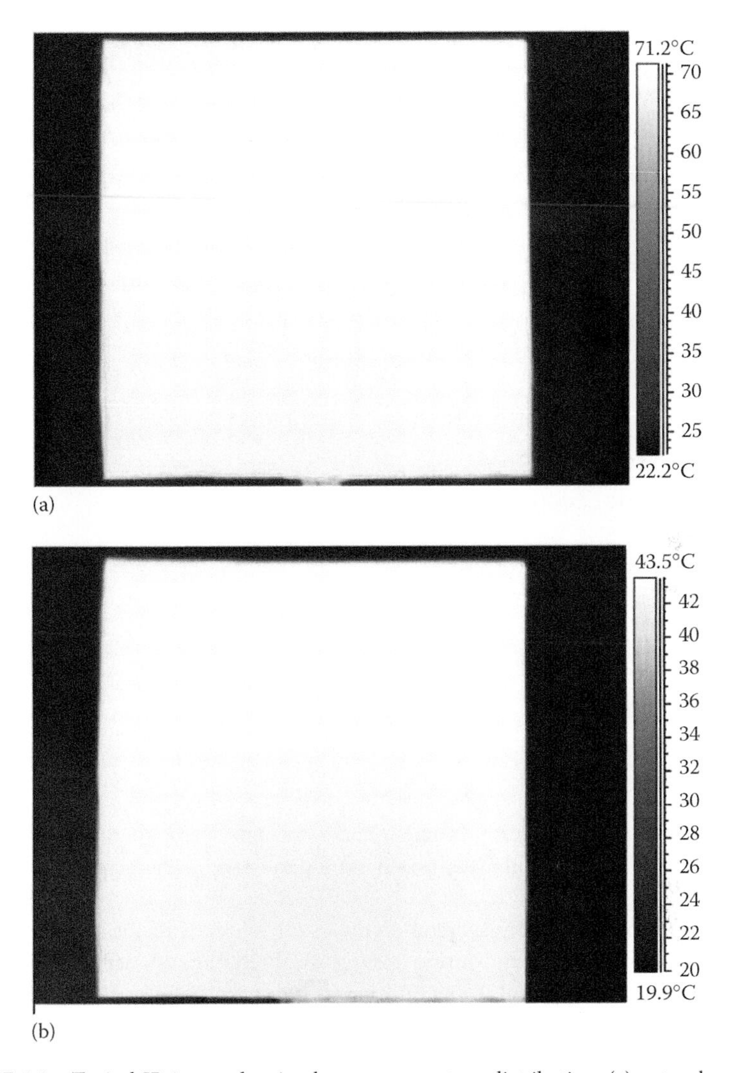

FIGURE 9.2 Typical IR image showing heater temperature distribution: (a) natural convection, 1.56 W heater power (maximum temperature = 71.2°C); (b) synthetic jet at 50 V_{rms} and 4500 Hz perpendicularly impinging on the heater center from a distance of 10 mm, 1.56 W heater power (maximum temperature = 43.5°C). (From Utturkar, Y. et al., *J. Heat Transfer*, 130, 062201, 2008. Reprinted with permission.)

temperatures, and heater-to-jet distance were all programmed into the data acquisition (DAQ) system, and a series of tests were performed for various jet operating conditions and geometrical variables. The data collected were then post-processed for the computation of thermal performance, jet power consumption, actuator disk displacement, and acoustic noise generated by the synthetic jet. The setup shown in Figure 9.3 was composed of

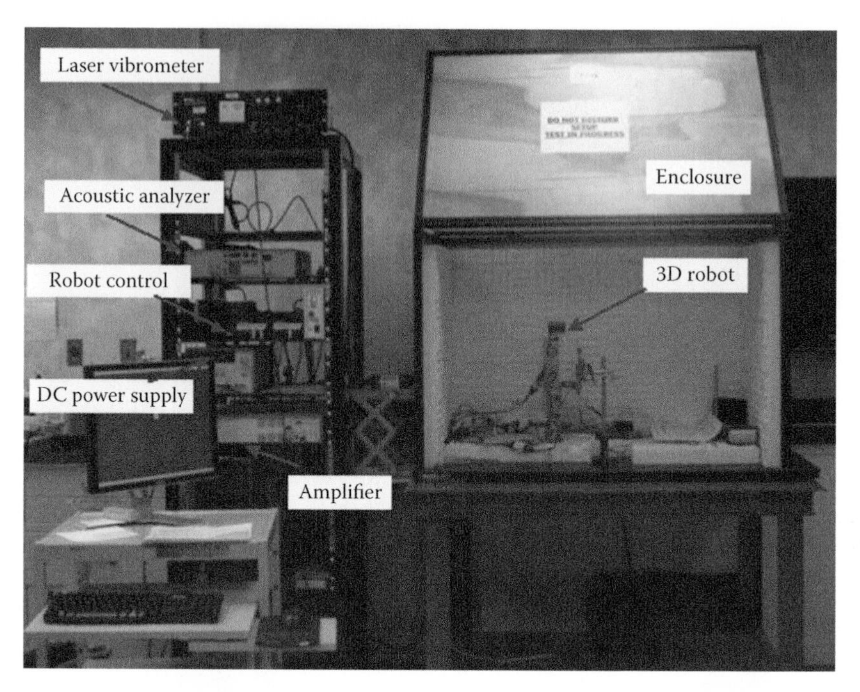

FIGURE 9.3 GE Global Research Center synthetic jet test rig. (From Arik, M. and T. Icoz. *J. Heat Transfer*, 134, 081901, 2012.)

enclosure, transverse mechanism, function generator, signal amplifier, laser vibrometer, and heat source, direct current (DC) power supply to energize the heater and the DAQ system.

The enclosure had acoustic insulation on the inside to isolate the heater and synthetic jet from environmental effects during testing. A three-dimensional (3D) transverse mechanism was used to control the jet position with respect to the heated surface, within an accuracy of ±0.5 mm. The synthetic jets were driven by the function generator with a low power sine wave, which then got amplified with the signal amplifier. Laser vibrometer was used to measure the piezoactuator displacement, which is an important parameter for the understanding of fluid–solid interactions. A lab windows DAQ program was used to control the equipment and the test conditions. Power to the resistive heater was provided from a DC power supply. Figure 9.4 shows the close-up picture of the synthetic jet placement and the heater.

Prior to conducting experiments, a tabulated list of operating voltages, frequencies, desired heater temperatures, and relative jet positions with respect to the heater was created. Using this list, the DAQ system automatically started the next test in the queue after recording data from prior test. Heater target temperature was set at 80°C for all tests, and it was controlled by a proportional-integral-derivative (PID) controller. The steady-state conditions were assumed to be satisfied when the standard deviation of surface temperature measurements for the last 30 seconds becomes smaller than 0.1°C.

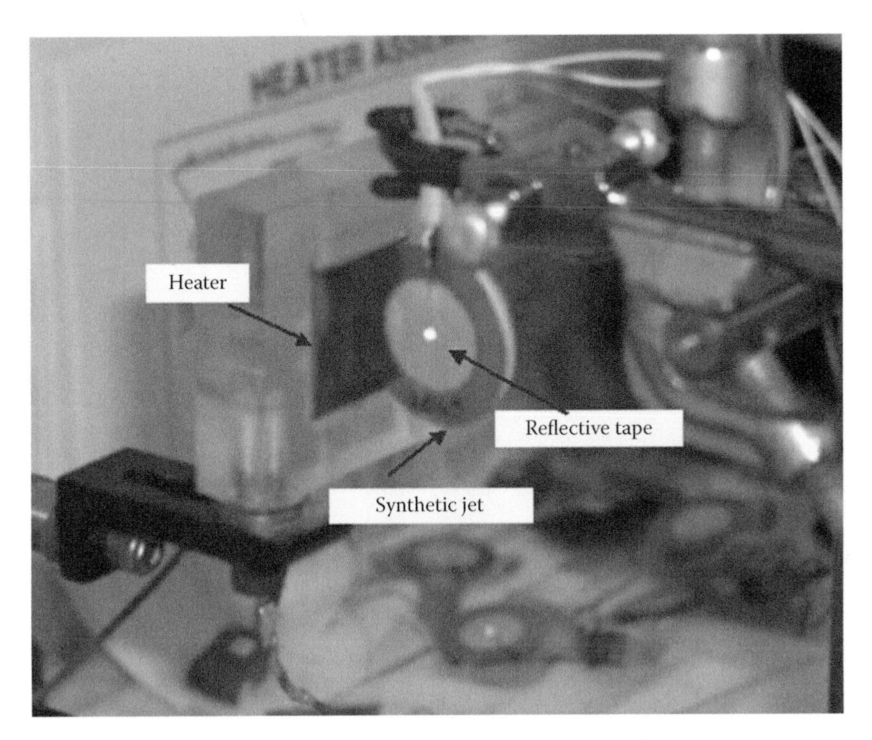

FIGURE 9.4 Close-up view of the synthetic jet and the vertical heater. (From Arik, M. and T. Icoz. *J. Heat Transfer*, 134, 081901, 2012.)

For the construction of the heat source, a 25 × 25 mm Kapton heater was placed at the backside of a 1.5-mm-thick copper plate, which was also equipped with three T-type thermocouples, as illustrated in Figure 9.5. A 50 × 50 mm Plexiglas substrate was used as the housing for the heater. Additionally, three T-type thermocouples were installed on this plastic substrate to quantify the heat losses. A series of tests were conducted on the heater assemble to measure the heat losses and develop a heat loss equation for the system. Heat losses can dominate the experimental measurements. Therefore, a thorough study with computational models and experimental validation has been performed and previously published. The heat loss test results (Figure 9.6) were inputted to the post-processing algorithm so that the net cooling effect of the synthetic jets can be accurately computed along with other parameters such as coefficient of performance (COP).

The impact of the driving frequency, voltage, and jet-to-heater spacing on the heat transfer and jet power consumption was studied. The resonance frequency (4.5 kHz) was previously determined from laser vibrometer measurements.

In the first set of heat transfer augmentation experiments conducted, the power to the heater was set to such a level that the heater temperature was in the vicinity of 80°C. The total power input to the heater for this setting was 1.56 W. The IR image of the heater under natural convection conditions is shown in Figure 9.2a. After the system reached the steady state under natural convection conditions, the IR temperature distributions were recorded. The axial distance between the jet and the heater was set to the desired location.

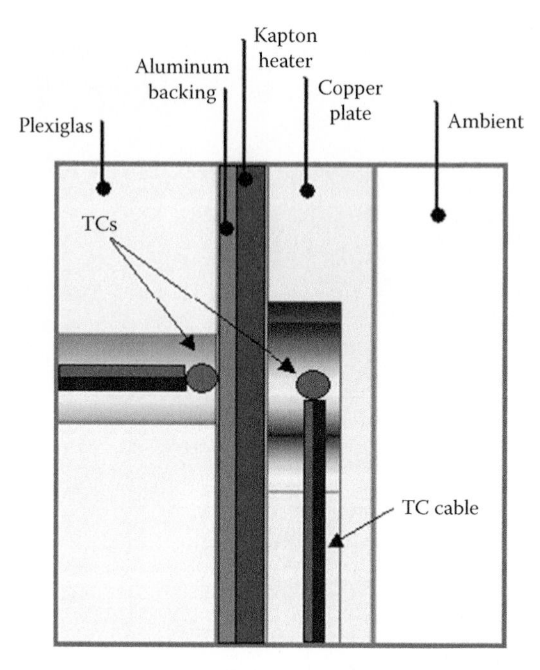

FIGURE 9.5 Heater construction. (From Arik, M. and T. Icoz. *J. Heat Transfer*, 134, 081901, 2012.) TC, thermocouple.

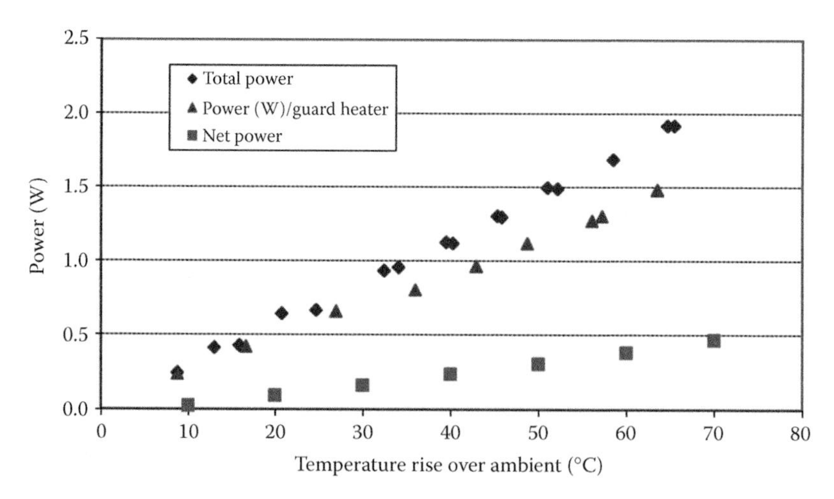

FIGURE 9.6 Heater temperature rise with heat rejection. (From Arik, M. and T. Icoz. *J. Heat Transfer*, 134, 081901, 2012.)

The synthetic jet was turned on with the desired driving frequency and voltage. The axial distance was varied between 5 and 50 mm.

Figures 9.7 and 9.8 present the effects of driving voltage and the axial distance over the heat transfer. As seen from Figure 9.7, the increased driving voltage does not increase the heat transfer at the same rate as the jet power consumption. This is particularly important because it might cause degradation of COP and coarsen the advantage of the higher EFs obtained at higher voltages.

An optimum placement of the jet from the heater is important. In an ideal system, the jet should not consume a large footprint area and volume. Therefore, it should be placed

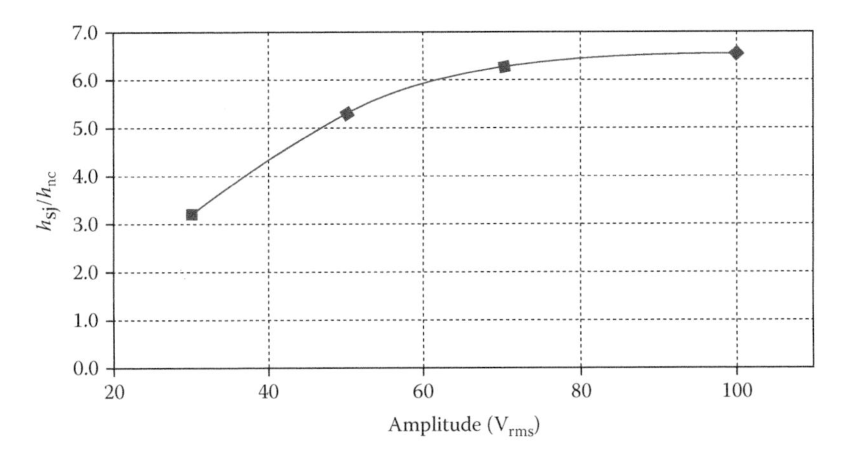

FIGURE 9.7 Effect of driving voltage on heat transfer enhancement. (From Utturkar, Y. et al., *J. Heat Transfer*, 130, 062201, 2008. Reprinted with permission.)

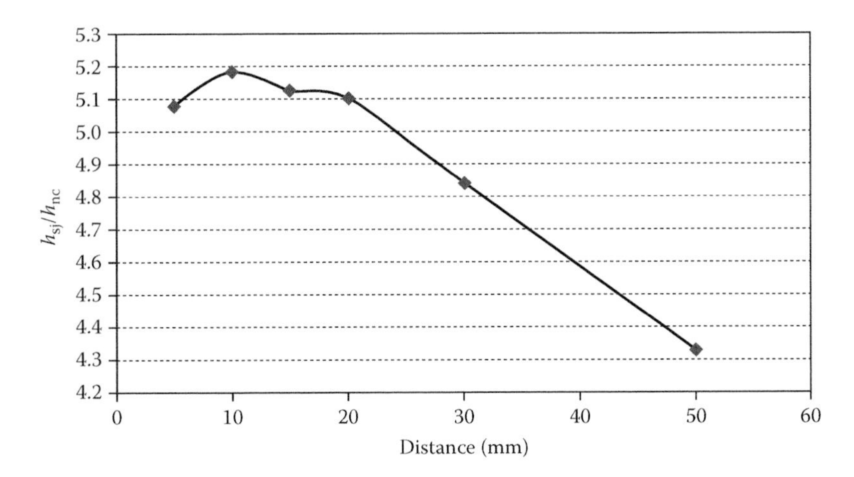

FIGURE 9.8 Effect of axial distance on heat transfer enhancement. (From Utturkar, Y. et al., *J. Heat Transfer*, 130, 062201, 2008. Reprinted with permission.)

as close as possible. As given in Figure 9.8, the jet can be placed as close as 5 mm with little performance degradation, though it gives a better performance at 10 mm. There are several reasons for an optimum spacing to provide a better heat transfer performance. First of all, at low spacing, the area of the heater being cooled by impingement is much smaller. This implies that the region of high heat transfer coefficient is smaller at smaller spacing, leading to a lower overall heat transfer coefficient. This in turn leads to higher average heater temperature. Another important aspect is the coolant temperature. The air exiting the jet during the exhaust cycle is hotter than the ambient, mainly because of the heat generated within the jet itself. When the jet is placed very close to the heat source, the temperature of the impinging air is close to the temperature of the air exiting from the jet. When the distance between the jet and the heater is increased, ambient air gets entrained, causing the temperature of the air impinging upon the heater to decrease substantially (Garg et al. 2005).

9.2.1.2 Velocity Measurements

The jet operates by periodic ejection and suction of the fluid. To capture both the suction and the ejection velocities, a hotwire probe was placed close to the jet orifice at a distance of approximately 0.5 mm (Figure 9.9a). The jet velocity is measured by means of a TSI IFA 100 hotwire anemometry system, which has 0.5-μm-thick probe wire. A computer code was used to separately average the ejection and suction peaks. A typical velocity profile for the synthetic jet is shown in Figure 9.9b. The average exhaust and average suction velocities were calculated. Because of the single-wire probe type, no directional output can be obtained; thus, we are able to tap only the local velocity magnitude.

The exhaust velocity was found to be increasing nonlinearly with the driving voltage as illustrated in Figure 9.10. Specifically, the peak exhaust jet velocity tends to increase at a lesser rate with driving voltage as the driving voltage is increased (Garg et al. 2005). Though the cause for this behavior has not been studied in detail, we expect it to be because of either flow/structural nonlinearities or air leakages occurring at the high voltages due to increase in strain.

Figure 9.11 presents the trend between the jet exit velocity and the heat transfer enhancement. The driving voltage was kept at 50 V, while the harmonic signal was swept between 2 and 5.5 kHz. The definition of EF is given in the following equation (Garg et al. 2005).

$$\mathrm{EF} = \frac{h_{\mathrm{jet}}}{h_{\mathrm{nc}}} = \left.\frac{q_{\mathrm{jet}}}{q_{\mathrm{nc}}}\right|_{(T_s - T_{\mathrm{air}})} \tag{9.1}$$

In Figure 9.11, an optimal air velocity of about 45 m/s is observed at 4.5 kHz, which is the natural frequency (or Helmholtz frequency) of the jet. Naturally, the best heat transfer performance (5.5× based on Equation 9.1) is also obtained at that resonant frequency.

9.2.1.3 Jet Power Consumption

Power consumption in an electronics cooling system is very important since it directly affects the COP of the cooling solution. Therefore, a careful study was carried out to obtain the true power consumption of the synthetic jets. Figure 9.12 shows the synthetic jet power

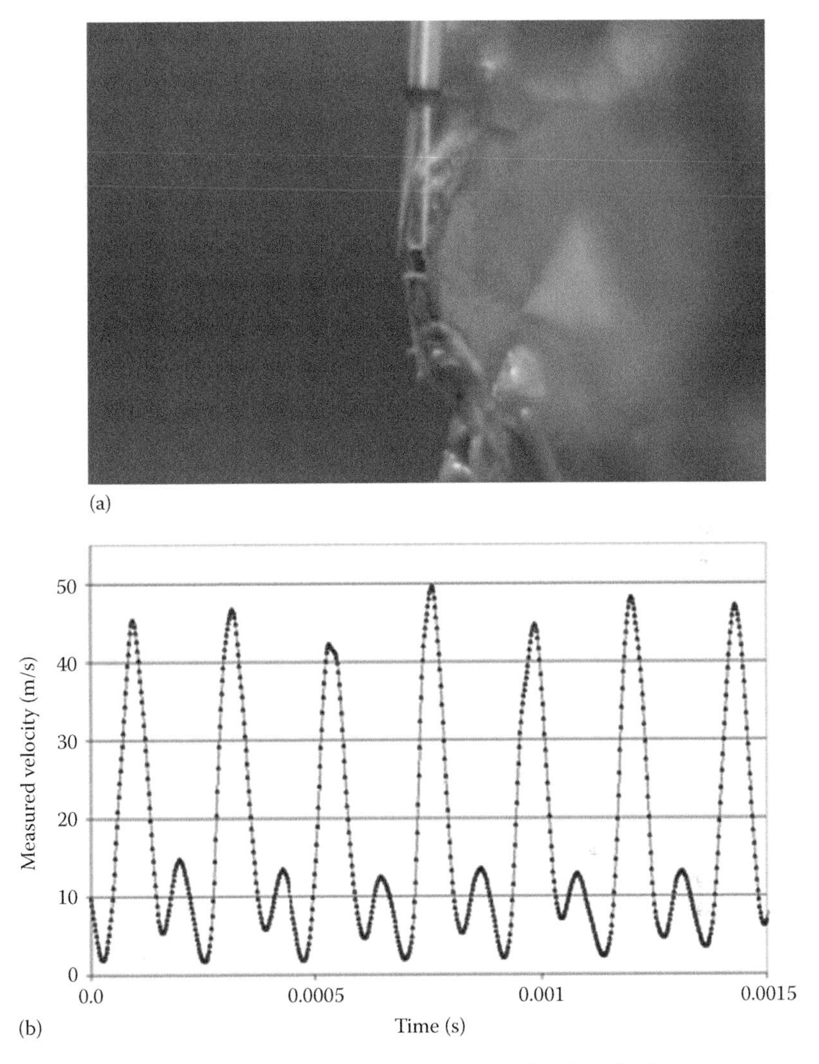

FIGURE 9.9 (a) Illustration of hotwire location near jet exit. (b) Typical velocity response tapped by the hotwire (50 V_{rms} and 4500 Hz). (From Utturkar, Y. et al., *J. Heat Transfer*, 130, 062201, 2008. Reprinted with permission.)

measurement setup. In addition to the basic synthetic jet heat transfer measurement setup, a 1 kOhm serial resistor was added to increase the accuracy of the current measurements. Voltage measurements were done before and after the resistance using a high-frequency differential probe (Tektronix 5205, 100 Hz, Tektronix, Inc., Beaverton, OR). By using a known resistance and voltage values, electrical current for the synthetic jet was calculated. Another high-frequency probe measured the exact driving voltage of the synthetic jets.

Effect of the driving voltage on the jet power consumption is depicted in Figure 9.12. It is interesting to note the steady increase of power consumption over 30 V of driving

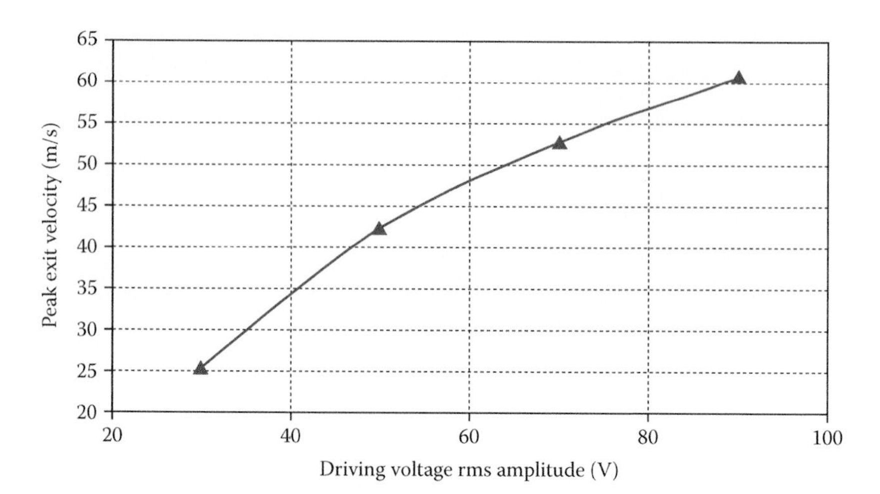

FIGURE 9.10 Variation of peak jet exit velocity with driving voltage amplitude. (From Utturkar, Y. et al., *J. Heat Transfer*, 130, 062201, 2008. Reprinted with permission.)

voltage. While the power consumption follows a steady trend, the jet exit velocity does not show the same steady trend over 30 V (Figure 9.10). This indicates that the jet may be less efficient at higher voltages (greater than 30 V).

The effect of the driving frequency over the jet power consumption is shown in Figure 9.13. The jet was driven at 50 V constant voltage supply, and the frequency was swept between 2000 and 6000 Hz. This range encompasses the structural and the Helmholtz natural frequencies of the jet. The jet showed maximum power consumption at the

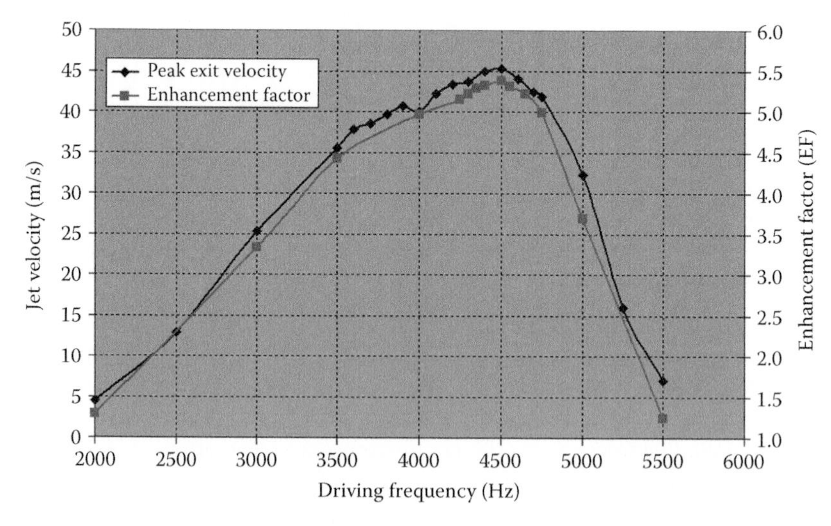

FIGURE 9.11 Effect of driving frequency on heat transfer enhancement and peak jet exit velocity. (From Utturkar, Y. et al., *J. Heat Transfer*, 130, 062201, 2008. Reprinted with permission.)

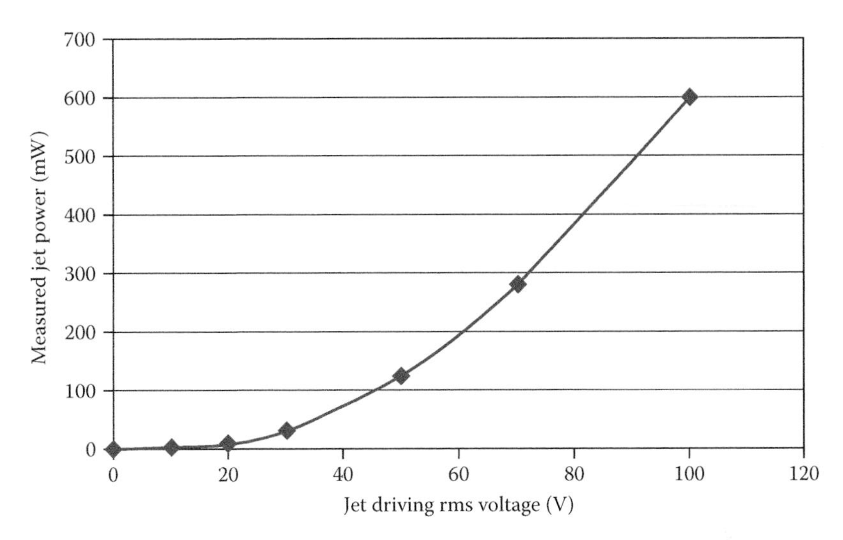

FIGURE 9.12 Synthetic power consumption with driving voltage. (From Utturkar, Y. et al., *J. Heat Transfer*, 130, 062201, 2008. Reprinted with permission.)

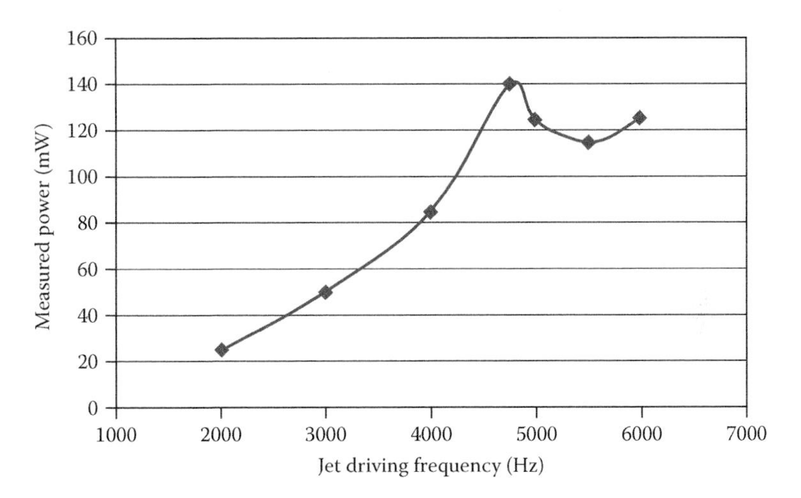

FIGURE 9.13 Synthetic jet power consumption with the frequency. (From Utturkar, Y. et al., *J. Heat Transfer*, 130, 062201, 2008. Reprinted with permission.)

resonance frequency of 4500 Hz. Before and after the resonance frequency, it showed lower power consumption. This is particularly significant because of the capacitive effect and the fluid-structural coupling of the jet.

Power consumption is a very important design constraint for the application of synthetic jets in electronics cooling applications. From Figures 9.11 and 9.13, we can comment that one can drive synthetic jets between 40 and 100 mW and get about 4–6 times heat transfer enhancement. Furthermore, based on the consumed jet power and the measured heater input power, a COP of about 10 is estimated for the synthetic jet actuators in this study.

9.2.2 Computational Models

9.2.2.1 Fluid Flow Model

The schematic of the two-dimensional (2D) synthetic jet model is depicted in Figure 9.14. The jet is modeled (Figure 9.14) by a 23-mm tall cavity. The motion of the side diaphragms is suitably modeled by a spatially varying velocity profile (Equations 9.2 and 9.3), as illustrated in Figure 9.14.

$$u_{\text{left}} = \frac{A}{2}\left[1 - \cos\left(\frac{2\pi y}{D}\right)\right]\sin(2\pi ft) \tag{9.2}$$

$$u_{\text{right}} = \frac{A}{2}\left[1 - \cos\left(\frac{2\pi y}{D}\right)\right]\sin(2\pi ft + \pi) \tag{9.3}$$

Here, A is the velocity amplitude, D is the diaphragm height, and f is the operating frequency. In this study, we assign $A = 3$ m/s and $f = 4500$ Hz. The above-mentioned profiles, besides yielding zero u-velocity, also produce a zero u-velocity gradient at the ends to imitate the clamping effect of the diaphragm (refer to Mittal et al. 2001; Utturkar et al. 2003). The width and height of the jet orifice are 1 mm. The chamber comprises of a 1″ long heater, which is modeled as a "thin" aluminum wall emitting 1.315 W of heat. All other walls are at 20.8°C (ambient). The distance of the heater from the jet orifice is

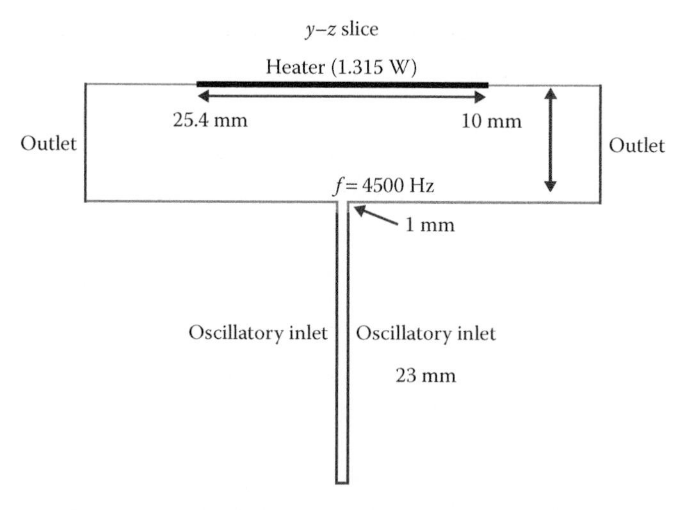

FIGURE 9.14 Schematic of a 2D synthetic jet model. (From Utturkar, Y. et al., *J. Heat Transfer*, 130, 062201, 2008. Reprinted with permission.)

chosen 10 mm based on a previously reported sensitivity analysis (Garg et al. 2005). All of the above-mentioned specifications are consistent with the experimental setup shown in Figure 9.1, and the simulation inputs are based on the experimental case corresponding to the peak performance point (Figure 9.11).

A commercially available computational fluid dynamics (CFD) software (Fluent, ANSYS, Inc., Canonsburg, PA) is employed to perform 2D, incompressible, time-dependent, Navier–Stokes computations on the above-mentioned geometry. The second-order upwind scheme is employed for the convective terms, and central differencing is used for the viscous terms. The energy equation is solved in addition to the mass (continuity) and momentum equations. Time stepping is performed via the first-order implicit Euler discretization. The commonly used, noniterative, operator-splitting technique, pressure implicit splitting of operators (PISO) (Issa 1985), is selected for the pressure–velocity coupling to expedite the computational time without sacrificing the formal accuracy.

The *Re* for the simulation, based on time-averaged mean velocity at the jet exit, is 2012. The Stokes number (*S*), which is a nondimensional form of the forcing frequency, is 42. The definitions of *Re* and *S* are mentioned below alongside other relevant parameters.

$$Re = \frac{V_{\text{inv}}^{\text{ave}} d}{\nu} \tag{9.4}$$

$$S = \sqrt{\frac{2\pi f d^2}{\nu}} \tag{9.5}$$

In Equation 9.4, $V_{\text{inv}}^{\text{ave}}$ represents the time-averaged (over half-cycle) and spatially averaged velocity in the jet orifice. Recently performed direct numerical simulations have demonstrated turbulent flow transitions for synthetic jets operating around $Re = 1150$ and $S = 17$ (Mahalingam and Glezer 2005). They also indicated the tendency of the flow to stabilize the laminar regime in the jet orifice. Thus, the flow regimes for synthetic jet actuators could range from being laminar in the orifice to turbulent in the external environment. Though use of Reynolds-averaged Navier–Stokes (RANS)-based turbulence models appears as an intuitive option, the resultant high effective viscosity may damp the small vorticity scales in the jet orifices and blur the flow physics. Considering this fact, we choose to perform laminar simulations in this study in conjunction with a finer mesh near the orifice vicinity (22 mesh points across the orifice).

9.2.2.2 Flow Simulation Results

In this section, the experimental data were employed to firstly develop and assess the computational model. Secondly, we extend the computational model/tool to predict the behavior of the jet under different alignment conditions.

The sensitivity of the model illustrated in Figure 9.14 is investigated by two grids: 16300 nodes (40 across the orifice) and 26800 nodes (60 across the orifice). The heater in the simulations is modeled by a constant heat flux boundary patch. Accounting the measured heat loss of 0.245 W for the given heater power of 1.56 W, a net heat flux of 1.315 W is applied over the 1″ long heater patch. Since the hotwire location, dimensions, and measurements are known, we can extract the velocity magnitude from the grid points at the hotwire location. The uncertainty in the hotwire length and location is within 0.25 mm. The time-dependent results in Figure 9.15 not only show that they are fairly grid-independent, but also illustrate a reasonable match with the experimental measurements in Figure 9.9b (given the above-mentioned hotwire uncertainties) for both mesh densities. This agreement instills confidence in the computational procedure.

The steady-state heater temperature obtained from IR images is averaged and compared with the average temperature on the 25.4-mm heat flux patch in the CFD model. For instance, the experimental heater average temperature mentioned in Table 9.1 is extracted from Figure 9.2b. As seen from Table 9.1, the average temperatures and average heat transfer coefficients, which are likewise calculated, agree within 5% between the experiments and simulations (Table 9.1). While the computational study only solved the fluid flow and heat transfer from heater surface, it did not consider the internal conduction effects in the thin aluminum plate. We assume that they are negligible.

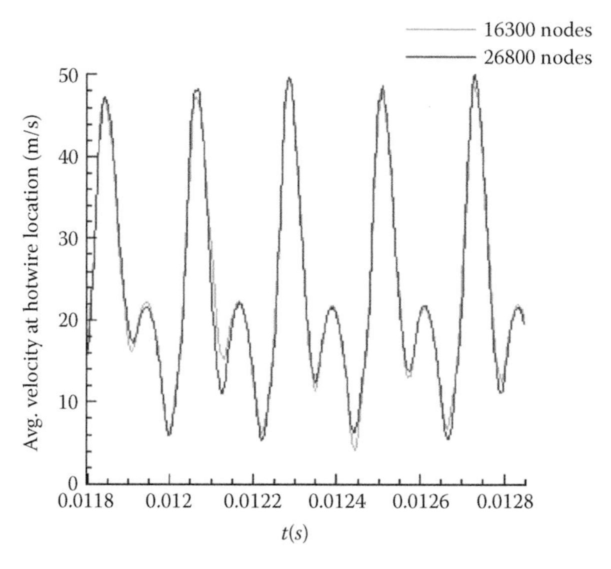

FIGURE 9.15 Time-dependent velocity magnitude extracted from computational domain at the hotwire location. (From Utturkar, Y. et al., *J. Heat Transfer*, 130, 062201, 2008. Reprinted with permission.)

TABLE 9.1 Comparison between CFD and Experiments (T_0 = 20.8°C)

Variable	Experiment	CFD	Difference (%)
$T_{average}$, °C	42.1	41.0	2.6
$h_{average}$, W/m^{2}K	95.6	100.0	4.6

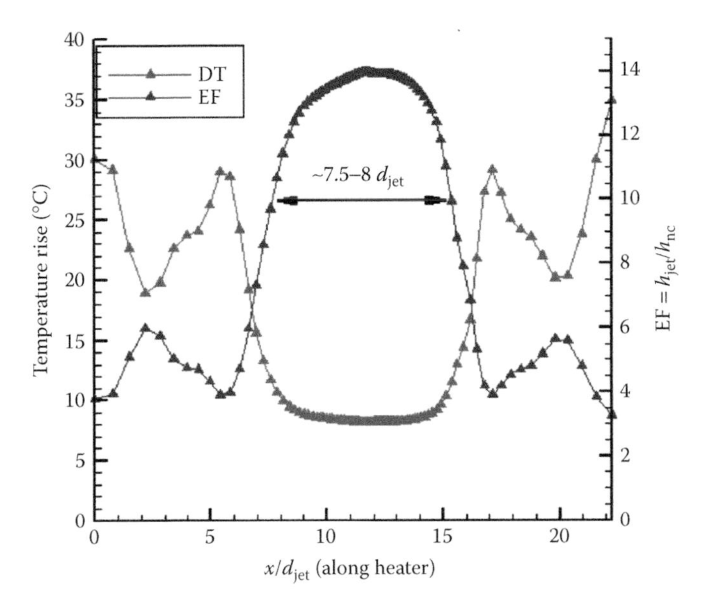

FIGURE 9.16 Variation of temperature rise and heat transfer enhancement along the heater surface. (From Utturkar, Y. et al., *J. Heat Transfer*, 130, 062201, 2008. Reprinted with permission.)

Our natural convection heat loss experiment yielded a natural convection heat transfer coefficient equal to 18 W/m^2K (it includes both natural convection and radiation) at 40°C heater temperature. Using this value, we could extract the EF (Equation 9.1) variation over the heater surface. Figure 9.16 depicts that though the overall enhancement on the heater is about 5× (indicated by the average heat transfer coefficient in Table 9.1), the middle portion of the heater is subjected to a 14× enhanced heat transfer. Furthermore, almost eight jet diameter wide area on the heater experiences at least 10× heat transfer. This information is highly useful in sizing the complete thermal system. For instance, for a given jet size, we are able to decide the maximum area that can be cooled with the jet, and vice versa, for a given heater size, it can indicate the minimum size of the jet to be used.

The instantaneous vorticity plots in Figure 9.17 reveal the relation between the heat transfer variation and the flow field. As observed, the jet periodically emanates vortex dipoles, which impinge on the central part of the heater and drive the heat away from the heater surface. Thus, we obtain the maximum cooling power at the exact impingement locations. The temperature difference plots in the same figure indicate the formation of hotspots at the heater ends, which is in line with the earlier discussion. The instantaneous vorticity plots are not able to clearly explain the formation of these hotspots, though they justify the heat transfer behavior at the heater center. As a result, we portray the time-averaged velocity field in Figure 9.18 to view from the perspective of a steady jet. The flow field is characterized by perpendicular impingement of the jet with a stagnation point at the heater center. In a time-averaged sense, the jet induces recirculation currents, which

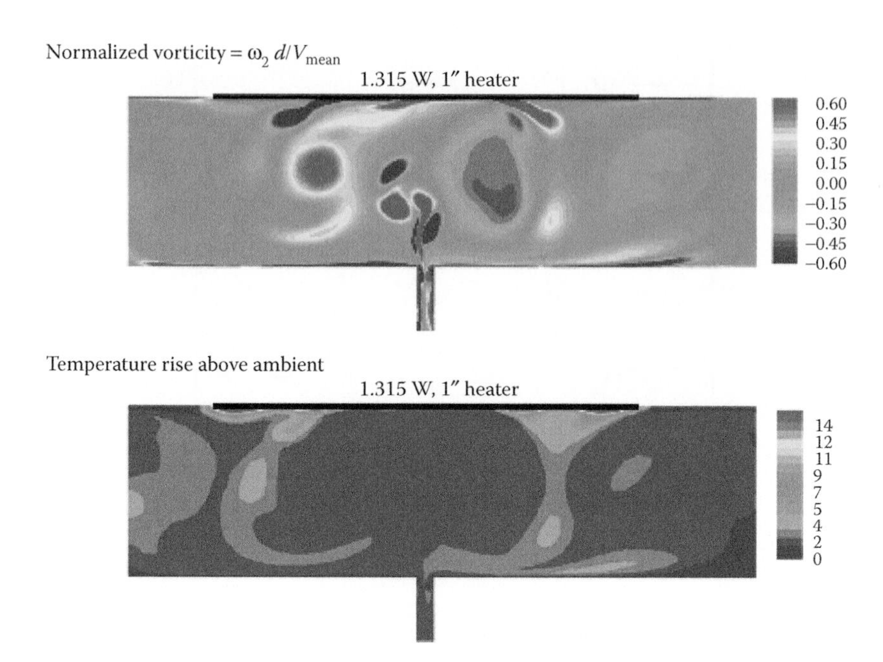

FIGURE 9.17 Normalized vorticity and temperature rise at the beginning of expulsion phase of the jet. (From Utturkar, Y. et al., *J. Heat Transfer*, 130, 062201, 2008. Reprinted with permission.)

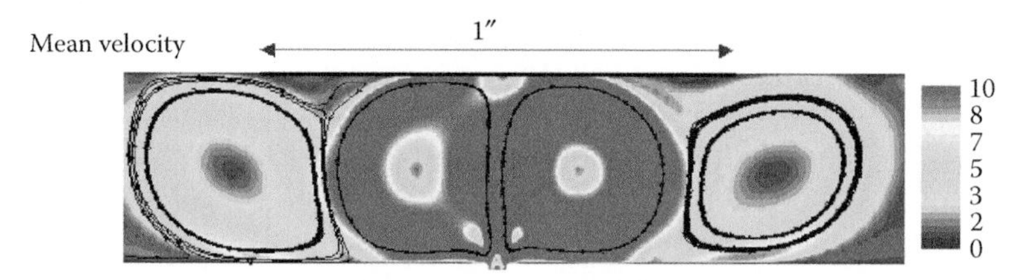

FIGURE 9.18 Time-averaged streamlines and synthetic jet velocity field. (From Utturkar, Y. et al., *J. Heat Transfer*, 130, 062201, 2008. Reprinted with permission.)

seemingly scale with the distance between the heater and the jet. Between two adjacent recirculation zones, there exist smaller secondary recirculation zones near the heater surface. Interestingly, the hotspots seen in Figures 9.16 and 9.17 develop near the location of these secondary vortices. This could be due to the fact that the secondary vortices stagnate the fluid within a smaller region and thus restrict the heat transfer. Furthermore, the velocity contours depict an upward jet movement at about 10 m/s. This value, which is the velocity actually facilitating the heat transfer, is fairly lower than the peak hotwire measurement (50 m/s). As a result, we believe that the hotwire measurement may be short in terms of gauging the heat transfer, and time-averaged results may be more valuable for this matter.

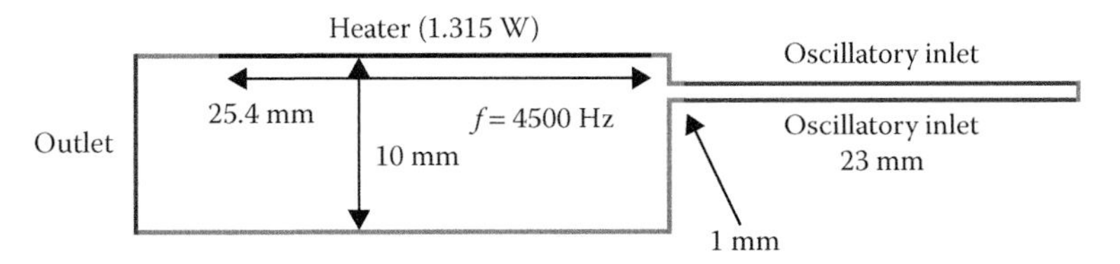

FIGURE 9.19 Schematic of the computational domain in the crossflow alignment. (From Utturkar, Y. et al., *J. Heat Transfer*, 130, 062201, 2008. Reprinted with permission.)

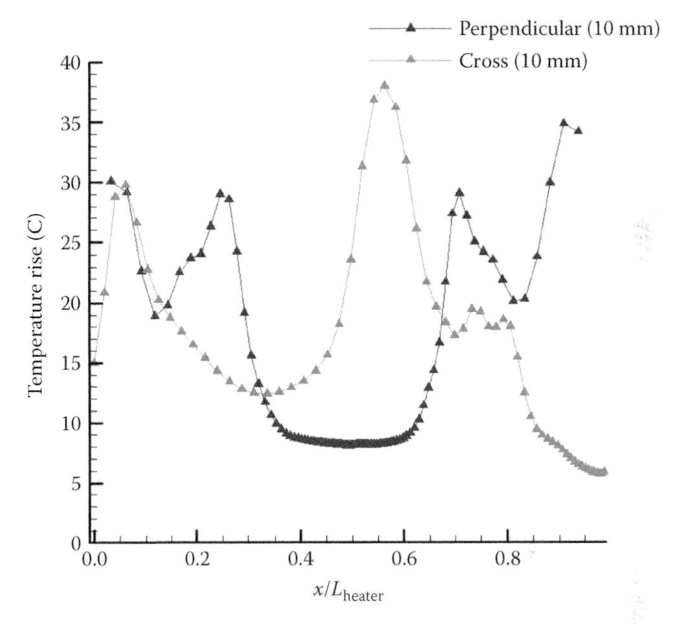

FIGURE 9.20 Temperature variation on the heater surface for the two flow alignments. (From Utturkar, Y. et al., *J. Heat Transfer*, 130, 062201, 2008. Reprinted with permission.)

We finally alter the jet alignment from the perpendicular impingement position to the crossflow orientation. This is motivated by packaging restrictions, which may arise due to space limitation between two circuit boards. Figure 9.19 shows the computational domain. Note that we just change the jet angle from 90° to 0° holding other computational parameters constant. Figure 9.20 compares the temperature rise on the heater surface between the perpendicular and crossflow positions. Though the heater temperature in the crossflow case is asymmetrical, its average value over the heater surface is within 0.1°C of the average value in the alternate case. Thus, the crossflow position works as good as the perpendicular impingement position. In case of the crossflow alignment, the heat transfer is facilitated through a boundary layer flow, as opposed to the perpendicular impingement. Though the flow structure near the heater surface is different in both cases, the heat transfer is less affected by it.

9.3 Forced Convection Heat Transfer with Synthetic Jets

Synthetic jets have been considered local cooling devices and for a majority of time they are thought for extending limits of natural convection. However, some researchers have considered for enhancing forced convection heat transfer. Utturkar et al. (2007) presented an experimental computational study to show the forced connection enhancement with synthetic jets.

9.3.1 Numerical Simulations

Figure 9.21 presents the computational domain used in this study. The synthetic jet is located 2 mm above and 1.5 mm away from the heater leading edge. The disk movement, as mentioned earlier, is simulated by moving boundary- and spring-based mesh smoothing techniques. Note that since the airflow surrounding the synthetic jet is solved as a part of the solution, the heat transfer condition on the synthetic jet disk does not need to be artificially stipulated and is coupled with the fluid flow solution. The upper and lower walls in the above model are adiabatic, and the heater is modeled as a heat flux patch. The displacement amplitude (A) is determined by converting the disk displacement, measured by laser vibrometer in the experimental setup, into an equivalent, volume-conserving, 2D amplitude. This corresponded to an experimental driving condition of 400 Hz and 90 V_{rms} for a 20-mm jet.

Figure 9.22a shows the temperature rise contours and the velocity vectors for cases with and without the jet. Naturally, the case without the jet shows the growth of a thermal boundary layer from the leading edge of the heater. The heat transfer between the hot wall and the cold ambient fluid occurs across the finite thickness of this thermal layer. As opposed to this condition, as seen from Figure 9.22b, the synthetic jet blows air with velocity of about 10–15 m/s in the direction of the bulk flow. Note that the jet velocity is about 10× the fan velocity. As observed from the temperature contours, this jet of air is

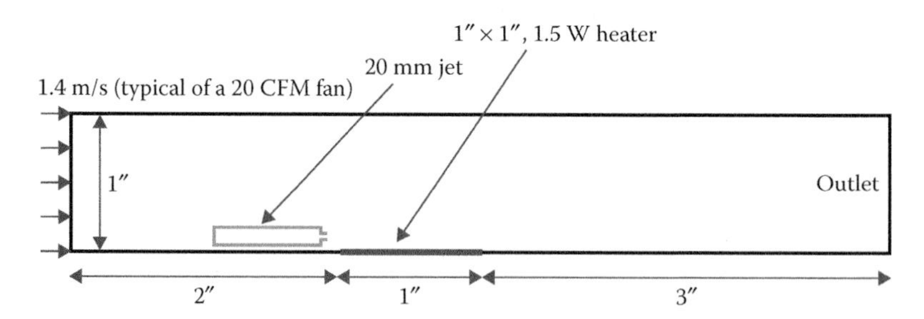

FIGURE 9.21 Schematic of the computational model. (From Utturkar, Y., M. Arik, and M. Gursoy, Assessment of Cooling Enhancement of Synthetic Jet in Conjunction with Forced Convection, *IMECE 2007*, Seattle, WA, IMECE2007-41446, 2007. Reprinted with permission.)

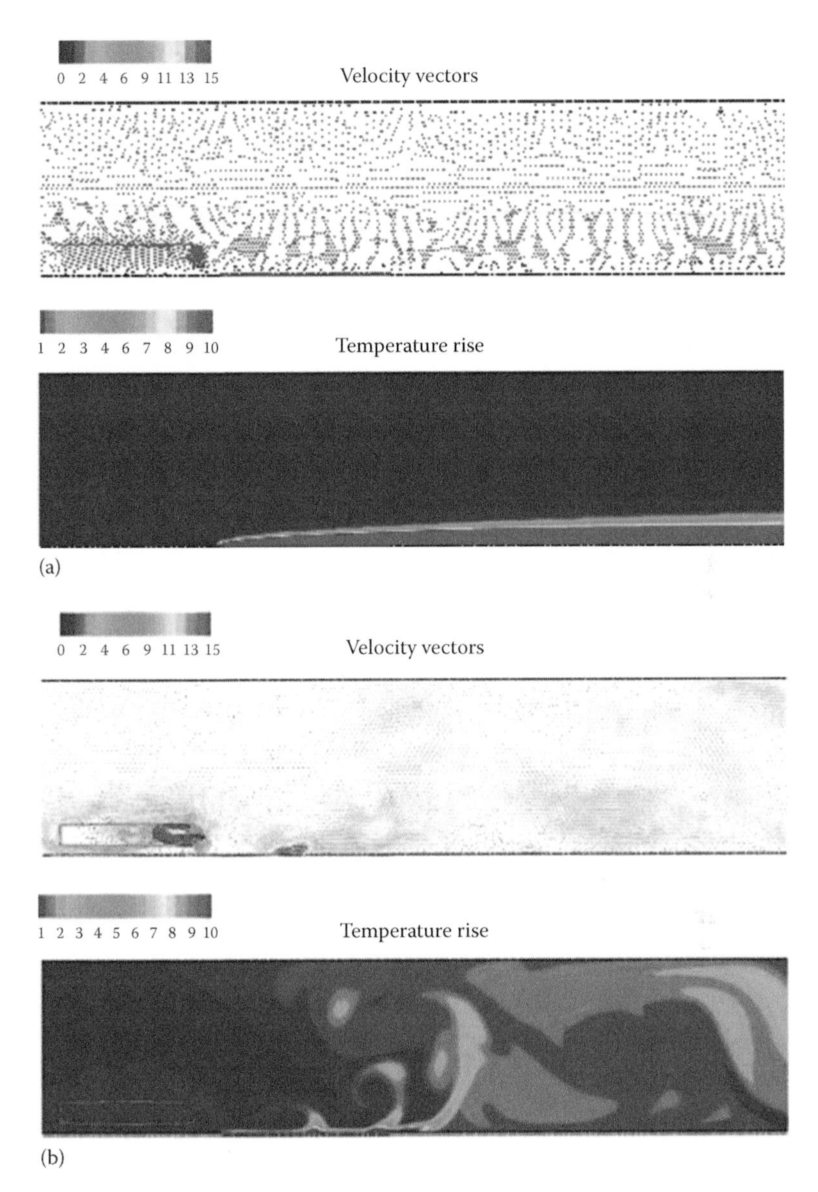

FIGURE 9.22 Velocity vectors and temperature rise: (a) no synthetic jet, only fan flow, inlet velocity = 1.4 m/s, interplate distance = 25.4 mm; (b) synthetic jet turned on, inputs derived from 400 Hz and 90 V operating condition. (From Utturkar, Y., M. Arik, and M. Gursoy, Assessment of Cooling Enhancement of Synthetic Jet in Conjunction with Forced Convection, *IMECE 2007*, Seattle, WA, IMECE2007-41446, 2007. Reprinted with permission.)

able to penetrate through the heater's thermal boundary layer, thus yielding a greater heat transfer rate. Additionally, the dissipated heat is convected away by the fan's bulk flow and is prevented from accumulating in the heater vicinity. As a result, the combination of a fan and the local effect of a synthetic jet is able to yield significant benefits over forced convection.

We further investigated the sensitivity of the configuration to changes to the inlet flow velocity and the distance between the upper and lower channel wall. Figure 9.23 illustrates

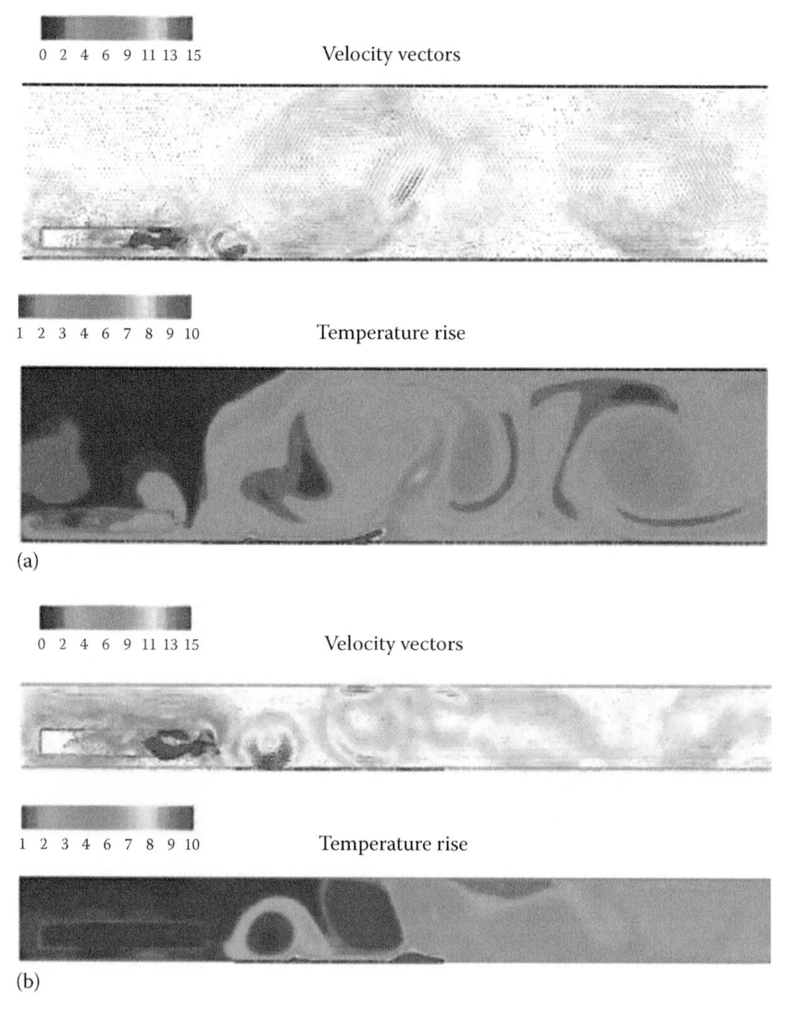

FIGURE 9.23 Velocity vectors and temperature rise: (a) inlet velocity = 0.5 m/s, interplate distance = 25.4 mm; (b) inlet velocity = 1.4 m/s, interplate distance = 10 mm. (From Utturkar, Y., M. Arik, and M. Gursoy, Assessment of Cooling Enhancement of Synthetic Jet in Conjunction with Forced Convection, *IMECE 2007*, Seattle, WA, IMECE2007-41446, 2007. Reprinted with permission.)

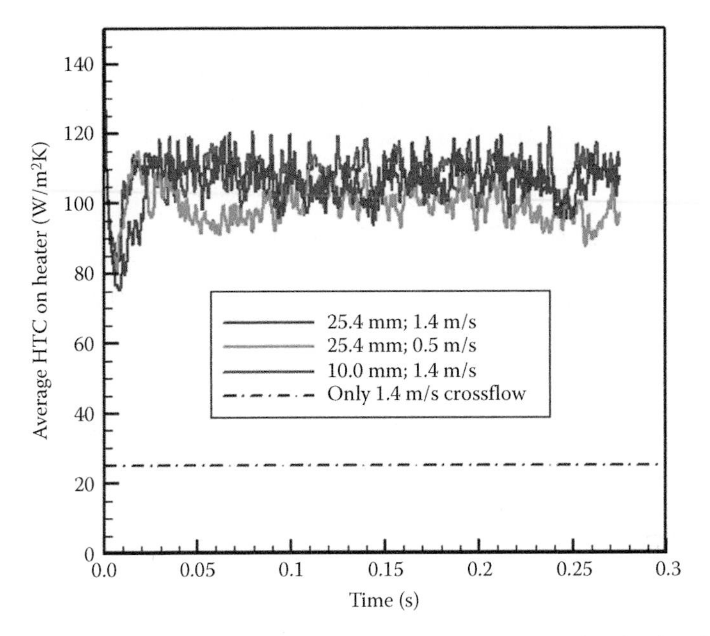

FIGURE 9.24 Comparison of heat transfer coefficient on the heater surface for four flow cases. (From Utturkar, Y., M. Arik, and M. Gursoy, Assessment of Cooling Enhancement of Synthetic Jet in Conjunction with Forced Convection, *IMECE 2007*, Seattle, WA, IMECE2007-41446, 2007. Reprinted with permission.)

the results for two cases with reduced inlet velocity and channel height. Clearly, both the cases apparently show greater heat accumulation downstream to the heater. This is because of the reduction in mass flow rate, which retards the heat advection. The differences between the four cases are quantified in Figure 9.24. Clearly, within the range of parameters investigated in this study, the synthetic jet seems to provide about 4× increase in the surface heat transfer coefficient over fan flow. Also, this enhancement is seemingly less sensitive to the variations explored in the study.

9.3.2 Forced Convection Experiments

An experimental test rig was designed to directly assess the CFD findings. This canonical experiment will show a benchtop demonstrator of the forced convection cooling by using conventional synthetic jets.

As shown in Figure 9.25, a small wind channel was designed with a fan and honeycomb straighter at one end. Channel walls were made up of 12 mm wall thickness. Channel width and height were both 100 mm, creating a square geometry. The fan was about 30 mm away from the heater surface. A square heater, 25.4 mm in length and 1 mm thickness, was aligned along one of the channel walls, and a synthetic jet was placed parallel to the wall such that its jet was in the direction of the fan flow. The jet was operated at different

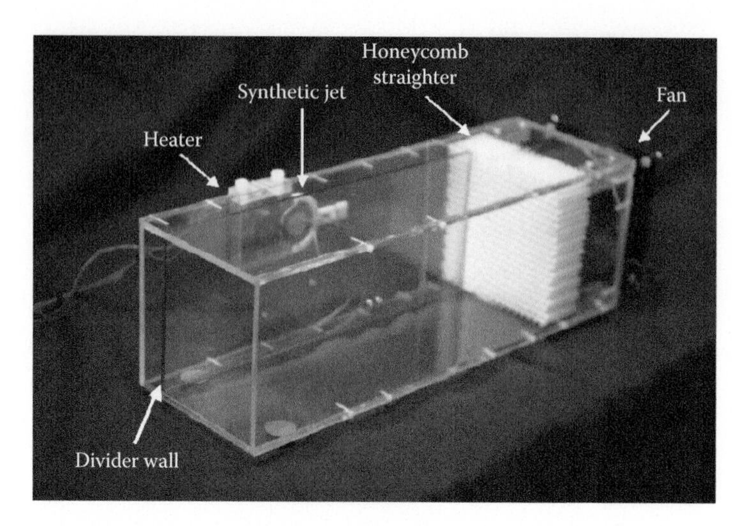

FIGURE 9.25 Experimental setup for investigating the impact of a synthetic jet on a flat heater in the presence of a fan. (From Utturkar, Y., M. Arik, and M. Gursoy, Assessment of Cooling Enhancement of Synthetic Jet in Conjunction with Forced Convection, *IMECE 2007*, Seattle, WA, IMECE2007-41446, 2007. Reprinted with permission; Arik, M. and T. Icoz. Predicting heat transfer from unsteady synthetic jets. *J. Heat Transfer*, 134, 081901, 2012.)

operating conditions, while the fan velocity was gradually increased. The flow straighter was enabled to avoid the nonuniform flow swirling effects, which are typically seen for a fan.

Figure 9.26 compares the experimental values of enhancement with a sample CFD data point. The EF is defined as

$$EF_{FC} = \left(\frac{Q_{FC+jet}}{Q_{FC}}\right)_{\Delta T} = \left(\frac{h_{FC+jet}}{h_{FC}}\right)_{\Delta T} \tag{9.6}$$

where:

Q is the heat removed from the jet-facing side of the heater
h is the heat transfer coefficient
ΔT is the temperature difference between the heater and ambient
The subscript "FC" means *Fan Condition*

The experimental enhancement levels of 4× are in line with the CFD predictions. As the fan speed increases, the enhancement levels drop because of the consequent increase in the force convection heat transfer. Furthermore, an increase in the jet driving frequency also seemingly has a significant impact on the performance. Higher driving frequency provided better cooling due to the strong vortex shedding behavior compared to lower frequency (500 Hz). At lower main flow speeds, *Re* numbers are lower and the weak effect of forced convection is dominated by synthetic jet flow. Due to the limited computational resources, only jet operating at resonance frequency was compared to computational results.

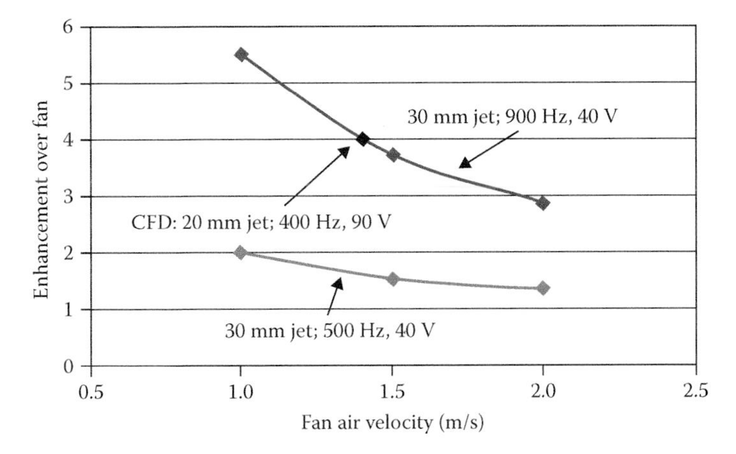

FIGURE 9.26 Enhancement of synthetic jet versus fan flow speed. (From Utturkar, Y., M. Arik, and M. Gursoy, Assessment of Cooling Enhancement of Synthetic Jet in Conjunction with Forced Convection, *IMECE 2007*, Seattle, WA, IMECE2007-41446, 2007. Reprinted with permission.)

Experiments were performed with a 30-mm diameter General Electric (GE) synthetic jet enclosed in a tunnel in parallel orientation to a heater. The heater power was measured at a given temperature drop with and without the jet to yield the jet's heat transfer performance enhancement. The fan flow velocity was varied to observe its impact on the enhancement values. In general, for fan velocities less than 1.5 m/s the jet is able to provide a 4× enhancement over forced convection. Naturally, the enhancement values reduce as the fan air speed is increased. This is because the difference between the fan velocities and the local jet velocities reduces with increasing fan speeds. Computational results are able to corroborate and clarify the findings pointed by the experiments. The jet's local flow is able to disrupt the thermal boundary layer over a heated body thus dramatically increasing the heat transfer rates. Furthermore, this effect seems quite robust to changes in geometry and inlet boundary conditions.

9.4 Synthetic Jets Cooling for Enhancing Natural Convection Heat Sinks

Flow through a small opening led people think that synthetic jets can be used only for local cooling. However, being able to have many openings and a number of jets can be synchronized enabled advancing heat sink cooling. A number of studies have been published in the open literature (Utturkar et al. 2007, 2009). The effects of number of jets, locations, arrangements, and operating ranges on heat transfer from a cylindrical heat sink are investigated both numerically and experimentally. CFD simulations are utilized to provide fundamental understanding to the fluid flow and airflow interaction with the heat sink fins, whereas experiments are conducted for validation of CFD models and to further investigate the effects of number of jets in the stack on thermal performance.

Test layout

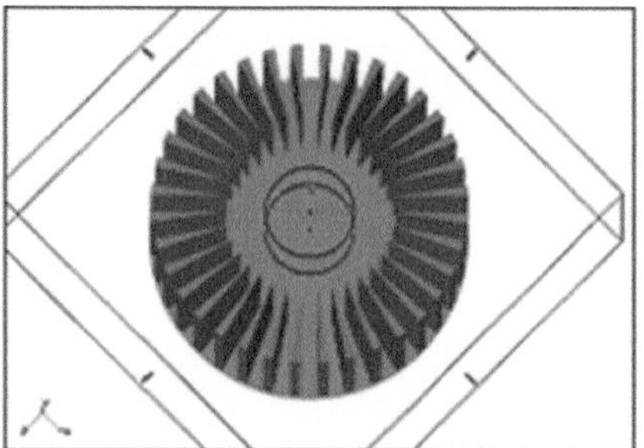

Model

Actual flow field:	Model:
• Highly unsteady	• Equivalent steady state
• Characterized by vortex dipoles	• Steady jet of air from the orifice
• Pulsed cooling	• Mimics a time-over aged cooling effect
• 3 days of CPU time required	• 2 hours of CPU time

FIGURE 9.27 Idealized CFD model and actual heat sink with synthetic jets.

A synthetic-embedded heat sink design was conceived and optimized using the design of experiments (DOE) approach and the equivalent steady-state computational strategy described in the Section 9.3 (Figure 9.27). The radius, fin height, and base plate dimensions of the sink are 50, 50, and 3 mm, respectively. Two synthetic jets were built in-house and mounted at the center of the heat sink equispaced along the height of the sink. Each jet has four orifices, and the jets are arranged such that they have a 45° angular offset. The heat sink has 36 triangular fins, which have been optimized for maximum heat removal. The base of heat sink is maintained at a constant temperature and the heat dissipated from the sink is calculated as a part of the solution. As a baseline to quantify the enhancement of the jet-embedded heat sink, a conventional heat sink with plate fins, having the same overall dimensions, is considered. Naturally, the fin area for the baseline heat sink is 57% more than the jet-embedded heat sink due to the availability of additional space that is occupied by the jets.

Experimental and CFD studies are performed for both the baseline and the jet-embedded heat sink to obtain the cooling enhancements. Commercially available software packages (Icepak, ANSYS, Inc., Canonsburg, PA, and Fluent) are used for the numerical computations. The jet movement is created using time-dependent velocity boundary conditions on the surfaces representing the jet disks. Second-order upwind scheme is used for the convection terms, while central differencing is employed for the diffusion terms. In general, 20 mesh points are employed to resolve the orifice flow and 10 mesh points are used for the fin gaps. The PISO methodology is employed for the pressure–velocity coupling and noniterative time advancement.

Figure 9.28 illustrates the heat dissipated by the baseline heat sink under natural convection at two different base temperatures. Note that the thermal resistance of the heat

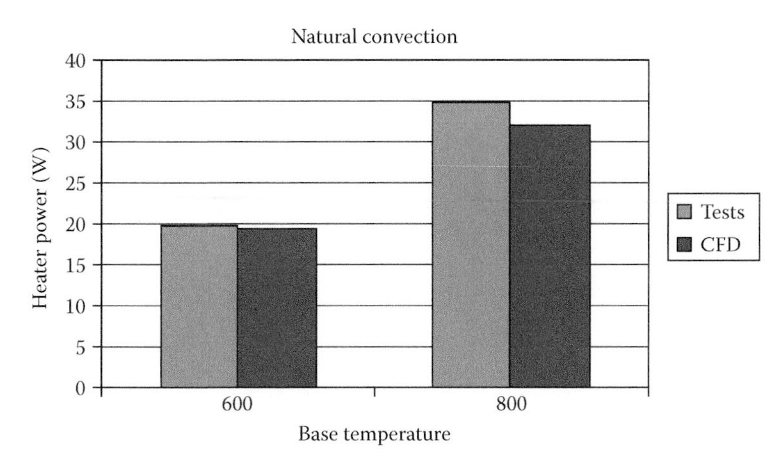

FIGURE 9.28 Performance of the heat sink, with the jets not operating, at two different base temperatures.

sink is equal to 2.0 C/W and 1.7 C/W at 60°C and 80°C base temperature, respectively. The average heat transfer coefficient (natural convection and radiation) over the fin surfaces, when the base is held at 80°C, is estimated from the computations as 6.8 W/m^2-K.

Following the natural convection calculations, computations are performed under the activated jet conditions over a frequency range of 500–800 Hz at 60 V. The surface heat transfer coefficient distribution over the fin surfaces is depicted in Figure 9.29 for the 700 Hz and 60 V condition. Note that the fins that lie exactly in the wake of the orifices show local enhancements of 4× the heat transfer coefficient. The overall enhancement of surface convection was noted around 2.6× for the chosen driving condition of the jet. This overall enhancement in heat transfer, while compensating for the loss of fin area due to the jets, additionally leads to an augmentation in cooling. The result is a thermal resistance of 0.8 C/W at 80°C base temperature, almost 2× better than the natural convection case.

An experimental investigation is carried out to validate single jet computational results and to evaluate thermal performance of multiple jets in the embedded heat sink design. For this purpose, an experimental setup is designed and fabricated. Various combinations of number of jets, number of orifices, and operating conditions are evaluated. Four- and 8-port jets are tested in stacks of 1 to 5 at 60 V operating voltage.

The heat sink of consideration here is 100 mm in diameter and made out of aluminum with 4 mm base thickness and 36 triangular fins, as shown in Figure 9.30. The fins are 50 mm high and only 27 mm long from the outer diameter of the heat sink, leaving a central cavity of 47.6 mm in diameter for synthetic jet integration. The geometry of fins and dimensions is determined following an exhaustive CFD analysis and reflects an optimal geometry for thermal performance.

The heat is applied to the heat sink via a Kapton film heater, which is mounted on the base of the heat sink. Two orthogonal grooves are opened at the base of the heat sink to allow thermocouple installation for base temperature measurements in various locations to ensure uniform heating condition is maintained throughout the testing. The backside of the heat sink is insulated with a 20-mm-thick insulation material.

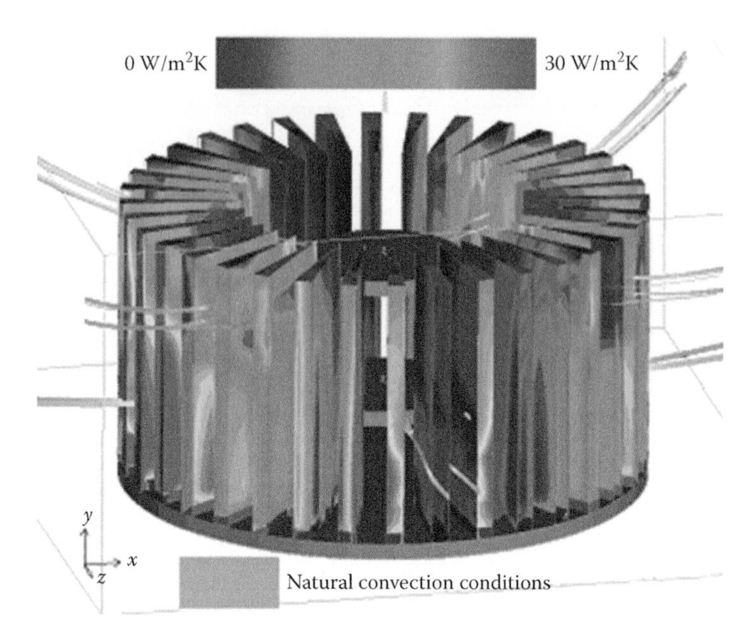

FIGURE 9.29 Heat transfer coefficient over synthetic jet-embedded heat sink fins (700 Hz and 60 V). (From Arik, M., M. Ozmusul, and Y. Utturkar, Effect of synthetic jets over natural convection heat sinks, IMECE 2008-68784, 2007. Reprinted with permission.)

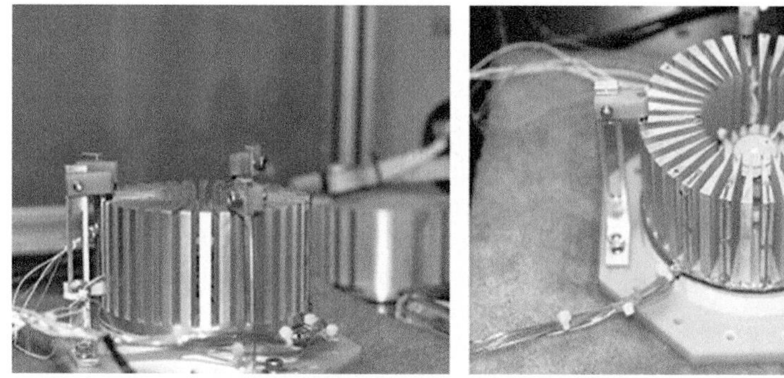

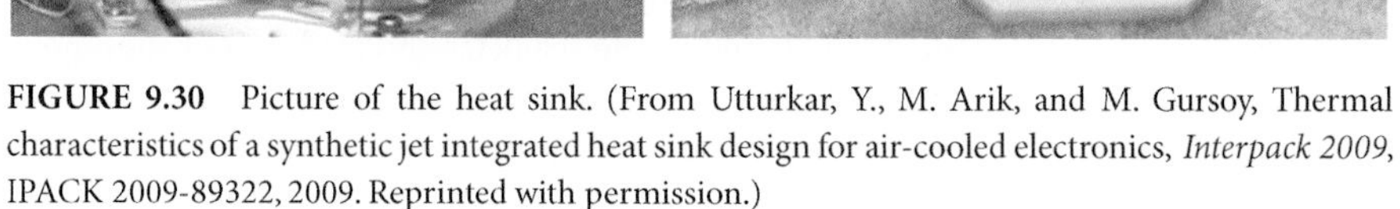

FIGURE 9.30 Picture of the heat sink. (From Utturkar, Y., M. Arik, and M. Gursoy, Thermal characteristics of a synthetic jet integrated heat sink design for air-cooled electronics, *Interpack 2009*, IPACK 2009-89322, 2009. Reprinted with permission.)

Heat loss measurements, using a guard heater, are conducted and results revealed that 8% of the total heat input was actually being lost through the insulation material. To minimize the heat loss, a secondary heater is then attached to the insulation material, as illustrated in Figure 9.31, to serve as a guard heater and create an adiabatic boundary condition for the backside of the heat sink. Two thermocouples are inserted into the insulation material,

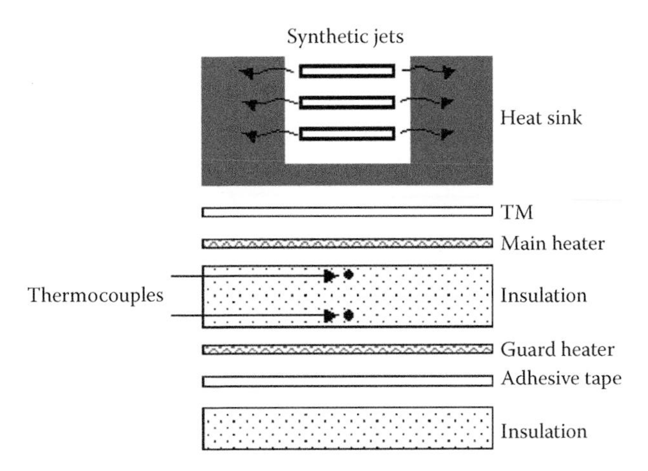

FIGURE 9.31 Heat sink and heater assembly. (From Utturkar, Y., M. Arik, and M. Gursoy, Thermal characteristics of a synthetic jet integrated heat sink design for air-cooled electronics, *Interpack 2009*, San Francisco, CA, IPACK 2009-89322, 2009. Reprinted with permission.)

separated by 10 mm to monitor the temperature gradient across the insulation material. This method is found to be very effective for minimizing the uncertainty in the heat load calculations. In addition, thermocouples are installed at the tips of five fins to measure the fin efficiencies and conduction losses in the heat sink. An uncertainty analysis on the final configuration is performed and found to be less than 4.2% for the overall thermal resistance results.

Synthetic jets are mounted onto the central cavity via plastic holders, which are attached to the insulating substrate. Special consideration is given to the design of these holders such that they would not affect the air jet exiting the jet orifices. Required driving conditions for the synthetic jets are controlled using a signal generator and an amplifier. Heat input and temperature data are recorded via Agilent data acquisition system.

For experimental investigation of multiple jets in the embedded heat sink concept, a series of experiments are performed on 4-port and 8-port synthetic jets.

For better comparison of two different jet designs, the total orifice sizes on both types of jets are kept identical. That is, the orifice size of 4-port jets is twice that of 8-port jets, where the orifices are evenly distributed around the circumference of the jets, as illustrated in Figure 9.32. Prior to the first use of the synthetic jets, natural frequencies and peak performance conditions of each jet are measured through characterization tests, which is performed at 60 V driving voltage.

The characterization tests revealed that natural frequencies of 4-port and 8-port jets are 600 Hz and 800 Hz, respectively. The thermal performance measurements summarized in this chapter are performed at natural frequencies of synthetic jets at 60 V condition.

As stated in Section 9.4, thermal performances of multiple jets embedded in the heat sink are evaluated in the experimental part of this study. The jets are evenly distributed and 10 mm apart from each other in the vertical direction, as depicted in Figure 9.33. Jet #1 is 5 mm above the heat sink base, whereas jet #5 is 5 mm below the fin tips.

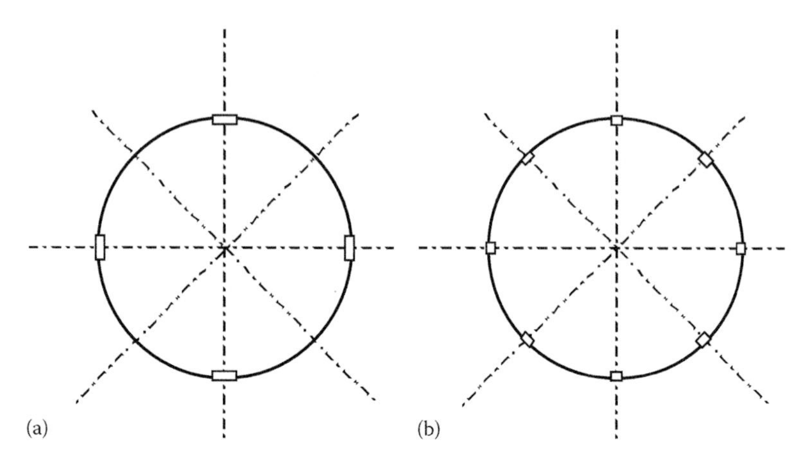

FIGURE 9.32 Layout of orifices: (a) 4-port jet with 3 mm orifices; (b) 8-port jet with 1.5 mm orifices. (From Utturkar, Y., M. Arik, and M. Gursoy, Thermal characteristics of a synthetic jet integrated heat sink design for air-cooled electronics, *Interpack 2009*, San Francisco, CA, IPACK 2009-89322, 2009. Reprinted with permission.)

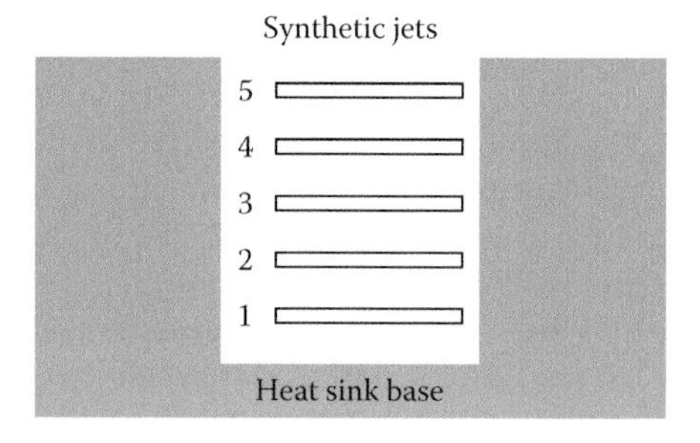

FIGURE 9.33 Layout of multi-jet stack. (From Utturkar, Y., M. Arik, and M. Gursoy, Thermal characteristics of a synthetic jet integrated heat sink design for air-cooled electronics, *Interpack 2009*, San Francisco, CA, IPACK 2009-89322, 2009. Reprinted with permission.)

First, thermal performances of 4- and 8-port jets with increased number of jets are compared at their natural frequencies, as shown in Figure 9.34. It is found that 8-port jets outperform 4-port jets by as high as 50%. This greater performance is attributed to the fact that the net effected fin surface area by the air jet created by the synthetic jets is larger for 8-port jets than that of 4-port jets. That is, there is more "wetted" fin area with 8-port jets than 4-port jets. It is further observed that as the number of synthetic jets increased from one to four, the resulting thermal resistance approaches asymptotic values. There is no additional performance gain by operating more jets. On the contrary, due to increased total power consumption by the jets, the COP drops significantly, as depicted in Figure 9.35. Although 8-port jets exhibit a higher COP, the decrease with the number of jets is found to be much more profound than 4-port jets.

$$\mathrm{COP} = \frac{Q_{\mathrm{SJ}} - Q_{\mathrm{freeconv}}}{P_{\mathrm{SJ}}} \tag{9.7}$$

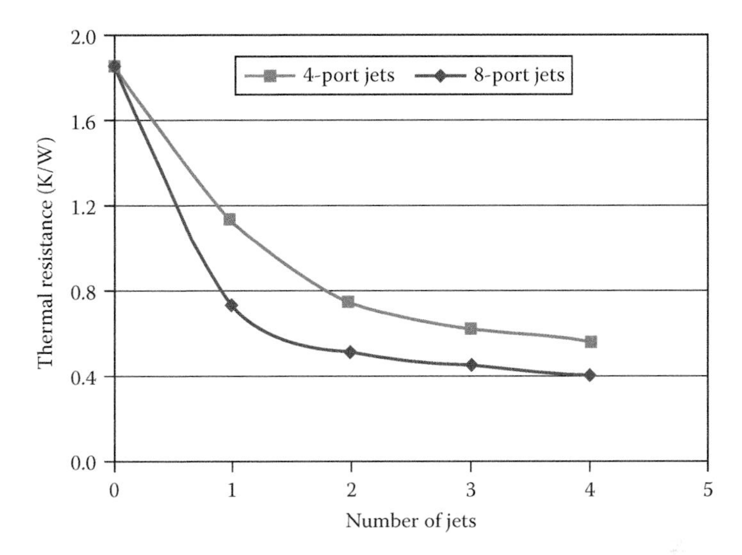

FIGURE 9.34 Thermal resistance comparison of 4- and 8-port jets. (From Utturkar, Y., M. Arik, and M. Gursoy, Thermal characteristics of a synthetic jet integrated heat sink design for air-cooled, electronics, *Interpack 2009*, San Francisco, CA, IPACK 2009-89322, 2009. Reprinted with permission.)

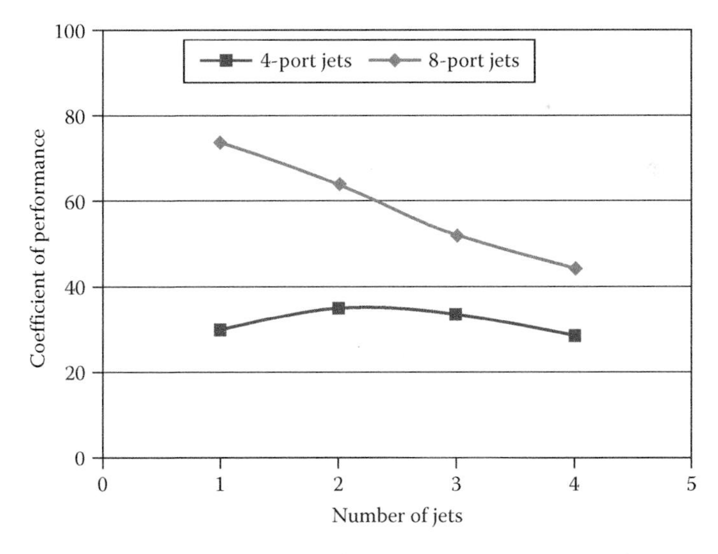

FIGURE 9.35 Coefficients of performance of 4- and 8-port jets. (From Utturkar, Y., M. Arik, and M. Gursoy, Thermal characteristics of a synthetic jet integrated heat sink design for air-cooled electronics, *Interpack 2009*, San Francisco, CA, IPACK 2009-89322, 2009. Reprinted with permission.)

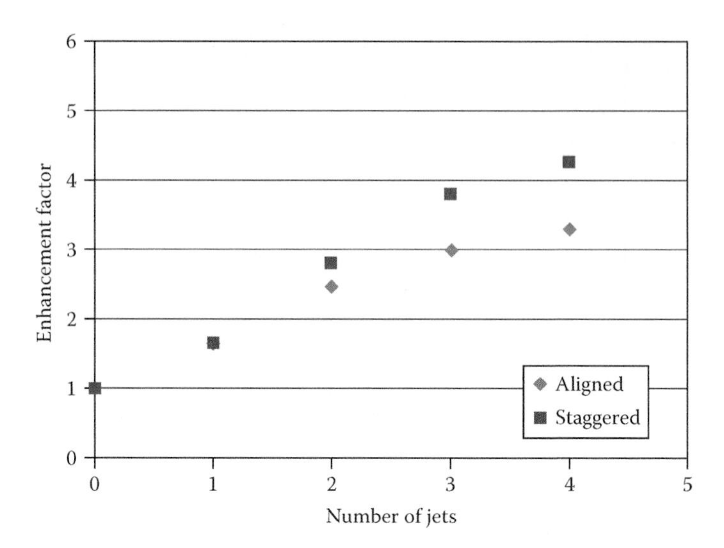

FIGURE 9.36 Enhancement factors measured with 4-port jets. (From Utturkar, Y., M. Arik, and M. Gursoy, Thermal characteristics of a synthetic jet integrated heat sink design for air-cooled electronics, *Interpack 2009*, San Francisco, CA, IPACK 2009-89322, 2009. Reprinted with permission.)

In Equation 9.7, Q represents the heat removed, and P denotes the power consumed by the jet.

The enhancements in the heat removal from the heat sink using synthetic jets are summarized in Figures 9.36 and 9.37 for 4-port and 8-port jets, respectively. The total cooling capability can be improved by a factor of 4.5 with respect to baseline, free convection heat sink. Furthermore, the effect of orientation of jet orifices on the EF is investigated. For this purpose, orifices of even-numbered jets are rotated by 45° for 4-port jets and 22.5° for 8-port jets. The motivation was simply maximizing the total wetted fin surface area. The results revealed that the staggered orientation of orifices increased the thermal performance by up to 30% on 4-port synthetic jets. However, the same effect could not be seen on 8-port jets. This suggests that the air stream from the eight orifices effectively distributes the flow across most of the fins such that staggered orifices make no further improvement in the thermal performance.

An experimental and computational study has been performed to justify the use of synthetic embedded heat sinks to meet the ever-rising cooling demands in the electronic industry. A baseline heat with plated fins has been compared to a heat sink embedded with two synthetic jets. The base of the heat sinks has been held at a constant temperature, and the power dissipated by the heat sink under varying operating conditions of the jet has been reported. The thermal resistance of the integrated heat sink is 0.8 C/W at 80°C base temperature, almost 2× better than the natural convection case. Furthermore, it is found that 8-port jets outperform 4-port jets by as high as 50% within the embedded heat sink. The staggering of orifices increased the thermal performance by up to 30% on 4-port synthetic jets. However, the same effect could not be seen on 8-port jets.

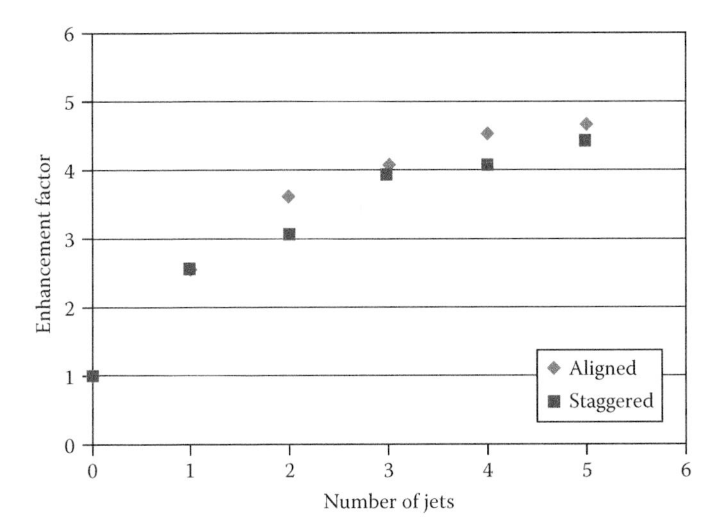

FIGURE 9.37 Enhancement factors measured with 8-port jets. (From Utturkar, Y., M. Arik, and M. Gursoy, Thermal characteristics of a synthetic jet integrated heat sink design for air-cooled electronics, *Interpack 2009*, San Francisco, CA, IPACK2009-89322, 2009. Reprinted with permission.)

9.5 Summary and Conclusions

Thermal management is currently one of the key limitations in the design of electronic systems. Parallel to the advancements in the electronics industry and increase in power dissipation, the development of effective, low-cost, compact heat removal solutions becomes extremely critical to ensure a fail-safe and reliable operation. While liquid cooling is poised to provide the cooling capability for next-generation electronics, its use in the present-day products is less prevalent due to the risks associated with condensation, leakage, and pumping power. Consequently, air-cooling strategies still continue to vie for near-term cooling needs in the electronic industry.

The need for new cooling technologies such as piezofans, ionic winds, and synthetic jets is obvious. This has led to many researchers and companies devoting their attention to new technologies. Synthetic jets, mesoscale cooling devices, have acquired noticeable attention and both academia and industry have devoted substantial efforts to these technologies. While synthetic jet technology is still progressing, we have summarized the state of the art for heat transfer applications. The conclusions drawn based on our findings are as follows:

1. The heat transfer application of synthetic jets was invented in early 1980s.
2. Various synthetic jet sizes, from tens of mm to hundreds of mm, are investigated and the results are presented.
3. Various actuation techniques are studied, and each actuation method has its own unique advantages and features.
4. Synthetic jets can enhance natural and forced convection heat transfer.
5. Driving voltage and frequency impact the performance of the jet, power consumption, and noise. A different design optimal point may exist for each application.

6. A number of computational models have been presented with especially 2D models, and there is a need for 3D models to better understand the flow scheme.
7. While there is some data about the reliability of the synthetic jets, further research is necessary before implementing in the actual applications.

Acknowledgments

This material is based upon the work supported by the Department of Energy National Energy Technology Laboratory under award number(s) DE-FC26-08NT01579.

The authors acknowledge the American Society of Mechanical Engineers (ASME) for granting them the permission to use the figures under their copyright for this chapter. They also thank the following people for their immense help in their synthetic jet research: Chuck Wolfe, Todd Wetzel, Tunc Icoz, Mustafa Gursoy, and Murat Ozmusul.

References

Arik, M. 2007. An investigation into feasibility of impingement heat transfer and acoustic abatement of meso scale synthetic jets. *Journal of Applied Thermal Engineering*, 27: 1483–1494.

Arik, M. and T. Icoz. 2012. Predicting heat transfer from unsteady synthetic jets. *Journal of Heat Transfer*, 134(8): 081901.

Coe, D. J., M. G. Allen, M. A. Trautman, and A. Glezer. 1994. Micromachined jet for manipulation of macro flow. *Proceedings of the Solid-State Sensor and Actuation Workshop*, Hilton Head Island, SC, June 13–16.

Crook, A., A. M. Sadri, and N. J. Wood. 1999. The development and implementation of synthetic jets for the control of separated flow. *Proceedings of the 37th Aerospace Science Meeting and Exhibits*, Reno, NV, June 11–14.

Erbas, N., M. Koklu, and O. Baysal. 2005. Synthetic jets for thermal management of microelectronic chips. *Proceedings of the ASME International Mechanical Engineering Congress and Exposition*, Orlando, FL, November 5–11.

Gallas, Q., R. Holman, R. Raju, R. Mittal, M. Sheplak, and L. Cattafesta. 2004. Low dimensional modeling of zero-net-mass-flux actuators. *The 2nd AIAA Flow Control Conference*, Portland, OR, June 28–July 1.

Garg, J., M. Arik, S. Weaver, T. Wetzel, and S. Saddoughi. 2005. Meso pulsating jet for electronics cooling. *Journal of Electronic Packaging*, 127(4): 503–551.

Gillespie, M. B., W. Z. Black, C. Rinehart, and A. Glezer. 2006. Local convective heat transfer from a constant heat flux flat plate cooled by synthetic air jets. *Journal of Heat Transfer*, 128: 990–1000.

Holman, R., Y. Utturkar, R. Mittal, B. Smith, and L. Cattafesta. 2005. Formation criterion for synthetic jets. *AIAA Journal*, 43(10): 2110–2116.

Issa, R. 1985. Solution of the implicitly discretized fluid flow equations by operator-splitting. *Journal of Computational Physics*, 62: 40–65.

Lee, C. Y. and D. B. Glodstein. 2001. DNS of micro jets for turbulent boundary layer control. *The 39th AIAA Aerospace Sciences Meeting and Exhibit, AIAA Paper* 2001-1013, Reno, NV, January 8–11.

Li, S. 2005. A numerical study of micro synthetic jet and its applications in thermal management. PhD thesis, Georgia Institute of Technology, Atlanta GA.

Mahalingam, R. and A. Glezer. 2005. Design and thermal characteristics of a synthetic jet ejector heat sink. *Transactions of the ASME*, 127: 172–177.

Mahalingam, R., N. Rumigny, and A. Glezer. 2004. Thermal management using synthetic jet ejectors. *IEEE Transactions on Components Packaging and Manufacturing Technology*, 27(3): 439–444.

Mittal, R. and P. Rampunggoon. 2002. On virtual aero-shaping effect of synthetic jets. *Physics of Fluids*, 14(4): 1533–1536.

Mittal, R., P. Rampunggoon, and H. S. Udaykumar. 2001. Interaction of a synthetic jet with a flat plate boundary layer. *The 31st AIAA Fluid Dynamics Conference and Exhibit*, Anaheim, CA, June 11–14.

Pavlova, A. and M. Amitay. 2006. Electronic cooling using synthetic jet impingement. *Journal of Heat Transfer*, 128(9): 897–907.

Raju, R., R. Mittal, Q. Gallas, and L. Cattafesta. 2005. Scaling of vorticity flux and entrance length effects in zero-net-mass-flux devices. *The 35th AIAA Fluid Dynamics Conference and Exhibit*, Toronto, Ontario, June 6–9.

Rathnasingham, R. and K. S. Breur. 1997. System identification and control of turbulent boundary layer. *Physics of Fluids*, 9(7): 1867–1869.

Smith, B. L. and A. Glezer. 2002. Jet vectoring using synthetic jets. *Journal of Fluid Mechanics*, 458: 1–34.

Smith, D., M. Amitay, V. Kibens, D. Parekh, and A. Glezer. 1998. Modification of lifting body aerodynamics using synthetic jet actuators. *The 36th AIAA Aerospace Sciences Meeting and Exhibit*, Reno, NV, January 12–15.

Timchenko, V., J. Reizes, and E. Leonardi. 2004. A numerical study of enhanced micro-channel cooling using a synthetic jet actuator. *The 15th Australian Fluid Mechanics Conference*, Sydney, Australia, December 13–17.

Utturkar, Y., M. Arik, and M. Gursoy. 2006. An experimental and computational sensitivity analysis of synthetic jet cooling performance. *Proceedings of the ASME International Mechanical Engineering Congress and Exposition*, Chicago, IL, November 5–10.

Utturkar, Y., M. Arik, and M. Gursoy. 2007. Assessment of cooling enhancement of synthetic jet in conjunction with forced convection. *Proceedings of the ASME International Mechanical Engineering Congress and Exposition*, Seattle, WA, November 11–15.

Utturkar, Y., M. Arik, and M. Gursoy. 2009. Thermal characteristics of a synthetic jet integrated heat sink design for air-cooled electronics. *InterPACK Conference Collocated with the ASME Summer Heat Transfer Conference*, San Francisco, CA, July 19–23.

Utturkar, Y., M. Arik, C. E. Seeley, and M. Gursoy. 2008. An experimental and computational heat transfer study of pulsating jets. *Journal of Heat Transfer*, 130(6): 062201.

Uturkar, Y., R. Mittal, P. Rampunggoon, and L. Cattafesta. 2003. Sensitivity of synthetic jets to the design of the jet cavity. *The 40th AIAA Aerospace Sciences Meeting and Exhibit*, Reno, NV, January 14–17.

Yassour, Y., J. Stricker, and M. Wolfshtein. 1986. Heat transfer from a small pulsating jet. *Proceedings of the 8th International Conference*, San Francisco, CA, August 17–22.

10

Application of Zero-Net Mass-Flux Actuators for Propulsion: Biology and Engineering

Mike Krieg
University of Florida

Kamran Mohseni
University of Florida

The jetting process used by jellyfish, squid, octopus, and other cephalopods to power locomotion is conceptually very similar to the mechanism used to create synthetic jets. These animals successively ingest and expel jets of water creating a net flow. Similar to synthetic jet actuators (SJAs) used to energize boundary layer flows in aeroshaping and flow control applications, the biological jetting mechanism adds both energy and momentum to the net flow, despite the zero-net mass-flux (ZNMF). Underwater thrusters inspired by this form of locomotion have gained attention recently because they have the capability to avoid a major limitation of the current underwater vehicle maneuvering technology.

Aside from gliders, typical autonomous underwater vehicles (AUVs) fall into one of the two categories. The first, which is termed the "torpedo" class vehicle, is characterized by a long slender streamlined body and control surfaces utilized for maneuvering forces.

The Woods Hole Oceanographic Institution (WHOI)'s Remote Environmental Monitoring UnitS (REMUS) is a good example of a torpedo class vehicle. This type of vehicle is very efficient at traveling long distances at high speeds. However, at low speeds the control surfaces provide little to no maneuvering forces and the vehicle cannot accurately control its trajectory. The use of ducted/gimbaled propellers provide some turning capabilities at low speeds, but the possible maneuver space is still limited to a narrow cone around the vehicle axis. The second category of vehicles is termed the "box" class. This type of vehicle is characterized by a bulky shape (low aspect ratio), with multiple thrusters positioned at several locations to provide the necessary control forces in any direction. These vehicles have much more accurate low-speed maneuvering, but have low top speeds and require significant energy to travel long distances. Remotely operated vehicles (ROVs) rely on a tether for power and control, which inherently limits mobility. For this reason, the majority of ROVs utilize a box design like WHOI's Jason or the Monterey Bay Aquarium Research Institute (MBARI)'s Ventana.

A hybrid class vehicle is desirable which has the long-range transit capabilities of the torpedo class and the maneuvering capabilities of the box class. This can be achieved by equipping a torpedo class vehicle body with a set of ZNMF thrusters inspired by squid and jellyfish within the vehicle hull. The submerged thrusters will have little to no protruding structure maintaining the streamlined shape, but provide maneuvering forces at low speeds, eliminating this setback in AUV design. This chapter describes the locomotion of jetting organisms and biomimicry intended to recreate this form of propulsion.

10.1 Cephalopod and Jellyfish Propulsion

Squid jet propulsion produces the fastest swimming velocities seen in aquatic invertebrates (O'Dor and Webber 1991; Anderson and Grosenbaugh 2005); but until recently, this method of underwater propulsion has been largely overlooked by the maritime community, and there is still much to be learned about the actual mechanics of generating thrust. While jetting is generally considered a less efficient form of locomotion than undulatory swimming (O'Dor and Webber 1991; Vogel 2003), squid morphology has evolved to fully exploit it. In fact propulsive efficiency was seen to rise as high as 78% in adult *Lactobacillus brevis* swimming at high velocities and averaged 87% (±6.5%) for paralarvae (Bartol et al. 2008), challenging the notion that a low-volume high-velocity jet inherently negates a high propulsive efficiency.

In general, locomotion begins when the squid inhales seawater through a set of vents or aperture behind the head, filling the mantle cavity. The mantle then contracts forcing seawater out through the funnel which rolls into a high-momentum vortex ring and imparts the necessary propulsive force (Anderson and Grosenbaugh 2005). Figure 10.1 shows the basic squid layout and anatomy used for jetting. Many members of the phylum Mollusca have evolved a specialized extension of the molluscan foot known as the siphon. The siphon is a flexible tube that allows fluid transport for various purposes including, but not limited to, locomotion, feeding, respiration, and even reproduction. In the Cephalopoda family, the siphon has evolved into the hyponome that is used primarily for generating propulsive jets (although the gills are also located near this region to enhance

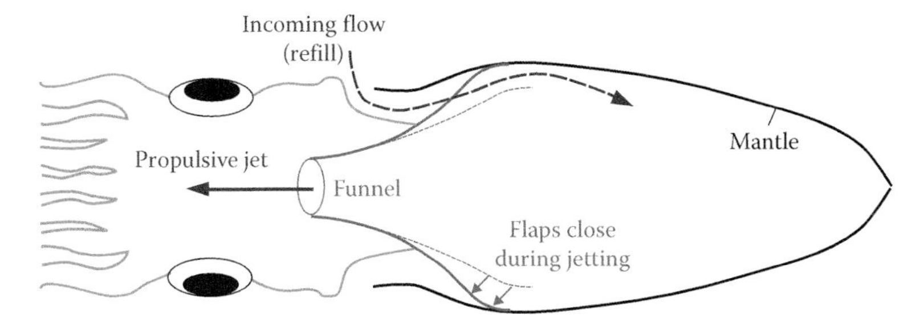

FIGURE 10.1 Diagram showing body layout of a squid focusing on the anatomy responsible for jetting locomotion.

respiration from the high fluid velocities). For most species, the hyponome is composed of a muscular tube (mantle) ending in a funnel, the exception being the nautilus, which has a rigid outer shell and the fluid motion is driven by an internal muscular flap mechanism rather than an outer mantle. Cephalopods have control over the shape and diameter of the funnel which can range from a straight cylindrical tube to a converging conical nozzle, and they are known to actively adjust this shape during jetting. The opening of this funnel can be generally thought of as an ellipse with a constant semimajor axis which runs horizontal to the ground (Anderson and DeMont 2000; Anderson and Grosenbaugh 2005). The semiminor axis can vary from being zero (completely closed funnel acts as valve) to equal to the semimajor axis at full extension. The funnel area was measured for adult squid in O'Dor (1988) and was observed to contract more than 30% during individual pulsations. However, it should be noted that the author admits a high lack of accuracy in these measurements due to the difficulty in imaging the funnel and that the predicted funnel diameters calculated from force calculations varied from the measured areas drastically.

The versatility of the squid locomotory system permits both low-speed steady swimming or cruising and fast impulsive escape jetting (Williamson 1965). During cruising, squid swim at nominal speed with a higher efficiency (Shevstov 1973), and two distinct gaits have been observed in steadily swimming squid as determined by the nature of the expelled jet (Bartol et al. 2009), those being above or below the jet formation number.* Escape jetting involves a hyperinflation of the mantle followed by a fast powerful contraction to impart significant acceleration (Gosline and DeMont 1985) at the cost of fluid dynamic losses, similar to the loss in efficiency seen in high-velocity jet locomotion of jellyfish (Sahin et al. 2009). Bartol et al. (2009) report cruising mode efficiency at 69% (±14%) averaged over several species and swimming speeds and 59% (±14%) for escape jetting.

The locomotion of jellyfish tends to be very similar to that of squid with some key differences, primarily that the refilling phase of the swimming jellyfish uses the same

* The formation number is defined as the formation time at which a jet reaches the circulation corresponding to the final resulting vortex ring (Gharib et al. 1998) and is described in great detail in Section 10.3.

bell opening as the jetting phase. However, despite the fact that squid do not use the funnel during refilling, the inlet vents are still on the anterior side of the mantle cavity (Figure 10.1), meaning that the locomotion for both organisms is quite different from the traditional continuous jet pumping mechanisms. Similar to the different gaits seen in squid locomotion, different species of jellyfish generally fall into two categories of swimmers based on the "quality" of vortex ring they produce. Oblate jellyfish have a very large bell opening, and the jetting is similar to a paddling motion; exact bell contractions for several species can be found in the work by Ford and Costello (2000). Prolate jellyfish have smaller bell openings with nozzle-like flaps at the velar cavity opening, have a much more distinct jetting motion (Ford and Costello 2000), and tend to have faster maximum velocities and faster acceleration (Colin and Costello 2002).

Jellyfish morphology during swimming has been digitally captured from experiment, and the body motions imported into numerical simulations to predict body forces on the swimming jellyfish, determining drastically different swimming efficiencies. Froude propulsive efficiency of jellyfish was calculated through this process by Sahin and Mohseni (2008, 2009) and Sahin et al. (2009) to be 37% for the oblate jellyfish *Aequorea victoria* and 17% for prolate *Sarsia tubulosa*. It should be noted that both species of jellyfish most likely do not use vortex generation for the sole purpose of locomotion. *Aequorea victoria* uses vortex generation for feeding as demonstrated through Lagrangian coherent structures (LCS) analysis (Lipinski and Mohseni 2009a, 2009b; Wilson et al. 2009), and *S. tubulosa* uses jetting as an escape mechanism, where survival supersedes the desire for efficient propulsion. Empirical data gathered by Dabiri et al. (2010) through digital particle image velocimetry (DPIV) measurements of several species show similar efficiency characteristics for the different swimming patterns.

Figure 10.2a shows stable and unstable LCS around a swimming jellyfish taken from simulations in Sahin et al. (2009) and Lipinski and Mohseni (2009b). At the instant in the jetting cycle captured in this figure, the velar opening is a converging conical nozzle, which creates the converging radial velocity in the expelled jet shown in Figure 10.2b. The squid funnel changes shape during pulsation and at times resembles a converging conical nozzle, and evidence of a converging radial velocity can be detected very close to the exit of the funnel in Figure 1.8 of Anderson and Grosenbaugh (2005), though this was not discussed in that study. This chapter will discuss the effect of these swimming behaviors, including the nonparallel jet flow, with respect to propulsive output.

10.2 Zero-Net Mass-Flux Thruster Design

A design for a ZNMF underwater thruster that was originally proposed by Mohseni (2004, 2006) and has been used extensively for underwater maneuvering (Krieg and Mohseni 2010; Krieg et al. 2011) is depicted schematically in Figure 10.3a, and a simple mock-up of the thruster constructed out of transparent plastic is shown in Figure 10.3b giving visual access to the different components. The device consists of a fluid cavity with an oscillating driving mechanism at one end that moves fluid in or out of the cavity and is open to the external fluid at the other end. The entire cavity is contained within the hull of the vehicle, so that the open end is flush with the vehicle surface. There are

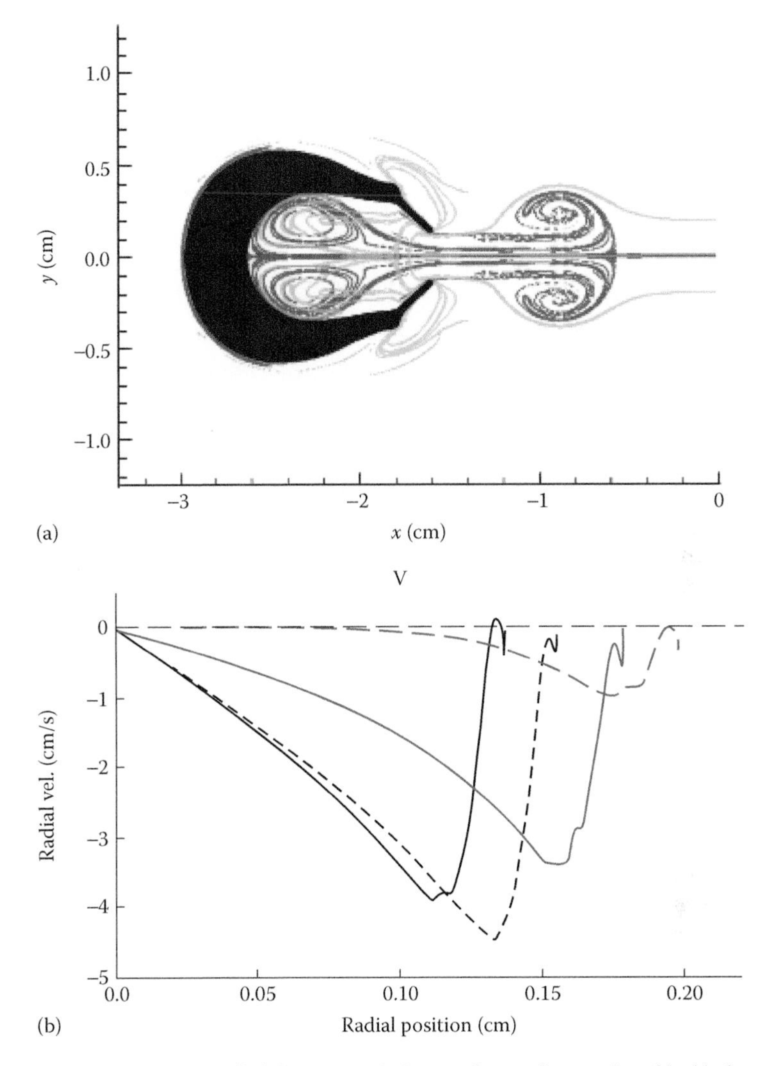

FIGURE 10.2 A swimming jellyfish, *Sarsia tubulosa*, is shown along with stable (dark gray) and unstable (light gray) material transport barriers around the body for a given instant during the jetting phase (a). Local radial velocity at the velar opening is plotted for several instances in (b). (Data reproduced from Sahin, M. et al., *J. Exp. Biol.*, 212, 2656–2667, 2009; Lipinski, D. and K. Mohseni, *J. Exp. Biol.*, 212, 2436–2447, 2009.)

several options for nozzles that can be attached to the cavity opening, but the two basic options are summarized in Figure 10.3a. An orifice nozzle consists of a flat plate with a central circular opening and a tube nozzle consists of a long cylindrical tube, generally with a very sharp taper angle at the outlet (γ), as shown in Figure 10.3a. Both the nozzle configurations create axisymmetric jets which aids in modeling of the thruster output, as is described in Section 10.3. Some of the fluid in the cavity must converge to the

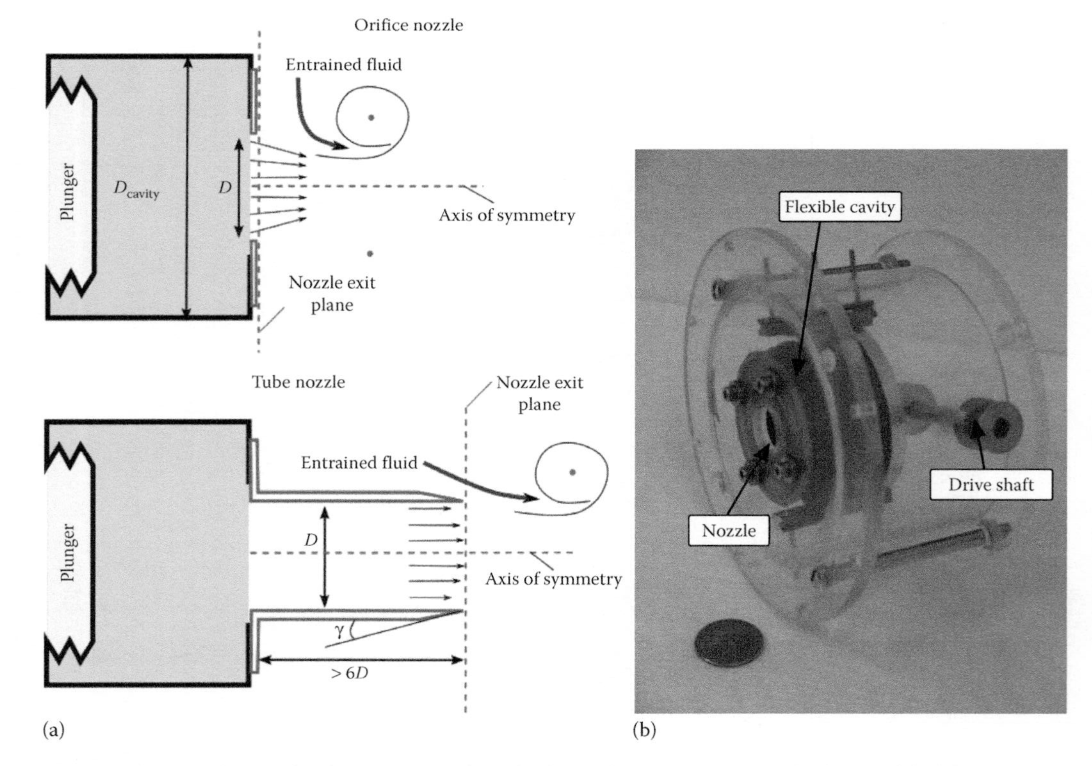

FIGURE 10.3 (a) Schematic of ZNMF thruster (Mohseni 2006) with multiple nozzle configurations and (b) a simplified thruster constructed in clear plastic for easy viewing. (From Mohseni, K., *Ocean Eng*., 33, 2209–2223, 2006.) The different nozzle configurations are tested in Krieg and Mohseni (2013) and the clear casing thruster is based on a design tested in Krieg and Mohseni (2010). (From Krieg, M. and K. Mohseni, *J. Fluid Mech*., 719, 488–526, 2013; Krieg, M. and K. Mohseni, *IEEE Trans. Robot*., 26, 542–554, 2010.)

centerline before passing through the orifice nozzle and the converging streamlines persist downstream beyond the nozzle exit plane. By contrast tube nozzles are generally designed to be sufficiently long so that the flow becomes parallel before exiting the nozzle. From a vehicle maneuvering standpoint, tube nozzles are undesirable since they protrude from the vehicle body inducing significant drag. Fortunately, as is discussed in Section 10.3, the converging radial velocity created by the orifice nozzles actually enhances the total impulse of the propulsive jet, making orifice nozzles an excellent option for maneuvering thrusters on underwater vehicles.

SJAs used for flow control in air are most often driven by piezoelectric actuation that provides a very high actuation frequency while the deflection of the piezodiaphragm is somewhat limited. In addition, acoustics plays a key role in SJAs used in air as described in Chapter 2, whereas actuators using water as a fluid medium are less affected by acoustics. Though ZNMF devices used to generate underwater propulsion utilize similar fluid principles, creating synthetic jets by asymmetrically oscillating fluid through a slit or orifice, the working medium is quite different, often requiring an alternative driving mechanism. One major difference comes from the fact that the load on the underwater devices is much higher (higher density fluid), which results in reduced deflection under increased loading when using solenoid actuators (Krieg et al. 2005). For propulsive devices, the deflection of the actuation method is critical because it is directly related to the total momentum transfer (larger jets carry more momentum). As a result, ZNMF devices used for underwater propulsion examined here are actuated by a mechanical system that guarantees the deflection and jet volume of each pulsation. Additionally, the pulsation frequency of these devices is in the orders of magnitude lower than that in air and limited by the introduction of cavitation inside the thruster at higher frequencies (Krieg and Mohseni 2008). The frequency limitation of these propulsors also lends well to mechanical actuation methods.

In Section 10.3, we will first examine general tendencies of axisymmetric starting jets, specifically identifying the effect that pulsation kinematics has on the final jet. These effects are then related to the thrust output of the device described in this section while operating in a ZNMF sense.

10.3 Starting Jet Dynamics

This section will focus on jets ejected through a circular opening, so that they can be considered axially symmetric. The position in this space will be defined by the cylindrical coordinates r, ϕ, and x. The velocity vector in the cylindrical coordinate system is $[u, v, w]^T$, but it is also assumed that there is no swirl ($w = 0$). This does not eliminate the possibility that propulsive devices can be made which utilize slit-type openings, but an in-depth discussion is unwarranted because of the lack of devices using such a configuration.

The study of short duration starting jets is nearly synonymous with the study of vortex ring formation, since starting jets ejected into a fluid reservoir roll into either a single vortex ring or a vortex ring with a trailing wake depending on the jet driving parameters (Gharib et al. 1998). When a jet of fluid is ejected into a similarly dense resting fluid, it forms a free shear tube/layer. The shear tube is unstable and spirals up at the free end (Pullin 1979; Pullin and Phillips 1981) forming a vortex ring at the leading edge of the starting jet.

The leading vortex ring continues to grow feeding off of the trailing shear layer until it reaches a critical "saturated" state driving the shear layer toward the centerline, and vorticity cancellation at the axis of symmetry causes the shear layer in this region to break, separating the primary vortex ring from the trailing shear layer (Gharib et al. 1998; Mohseni et al. 2001). The free end of the trailing shear layer rolls into a secondary vortex ring, and the primary vortex ring settles upon a stable arrangement. The primary vortex ring quickly travels downstream out of range of the influence of the secondary slower moving vortex ring, and the evolution of the ring becomes only dependent on viscosity (refer to the work of Maxworthy 1972, 1977). Gharib et al. (1998) defined a nondimensional timescale that is closely related to the vortex formation process. The timescale, called *formation time*, is the amount of time at which the pulsation is actively scaled by the piston velocity (u_P) and nozzle diameter (D).

$$t^{\star} = \frac{\int_0^t u_P(\tau)\,d\tau}{D} \tag{10.1}$$

The piston velocity is the jet volume flux divided by the nozzle area, in reference to cylinder piston vortex generators (which have historically been used to generate this kind of jet flow). The *formation number* is defined as the formation time at which the jet first reaches the circulation of the final vortex ring, and it was observed that the jets ejected with a wide variety of piston velocity programs all had a "universal" formation number falling between 3.6 and 4.2 (Gharib et al. 1998). Mohseni and Gharib (1998) showed that the formation number of vortex rings generated from uniform parallel starting flow should be close to 4 by equating the resulting bulk flow quantities to a known family of "standard" vortex rings. This means, in a loose sense, that jets ejected with a stroke ratio (L/D) below the formation number will form a single vortex ring, and jets with stroke ratio above the formation number will form a leading vortex ring with a trailing wake. This is represented graphically in Figure 10.4, which is reproduced from Gharib et al. (1998). Figure 10.4a shows a jet created with an impulsive velocity program and a stroke ratio of $L/D = 2$, and the entire jet has rolled into the leading vortex ring; a jet with $L/D = 3.8$ (very close to the formation number) is shown in Figure 10.4b where most of the jet is contained within the leading ring, but a slender trailing wake has started to form. The large stroke ratio jet ($L/D = 14.5$) shown in Figure 10.4c has a substantial trailing jet with multiple trailing vortex rings.

Though the universal formation number suggests a critical state of the vortex ring, it has been observed by multiple investigators (Gharib et al. 1998; Mohseni and Gharib 1998; Mohseni et al. 2001) that the final state of these vortex rings does not actually correspond to the hypothetical maximum ring "thickness," which is Hill's spherical vortex (Hill et al. 1894). As the vortex grows, its induced velocity also grows, and when this velocity becomes large enough, it destabilizes the feeding shear layer and separates from the starting flow. Figure 10.5, which is reproduced from Mohseni et al. (2001), shows the evolution of a vortex ring generated through simulation. This particular run was created with long-duration forcing, which is the equivalent of a large stroke ratio starting jet, so pinch-off is readily observable. The dimensionless vorticity contours as well as the axial velocity and the vorticity profiles extending through the toroidal plane are shown for the vortex ring at several representative downstream locations. It can be seen that while the

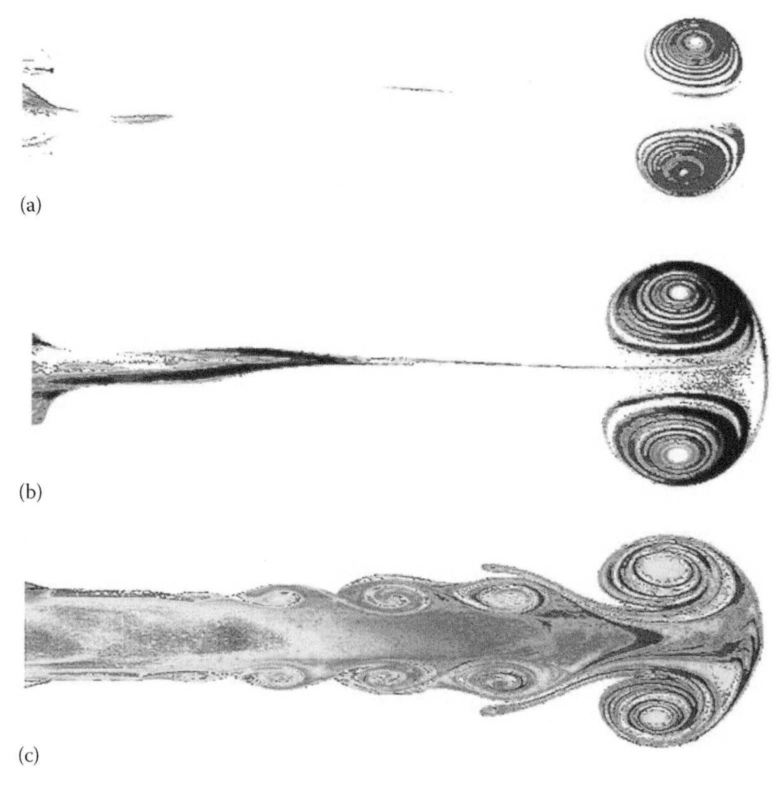

FIGURE 10.4 Visualization of vortex rings at $X/D \approx 9$ for (a) $L/D = 2, Re \approx \Gamma/\nu \approx 2800$; (b) $L/D = 3.8, Re \approx 6000$; and (c) $L/D = 14.5$ (picture taken at $t^\star = 8$). All three cases were generated by an impulsive piston velocity. (From Gharib, M., E. Rambod, and K. Shariff, *J. Fluid Mech.*, 360, 121–140, 1998; Figure 1.3.)

vortex ring is still growing, represented by the location $x \approx 6$ in Figure 10.5, the vorticity profile has two distinct peaks corresponding to the radial locations of the centroid of the growing vortex ring and the feeding shear layer. At the pinch-off location ($x \approx 12.5$), the feeding shear layer is driven to the axis of symmetry where vorticity cancellation provides a physical mechanism for separation. It was further hypothesized by Mohseni et al. (2001) that this process could be delayed by accelerating the shear layer or expanding the shear layer diameter during pulsation to account for the vortex growth. Subsequent simulations using these driving conditions showed that a final vortex ring could be created with increased dimensionless circulation and decreased dimensionless energy corresponding to a "thicker" vortex ring closer to Hills spherical vortex.

Starting jets, as well as the vortex rings they produce, are often modeled in terms of bulk flow quantities such as circulation, hydrodynamic impulse, and kinetic energy. For jet flows with large enough Reynolds number (Re), the effects of viscous dissipation become negligible, so that these bulk quantities can be considered invariants of steady flow describing a unique stable configuration and can be determined from flux terms

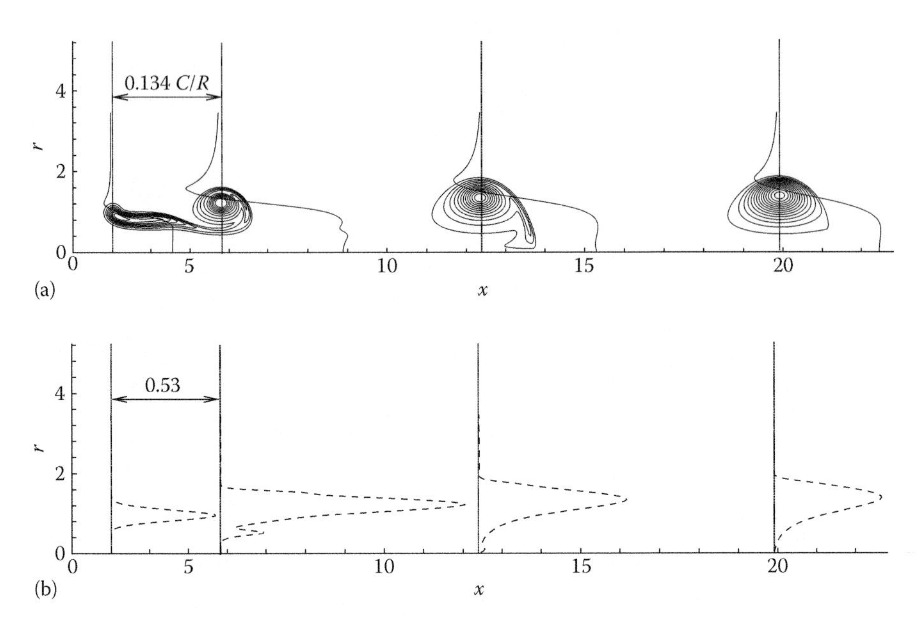

FIGURE 10.5 The evolution of a vortex ring created from a large stroke ratio jet is shown at multiple characteristic downstream locations. (Reproduced from Mohseni, K. et al., *J. Fluid Mech.*, 430, 267–282, 2001. With permission.) The nondimensional vorticity contours and axial velocity profiles are depicted in (a) and the vorticity profiles are shown in (b).

along the boundary of the fluid domain. It has been observed in several studies that the circulation of a starting jet ejected through an orifice nozzle is substantially greater than the circulation of a starting jet ejected through a tubular nozzle with identical jet velocity and nozzle diameter (Didden et al. 1979; Krueger and Gharib 2005; Rosenfeld et al. 2009). This might be counterintuitive since there is more boundary layer development on the surface of the tube nozzle. Rosenfeld et al. (2009) attributed the added circulation to the existence of the gradient $\partial v/\partial x$ at the nozzle exit plane, which increases the vorticity flux across that surface. This gradient is nonzero for jets ejected through an orifice nozzle because of the enduring converging streamlines downstream. However, by conservation of mass in an axisymmetric jet, the converging radial velocity at the nozzle must decrease in the downstream direction with a coupled axial velocity acceleration.

Krueger observed an increase in both circulation and impulse of starting jets over the one-dimensional (1D) slug model, when the jet stroke length was at or below the formation number, for both tubular (Krueger and Gharib 2005) and orifice nozzles (Krueger 2008). This increase was attributed to a phenomena called "overpressure," where the interaction of the jet flow with the formation of the primary vortex ring induces a pressure at the nozzle exit plane above the ambient pressure. Using a potential flow model, Krueger and Gharib (2005) identified a correlation between overpressure at the nozzle and the gradient $\partial v/\partial x$ and attributed the increased impulse to the acceleration coupled with that gradient.

Krieg and Mohseni performed a control volume analysis on nonparallel starting jet flows, meaning that the jet flow entering the control volume is not restricted to flows with no radial velocity profiles [$v(r)$]. This analysis assumed a perfectly inviscid, axisymmetric volume of fluid, starting from rest with no swirl, so that the total rate of change of circulation (Γ), impulse (I), and kinetic energy (E) of the control volume can be calculated according to surface integrals at the entrance boundary.

It was shown by Lamb (1945) (and in vector form by Saffman 1992) that in an unbounded fluid with confined vorticity, the total nonconservative body forces acting on the fluid are equal to the rate of change of a quantity known as the hydrodynamic impulse, which is defined as

$$\vec{I} \equiv \frac{1}{2} \int_{CV} \vec{x} \times \vec{\omega} \, \mathrm{d}\vec{x} \tag{10.2}$$

Therefore, the quantity which is directly related to the propulsive output of the jet is the hydrodynamic impulse. It should be noted that the total momentum of the control volume ($\vec{H} = \int \vec{u} \mathrm{d}\vec{x}$) is not in general equal to the hydrodynamic impulse. Classical analysis by Cantwell (1986) has shown that in a spherical control volume, extended to infinity, with vorticity localized to a confined central region, the relationship is $H = (2/3)I$. The complete derivation of circulation, hydrodynamic impulse, and kinetic energy of a generic starting jet is given by Krieg and Mohseni (2013), which will be briefly summarized here. Following a method similar to that used in Saffman (1992), the rate of change of hydrodynamic impulse can be found by calculating the time derivative of the impulse definition (Equation 10.2) and inserting the vorticity transport equation to define the dynamics, $\partial\omega/\partial t$, where ω is the local vorticity.

$$\frac{\mathrm{d}I}{\mathrm{d}t} = \pi \int_0^\infty \left(2u^2 r + u \frac{\mathrm{d}v}{\mathrm{d}x} r^2 - v^2 r \right) \mathrm{d}r \tag{10.3}$$

The rate of circulation added to the system is just the flux of vorticity across the nozzle exit plane since circulation is an invariant of motion for inviscid fluids (Didden 1979). As was noted by Rosenfeld et al. (1998), this flux can be separated into a term that is only dependent on the centerline velocity (u_0) and a simple surface integral.

$$\frac{\mathrm{d}\Gamma}{\mathrm{d}t} = \frac{1}{2} u_0^2 + \int_0^\infty u \frac{\mathrm{d}v}{\mathrm{d}x} \mathrm{d}r \tag{10.4}$$

Multiple classic texts (Lamb 1945; Landau and Lifshitz 1959) have derived the rate of change of kinetic energy for an incompressible inviscid control volume that reduces to the sum of the flux of kinetic energy and pressure work across all the domain boundaries.

$$\frac{\mathrm{d}E}{\mathrm{d}t} = \pi \int_0^\infty \left(u^2 + v^2 + 2P \right) ur\mathrm{d}r \tag{10.5}$$

In Equations 10.3 through 10.5, the surface integrals are all evaluated at the nozzle exit plane. To calculate the rate of change of kinetic energy of the volume, the pressure profile must also be known along the nozzle exit plane, which determines the pressure work done on this boundary. Krieg and Mohseni (2013) took advantage of the axial symmetry of the problem and utilized potential flow analysis along the centerline to recover the pressure at the center of the nozzle exit plane in terms of the farfield stagnation pressure (P_∞) and the line integral of velocity along the centerline. By the definition of circulation, the line integral can be related to the total circulation and a line integral of velocity along the nozzle exit plane (which is presumably the known boundary condition). Finally, the total pressure distribution was then recovered by integrating the momentum equation in the radial direction.

$$P(r) = P_\infty - \frac{1}{2}v(r)^2 + \int_r^{R_\infty} u\frac{\partial v}{\partial x} + \frac{\partial v}{\partial t}\,\mathrm{d}r \tag{10.6}$$

It can readily be seen from the above equation that in the absence of any radial velocity, or radial velocity gradient, the pressure along the nozzle exit plane is the farfield stagnation pressure, meaning that overpressure is impossible without any radial velocity content (or vice versa, depending on how you approach the problem). The overpressure observed by Krueger for low stroke ratio jets ejected through tube nozzles is due to the fact that at the onset of pulsation, the free shear layer rolls into a vortex ring very close to the nozzle exit, and the close proximity of the forming vortex ring induces a discernible converging radial velocity on the jet flow entering the domain, which was observed experimentally for parallel jet flows by Krieg and Mohseni (2013). As the shear layer continues to feed the vortex ring, it grows in strength but is still bounded to the shear layer and remains fairly close to the nozzle exit plane, and the stronger vortex ring induces a greater radial velocity on the emanating jet flow. As the jet stroke length surpasses the formation number, the primary vortex ring "pinches-off" of the trailing shear flow and advects downstream. As the vortex ring translates downstream, its influence at the nozzle exit plane diminishes and the total effect of induced converging velocity becomes negligible over the entire pulsation for very large stroke ratios. This explains why Krueger observed that the relative effect of overpressure was maximized for jets with stroke ratio near the formation number and diminished significantly for large stroke ratios.

The exact velocity profiles of experimentally generated jet flows are heavily dependent on the geometry of the nozzle. The interested reader should refer to Krieg and Mohseni (2013) for a parameterization of velocity profiles for both tube and orifice nozzles. However, by far the simplest and most common approximation for velocity profiles of starting jet flows is the "1D slug model" approximation, which assumes that the fluid is ejected with a uniform axial velocity profile [$u(r) = u_P$] and no radial velocity profile [$v(r) = 0$]. Under these approximations, Equations 10.3 through 10.5 can be drastically simplified as

$$\frac{\mathrm{d}\Gamma_{1D}}{\mathrm{d}t} = \frac{1}{2}u_P^2, \qquad \frac{\mathrm{d}I_{1D}}{\mathrm{d}t} = \pi u_P^2 R^2, \qquad \frac{\mathrm{d}E_{1D}}{\mathrm{d}t} = \frac{\pi}{2}u_P^3 R^2 \tag{10.7}$$

The 1D slug model was used to predict total jet circulation in Glezer (1988), Shariff and Leonard (1992), and Gharib et al. (1998) and was used to predict circulation, impulse, and energy of the jet in Mohseni and Gharib (1998).

The circulation, impulse, and energy of a starting jet ejected through an orifice nozzle with a stroke ratio of $L/D = 6.9$, as measured in Krieg and Mohseni (2013), are reproduced

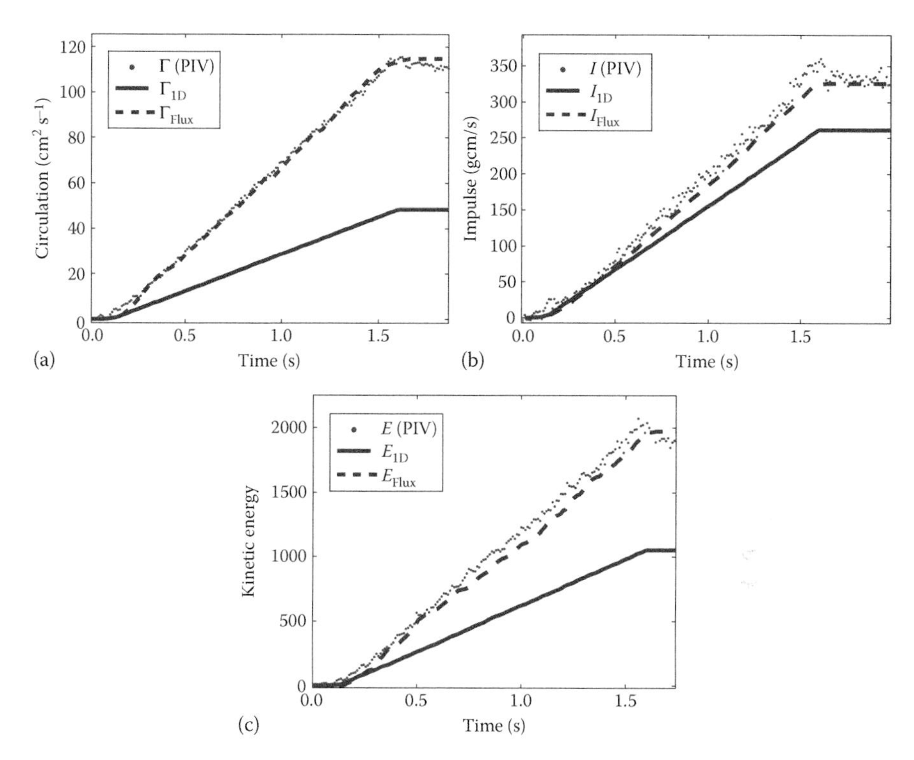

FIGURE 10.6 (a) Evolution of circulation, (b) hydrodynamic impulse, and (c) kinetic energy of a converging starting jet throughout pulsation. The jet was created with a 0.93-cm radius orifice nozzle and a piston velocity of $u_P \approx 7$ cm/s. In each of these figures, the subscript "1D" refers to quantities calculated from the 1D slug model (Equation 10.7) and the subscript "Flux" refers to quantities calculated from the exact flux (Equations 10.3 through 10.5). (Data reproduced from Krieg, M. and K. Mohseni, *J. Fluid Mech.*, 719, 488–526, 2013.)

in Figure 10.6. This figure not only shows the substantial effect that a converging radial velocity has on these quantities but also shows the validity of using Equations 10.3 through 10.6 to predict the bulk quantities from known velocity profiles at the jet source, as opposed to the 1D slug model (Equation 10.7).

Section 10.3 discussed starting jet dynamics as they depend on driving parameters assuming that the jet is ejected into a resting fluid environment. Invariably if this technology is used for vehicle maneuvering, then the thruster will be required to operate continuously. A thruster operating continuously (high-frequency operation) does not eject a jet of fluid into a resting reservoir, but rather a domain where previous jets have already been established, so the propulsive jet will be influenced by the previous actuation cycles. In addition, the fluid that is ejected to create the propulsive jet is also influenced by the refill dynamics. Though both of these features are difficult to model in a general sense, an empirical characterization of the thruster output while operating continuously is provided in Section 10.4.1.

10.4 High-Frequency Operation

This section describes the average thrust of a prototype thruster pulsating at high frequencies and shows that the average thrust is closely related to vortex ring pinch-off dynamics. The average thrust is defined $\overline{T} = If$, where I is the total impulse of each pulsation, including refilling, and f is the frequency of actuation.

In between pulsations, the prototype thruster must refill the internal cavity in preparation for the next jet (hence ZNMF). The prototype thruster discussed here ingests fluid through the same aperture through which the jet is expelled. Alternatively, the refilling could be supplied from a separate vent, like squid jetting, but this option requires complicated valving mechanisms and introduces additional structure to vehicle frameworks, both of which should be avoided if at all possible. A special type of orifice nozzle has been investigated that works like a standard orifice nozzle when closed, but flaps open to allow greater nozzle area during refilling and is shown in Figure 10.7. However, initial testing showed that the passive opening and closing mechanism of this nozzle was insufficient to guarantee proper configuration during jetting, and an active control mechanism was not pursued due to complexity. Additionally, jellyfish (which loosely inspired this technology) refill the velar cavity through the same opening, and although squid refill through separate vents behind the head, they are still on the anterior side of the animal (Figure 10.1). Therefore, losses associated with this type of refilling mechanism might be minimized for certain conditions.

Figure 10.8 shows two control volumes that help to describe the total jetting cycle and thrust output. The first control volume, surrounded by the boundary δ_{CV1}, includes all fluid downstream of the thruster nozzle plane. The rate at which the total circulation, hydrodynamic impulse, and kinetic energy of this control volume are increased is given in terms of the jet kinematics in Equations 10.3 through 10.5, and there are no nonconservative external forces being applied. The second control volume includes all the fluid surrounding the vehicle. The boundary of this control volume, δ_{CV2} in Figure 10.8, extends around the vehicle surface and into the thruster cavity so that there is no flux of fluid across this boundary. The net external force applied to this volume, which is the jetting force, is

FIGURE 10.7 An orifice nozzle mechanism that allows flaps to passively open and allow fluid to enter through entire nozzle area during refilling and close again during jetting. Nozzle shown in closed position in (a) and in open position in (b).

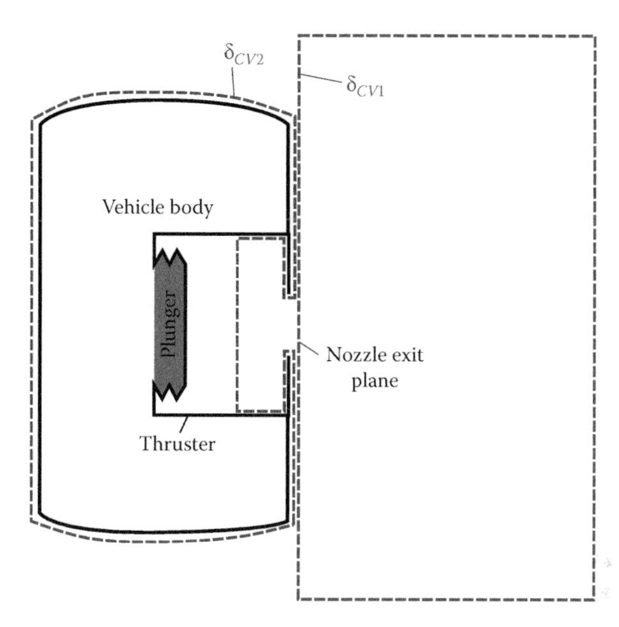

FIGURE 10.8 Two representative control volumes that help to analyze thruster output.

equal to the rate of change of that volume's hydrodynamic impulse (Lamb 1945; Saffman 1992). During the jetting phase, the rate of impulse created in $CV2$ is equal to the rate of impulse advected into $CV1$, and the jetting thrust can be determined from Equation 10.3. There are forces associated with accelerating the fluid inside the cavity before it crosses the nozzle exit plane, but these forces are similar to the forces required to create the plunger acceleration. These internal thruster forces are inherently cyclical, meaning that the force required to accelerate the plunger/fluid are both equal and opposite to the force required to decelerate the plunger/fluid at the termination of pulsation. Therefore, the fluid is not considered part of the jet until it crosses the nozzle plane. This is why the boundary δ_{CV2}, drawn in Figure 10.8, excludes some of the fluid that never leaves the thruster cavity.

During the refilling phase, the impulse transferred to the internal cavity fluid cannot be modeled by the starting jet Equation 10.3, because the cavity violates the assumption that the only unbounded surface is the nozzle plane, and the interaction of the incoming fluid and the internal cavity walls is significant. Due to the wall interactions, the fluid drawn into the cavity must come to rest after the refilling is terminated, and if the impulse of the jet (I_{CV1}) is unaffected, then there is no net change in impulse during refilling. Again there are forces associated with accelerating the fluid to bring it into the cavity which are negated by the forces bringing the fluid to rest inside the cavity, so as a first-order approximation the net impulse of jet refilling is zero.

10.4.1 Average Thrust Characteristics

Krieg and Mohseni (2008) characterized the thrust output of a prototype thruster operating at high frequencies with a sinusoidal jet velocity program. The prototype thruster was equipped with a set of orifice nozzles that varied in diameter allowing jets to be created with stroke ratios ranging from $L/D = 2$ to $L/D = 14$, independent of actuation frequency. This study directly measured the thrust over the range of stroke ratios, and a frequency range from approximately 3 to 20 Hz, limited at the upper end by the formation of cavitation bubbles within the cavity.

In order to show the relative effect of the actuation frequency and stroke length on thruster output, the average thrust at each test case ($\overline{T}$) is normalized by the average thrust predicted by the 1D slug model ($\overline{T}_{1D}$); therefore, $\sigma = \overline{T}/\overline{T}_{1D}$. The prototype thruster uses a sinusoidal piston velocity program, which reduces the loads on the mechanical driving system at high actuation frequencies. Integrating the impulse according to the 1D slug model (Equation 10.7) with a sinusoidal program yields,

$$\overline{T}_{1D} = \rho \frac{\pi^3}{16} D^4 \left(\frac{L}{D}\right)^2 f^2 \tag{10.8}$$

Therefore, as was observed in Mohseni (2006), the 1D slug model predicts that the average thrust output is proportional to both the square of the stroke ratio and the square of the actuation frequency, further indicating that high pulsation frequencies will be required to provide usable control forces.

The thrust response of the actuator operating with stroke ratio below the formation number is shown in Figure 10.9a. In the low-frequency regime, the average thrust is higher than that predicted by the 1D slug model due to the converging radial velocity at the nozzle exit plane (as discussed in Section 10.3), and there is no net momentum transfer during the refill phase. However, as the frequency increases the total thrust settles on the value predicted by the 1D slug model, meaning that in this frequency range the increased jet impulse due to converging radial velocity during expulsion, which is also referred to as the impulse due to overpressure, is negated by a loss of impulse encountered during refilling, which is on the same order of magnitude. As a result, the average thrust output in this frequency range is reasonably well predicted by the 1D slug model.

When the thruster is operating above the formation number (Figure 10.9b), the low-frequency ranges again exhibit an enhanced impulse due to the radial velocity effects, but the high-frequency range exhibits a relative loss in thrust with respect to the 1D slug model prediction. This loss increases monotonically with both actuation frequency and stroke ratio and suggests that the assumptions made concerning control volume impulse during refilling are no longer valid at increased frequencies. The net change in hydrodynamic impulse of the control volume $CV2$ (see Figure 10.8) during refilling is equal to zero if it is assumed that the jet is unaffected by the intake, which is valid for frequencies below ≈ 3 Hz. As the pulsation frequency increases, the refilling phase begins to decrease the jet impulse by taking in some of the trailing wake and bringing it to rest inside the cavity, and large stroke ratio jets are observed to be disproportionately affected. When a jet is ejected with a stroke ratio above the formation number, a substantial amount of the shear flow is left behind in the trailing wake of the leading vortex ring. The trailing wake has a lower momentum than the leading vortex ring and travels at a much lower velocity, but still has a momentum substantially larger than the surrounding resting fluid. The lower velocity of the trailing wake makes it susceptible to being reingested, and the fact that more of the jet ends up in the trailing wake as stroke ratio is increased above the formation number

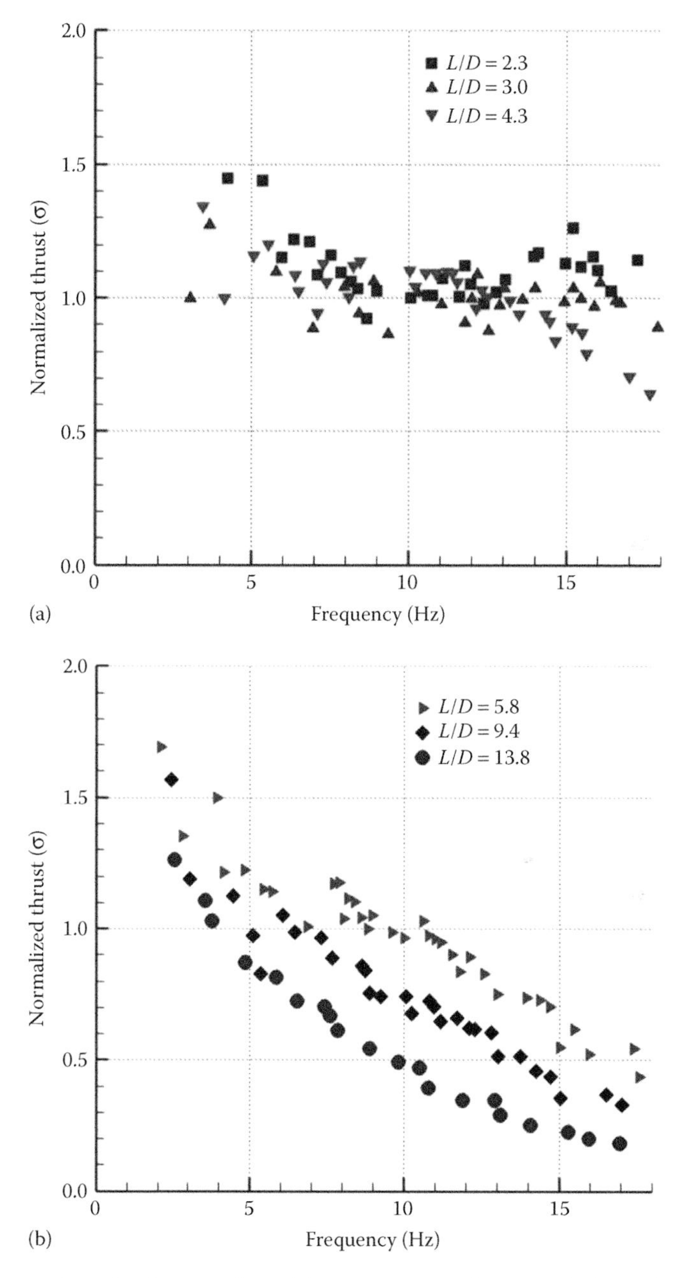

FIGURE 10.9 Normalized average thrust (σ) versus actuation frequency for (a) stroke ratios below the formation number and (b) stroke ratios above the formation number. (Reproduced from Krieg, M. and K. Mohseni, *IEEE J. Ocean. Eng.*, 33, 123–132, 2008. With permission.)

explains the increased relative loss with stroke ratio. Much more of the jet ends up in the primary vortex ring for the low stroke ratio cases, and the faster propagation velocity of the vortex ring makes it less affected by the refilling. In fact the $L/D = 2$ case, where the entire jet rolls into the leading vortex ring, shows enhanced impulse from contracting radial velocity even at very high pulsation frequencies.

10.4.2 Transient Thrust

Propeller-type thrusters experience a time delay associated with reaching a steady thrust, which is inversely proportional to that thrust level (Yoerger et al. 1990; Fossen 1991). Similar to propellers, jet actuators used for propulsion have time delays that are inversely proportional to the desired level of thrust. However, it should be noted that this type of thruster has rise times on the order of fractions of a second, whereas typical propeller-type thrusters experience rise times on the order of several seconds (Yoerger et al. 1990; Fossen 1991). Though the thrust produced by jet propulsors is highly dynamic with large oscillations aligned with the pulsation frequency of the thruster, the rise time of the average thrust can be examined by fitting the thrust curve to the form of a linear damper. This was done in Krieg and Mohseni (2010), and the resulting thrust curves are reproduced in Figure 10.10 showing the general trends in rise time.

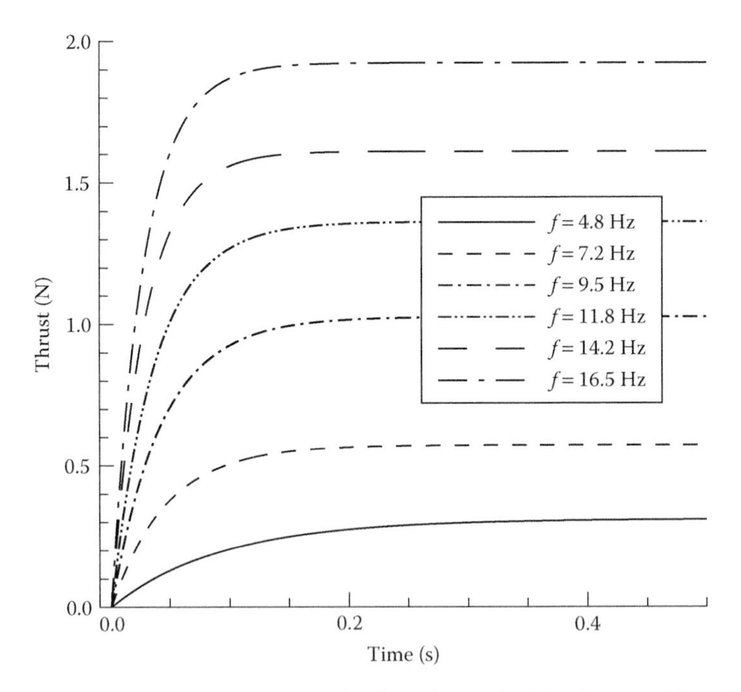

FIGURE 10.10 Thruster transient response fitted to a first-order delay. (Recreated from Krieg, M. and K. Mohseni, *IEEE Trans. Robot.*, 26, 542–554, 2010.)

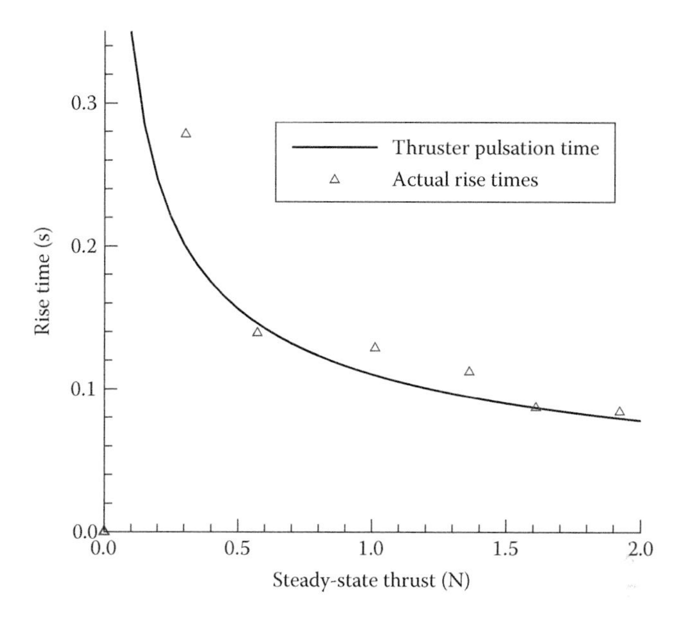

FIGURE 10.11 Thrust rise time as a function of steady-state thrust level for the thruster tested in Krieg and Mohseni (2010), which had a consistent jet volume of 33 ml and achieved the steady-state thrust levels reported by adjusting the pulsation frequency between 5 and 17 Hz. (From Krieg, M. and K. Mohseni, *IEEE Trans. Robot.*, 26, 542–554, 2010.)

Dimensional analysis suggests that the rise time should converge if the thrust and timescales are normalized appropriately (Krieg and Mohseni 2010). The natural choice for characteristic thrust and timescales are the steady thrust level and pulsation period, respectively, indicating that the rise time is proportional to the period of a single jet cycle. In fact the rise time is very close to the pulsation period as might be expected, which are both shown in Figure 10.11.

10.5 Biomimetic Vehicles and Thruster Integration

In the introduction of this chapter, we discussed the existence of two general classes of unmanned underwater vehicles. The abundance of remote marine research sites requiring high positioning accuracy for inspection, as well as the desire to create fully autonomous vehicle sensor networks, has inspired significant research in a hybrid class of vehicles with the efficient cruising characteristics of the torpedo class and the high positioning accuracy of the box class. Vehicles have been designed utilizing tunnel thrusters that run through the hull of the vehicle to give low-speed maneuvering capabilities to vehicles without compromising the forward drag profile (Mclean 1991; Torsiello 1994), and some researchers have moved the thrusters into the fins themselves (Dunbabin 2005). However, tunnel thrusters have been determined to be less effective when a crossflow is present and have been observed to continue producing a force even after being terminated (Mclean 1991), in addition to the large rise times discussed in Section 10.4. Other researchers observe that marine animals have a healthy balance of long distance endurance and high-accuracy low-speed maneuvering. To this end, vehicles have been designed to use fins for both high-speed maneuvering as well as mimic the low-speed flapping of turtles and marine mammals (Licht et al. 2004; Licht 2008), and some use tail fins as a primary means of propulsion (Barrett et al. 1999; Triantafyllou et al. 2000). Another option is to submerge the ZNMF devices described in Section 10.2 within the vehicle and use these thrusters to generate maneuvering forces without compromising vehicle drag.

Though much less prevalent than propeller-based designs, there are a few underwater vehicles that utilize ZNMF devices for underwater propulsion. The first generations of vehicles using this type of thruster were developed at the University of Colorado (Mohseni 2004; Krieg et al. 2005; Mohseni 2006). One vehicle to come out of this study (dubbed KRAKEN) had accurate low-speed maneuvering capabilities, and many of the maneuvers could be performed autonomously as directed by signals from an image recognition system (Clark et al. 2009). The forward propulsion for this vehicle was still provided by a propeller thruster to take advantage of the high propulsive efficiency capable at high velocity steady transit. This vehicle was mostly designed as a technology demonstrator and received high acclaim at the 2008 Association for Unmanned Vehicle Systems International (AUVSI) unmanned underwater vehicle competition, receiving the award for "Best New Entry." Figure 10.12 shows this vehicle performing a simulated parallel parking maneuver autonomously, guided by markers on the bottom of the pool.

More recently, another vehicle, named RoboSquid, was developed at Southern Methodist University (SMU) and utilized a single pulsatile jet thruster for forward propulsion (Moslemi and Krueger 2009). This vehicle was primarily designed to assess the propulsive efficiency of pulsed jet propulsion, and thus, the vehicle did not possess full operational capabilities and its motion was restricted to a single degree of freedom by suspending it from an air track. In this setup, RoboSquid was used to show that the propulsive efficiency of pulsed-jet devices increases with decreasing stroke ratio, and more significantly that the propulsive efficiency can surpass that of continuous jet propulsion for low enough stroke ratio and high enough frequency (Moslemi and Krueger 2010). Moslemi and Krueger (2011) also showed that the ratio of propulsive efficiency of pulsed jets to continuous jets increases with decreasing Reynolds number, suggesting a clear advantage for low Reynolds number vehicles.

The lessons learned from KRAKEN led to the development of a second-generation vehicle with improved sensing and reduced electronic footprint as summarized in Krieg et al. (2011). The AUV, named CephaloBot, is displayed in Figure 10.13. This vehicle is equipped with a set of ZNMF thrusters giving it accurate maneuvering capabilities in the horizontal plane (surge, sway, and yaw), whereas motion in the vertical plane (heave and pitch) is controlled by a set of ballast chambers. Though the buoyancy-driven maneuvering forces are weaker and slower acting, the minimal energetic cost of this type of forcing was decided to outweigh the speed limitations. The vehicle is controlled by a custom-designed, compact-embedded system platform, which includes several standard navigational sensors. An image recognition system and an acoustic sensing system are also included in the vehicle sensor suite, and the acoustic system is designed to provide communication and localization simultaneously. The vehicle is also equipped with a radio

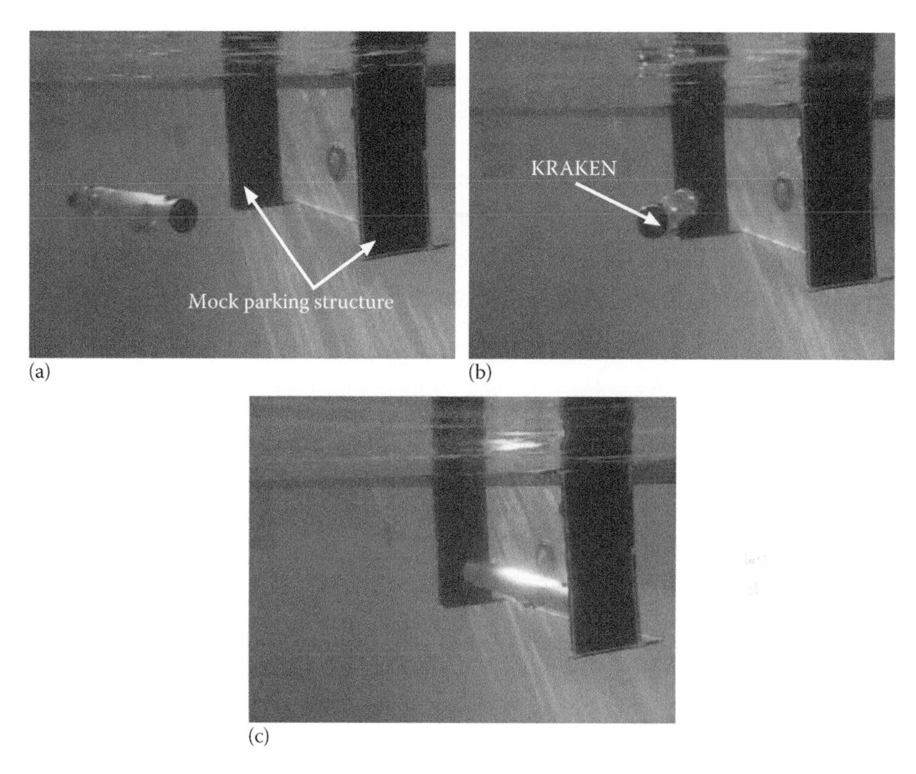

FIGURE 10.12 The KRAKEN prototype vehicle performing a simulated parallel parking sequence to demonstrate the maneuvering capabilities of this type of propulsion. The maneuver was filmed on October 8, 2008, during an unused discovery channel shoot, and the vehicle systems are described in Clark et al. All vehicle actions were performed autonomously according to the relative orientation of markers on the pool floor. (a) Vehicle stops near desired spot; (b) vehicle rotates to correct orientation and translates into slot; (c) vehicle in docking position. (From Clark, T. et al., Kraken: Kinematically roving autonomously kontrolled electro-nautic. *The 47th AIAA Aerospace Sciences Meeting*, Orlando, FL, January 5–8, 2009.)

FIGURE 10.13 The most recent, and most capable prototype vehicle utilizing ZNMF thrusters to generate maneuvering forces. The vehicle named CephaloBot is described in great detail in Shaw and Mohseni. (From Krieg, M., P. Klein, R. Hodgkinson, and K. Mohseni, *Mar. Technol. Soc. J.*, 45, 153–164, 2011.)

frequency (RF) antenna to provide more long-range communication and localization when operating at the surface. Therefore, CephaloBot represents the most complete vehicle to date utilizing ZNMF thrusters and is nearly fully operational.

10.5.1 Thruster Transfer Function Approximation

AUVs are regularly required to operate in chaotic environments with rapidly oscillating currents, further hindering accurate maneuvering capabilities. The energy of turbulent marine environments is well defined in the spectral domain (Pierson and Moskowitz 1964). A linear time invariant (LTI) transfer function model of the thruster dynamics is desirable since it allows the thruster parameters to be selected with respect to the mission-specific environmental dynamics. Krieg and Mohseni (2010) derived and experimentally validated an approximate LTI transfer function to represent the dynamics of a hypothetical underwater vehicle that uses the ZNMF thruster to generate maneuvering forces. This study assumed the vehicle to be a simple cylinder constrained to move in a single direction orthogonal to the cylinder axis; the drag coefficient is approximated as the drag of an infinitely long cylinder in crossflow.

Consider the LTI systems summarized in Figure 10.14, where the block labeled G_{Thruster} represents a thruster plant that converts a frequency input to a thrust output, the block labeled G_{Sub} represents the linearized dynamics of a cylindrical vehicle that is constrained to a single direction, X is the trajectory of the vehicle, u_f is the local fluid velocity, $\tilde{X}$ is the desired trajectory, and K represents the feedback error gains. This section will discuss a method to linearize the thruster plant (G_{Thruster}) and the vehicle plant (G_{Sub}) and then examine the accuracy of this modeling with respect to the thruster-vehicle system frequency response for both open- and closed-loop controllers in the absence of velocity disturbances.

Equation 10.8 calculates the average thrust output of the thruster with a sinusoidal velocity profile and actuation frequency (f) as predicted by the 1D slug model. This average thrust equation can be considered fairly accurate provided that the jet stroke ratio is below the formation number; however, if higher accuracy is desired, the average thrust

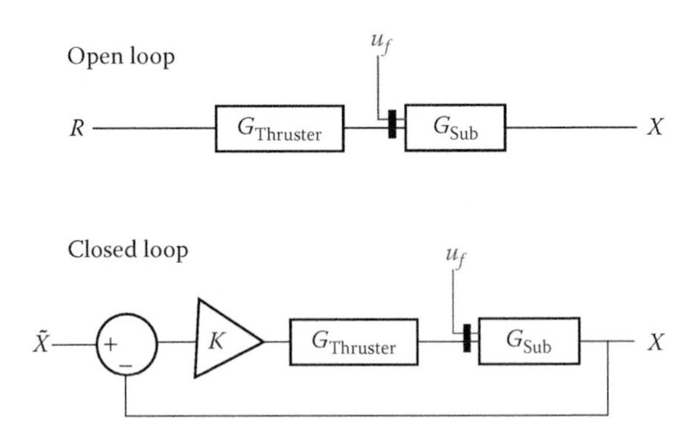

FIGURE 10.14 Open- and closed-loop block diagrams representing linearized dynamics of the thruster-vehicle system.

equation can be multiplied by the normalized thrust (σ) depicted in Figure 10.9. The exact, time-dependent thrust has large oscillations aligned with the thruster pulsation frequency and can also be approximated by the 1D slug model. If the approximate time-dependent thrust output is mapped into the spectral domain through a Laplace transform, it becomes a nonlinear integral function of the frequency input that can be linearized by setting the input function, which in this case is the frequency, equal to a weighted Heaviside function $f(t) = f_0 \int_0^t \delta(\tau)d\tau$ trimmed at some nominal pulsation frequency f_0. The linearized thrust equation in the spectral domain is

$$\hat{T}(f_0, s) \approx C_t \left(\frac{f_0^2}{s} - \frac{f_0^2}{s + f_0/\tau^\star} + \frac{\eta 2\pi f_0^3}{s^2 + 4\pi^2 f_0^2} \right) \tag{10.9}$$

where:

η is the ratio of thrust oscillation amplitude to average thrust
s is the maneuvering frequency (input)
$\tau^\star$ is a characteristic time scale

In the above equation, C_t is a coefficient associated with the thruster configuration ($C_t = \rho(\pi/16)D^4(L/D)^2$). The thruster plant is the ratio $\hat{T}(s)/F(s)$, where $F(s)$ is the Laplace transform of the input function $F(s) = \int_0^\infty f(t)dt = f_0/s$. This should be an accurate approximation for the thruster plant so long as the rate of change of $f(t)$ is slow with respect to the rate of change of $X(t)$

$$G_{\text{Thruster}}(s) = C_t \left(\frac{f_0^2}{\tau^\star s + f_0} + \frac{\eta 2\pi f_0^2 s}{s^2 + 4\pi^2 f_0^2} \right) \tag{10.10}$$

Therefore, the thruster plant describes the linearized dynamics between the input actuation frequency and output thrust and is purely a function of the trim frequency f_0.

The drag force on the simple one degree of freedom cylindrical vehicle is equated to the drag on a long cylinder in crossflow, which is a nonlinear function of the vehicle velocity and the Reynolds number of the flow. The dynamics of the vehicle can be described by Newton's second law where the force acting on the body is the sum of the thruster output and the drag force. The vehicle dynamics must be linearized about some trim velocity ($\dot{X}_{\text{trim}}$) to be transformed into the spectral domain that results in the approximate vehicle plant:

$$G_{\text{Sub}}(s) = \frac{1}{ms^2 + Cs}, \qquad C = \frac{1}{2}\rho C_D(Re_{\text{trim}})\dot{X}_{\text{trim}} \tag{10.11}$$

where:

m is the vehicle mass
C_D is the coefficient of drag

A more detailed derivation of the approximate thruster and vehicle plants, as well as a discussion of appropriate trim conditions for both plants, is provided in Krieg and Mohseni (2010).

The accuracy of this linearized transfer function approximation was tested by Krieg and Mohseni (2010) by operating the thruster in a hybrid vehicle simulation and recording the system frequency response. In this simulation, the behavior of a virtual vehicle is simulated incorporating the thrust measured empirically from a prototype thruster in a controlled static setup. Using this procedure, the validity of the thruster model can be tested with respect to a "pure" vehicle that acts predictably according to the full nonlinear cylinder drag equations. Also approximation/modeling errors may be determined independently from inconsistencies due to environment unpredictability. A driving signal generated by the vehicle controller is sent to the prototype thruster. The corresponding force is measured directly using a load cell in the static setup. The unfiltered thrust is then fed into the virtual algorithm, and the vehicle motion is integrated from rest according to the exact vehicle governing equation. In real time, the control algorithm drives the virtual vehicle using the actual forces generated by the thruster.

To demonstrate the accuracy of the LTI approximation, the hybrid simulation must be tested for a range of maneuvers with fundamentally different scaling. A quantity termed the "scale factor" was introduced to quantify the different maneuvering regimes (Krieg and Mohseni 2010). The scale factor is defined $A^\star = A/d$, where A is the maneuvering amplitude and d is the vehicle diameter (or characteristic size). The hybrid simulation was tested for three different maneuvering regimes: the *cruising regime* (maneuver larger than vehicle size, $A^\star = 3$), the *docking regime* (maneuver smaller than vehicle size, $A^\star = 0.5$), and the *transition regime* (the transition between cruising and docking, $A^\star = 1$). A more complete description of how the maneuvering regime affects the appropriate trim conditions in the LTI model is found in the work by Krieg and Mohseni (2010).

10.5.2 Frequency Response

Within the hybrid simulation, the open-loop frequency response was determined for the thruster-vehicle system for all three maneuvering regimes. The frequency response curves of the open-loop system are shown in Figure 10.15a. It can be seen from this graph that several features of the frequency response are accurately modeled assuming constant pulsation frequency including the cutoff frequency and the convergence of different maneuvering regime response curves near the cutoff frequency.

In the low-frequency maneuvers, the spread between the maneuvering regimes is more drastic than at the corner frequency of the system, which happens to be nearly identical for all three maneuvering regimes. At low maneuvering frequencies, the thruster is required to deliver smaller forces, which in turn result in lower actuation frequencies, so that at this level the vehicle experiences individual pulsations. In the docking regime (low amplitude maneuver), this results in a higher gain since pulsations enact an acceleration before drag forces take effect. In the transit regime, however, this results in less gain since the drag terms dominate in between pulsations. This trend is accurately captured by the approximated LTI model.

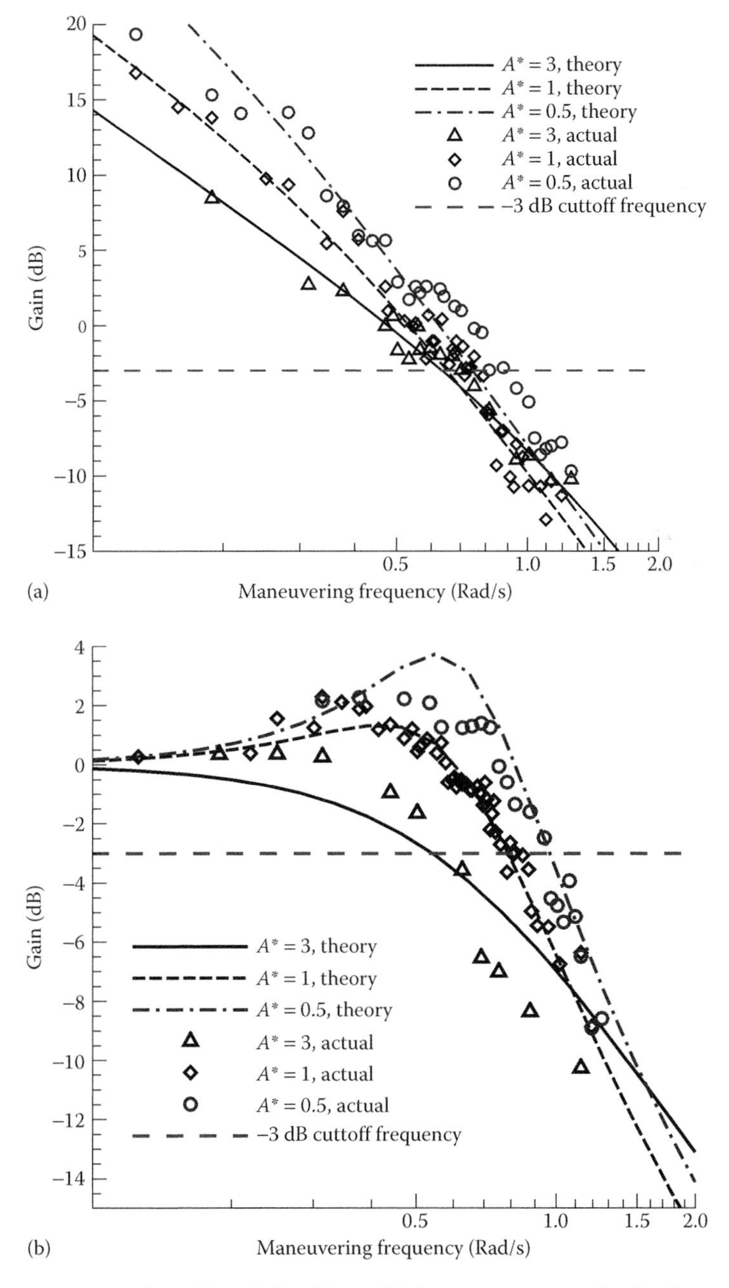

FIGURE 10.15 Open-loop (a) and closed-loop (b) frequency response for the thruster-vehicle system; cruising regime shown by $A^\star = 3$, transition regime shown by $A^\star = 1$, docking regime shown by $A^\star = 0.5$. Theoretical response modeled assuming $f_0 = 20$, 9, and 5 Hz, respectively. (Recreated from Krieg, M. and K. Mohseni, *IEEE Trans. Robot.*, 26, 542–554, 2010.)

10.5.3 Feedback Control

A simple proportional derivative (PD) feedback algorithm was then implemented in the hybrid simulation providing a control signal proportional to the error between desired and actual vehicle position and velocity. The important parameters which drive the controller gain selection are motivated by different goals for the different maneuvering regimes. The docking regime requires very accurate tracking with minimal overshoot; whereas the cruising regime is generally indifferent to overshoot and is much more concerned with a fast approach time (so that the vehicle can move to a site of interest before the phenomena of interest dissipate). For this study, the feedback gain was set to 4 that was chosen to keep the required thrust within the thruster capacity; and the derivative gain ratio was set to 0.75 to keep the position overshoot within acceptable bounds in the docking regime.

The system closed-loop frequency response was determined over a similar frequency range for the same maneuvering scale factors as the open-loop response, with identical trim conditions. The LTI model is seen in Figure 10.15b to approximate the closed-loop behavior of the thruster-vehicle system sufficiently well including the cutoff frequency. Though the transfer function model for the cruising regime comes close to approaching the actual cutoff seen in the system, it incorrectly predicts the gain on either side of the cutoff frequency. This is due to the fact that maneuvers in the cruising regime experience drastically variable drag forces, since the drag force is nonlinear and the vehicle has a larger velocity range in the cruising regime. Because of these nonlinearities and the sizable velocity range, the linearization about a single trim velocity predicts a drag that is too large for the low-frequency maneuvers and similarly predicts a drag that is too low for high-frequency maneuvers. Therefore, the accuracy of the drag approximation will decrease as the maneuvering scale increases and the large velocity range in the cruising regime requires a sliding model to accurately predict system frequency response. Conversely, the small-scale maneuvers are well approximated by a single trim velocity.

Additionally the model has no limitation on thrust level, and in the cruising regime, it drives the thruster beyond its physical capacity (even with relatively low gains). Though this is an unmodeled nonlinear effect, it also addresses an interesting design consideration. This analysis demonstrates that a high-accuracy ZNMF thruster may not have a large enough range to be completely effective in the cruising regime. The thruster could be designed with a larger output but this would reduce the accuracy of the system in the docking regime. Fortunately, maneuvers in the high-frequency cruising regime are also generally coupled with significant forward vehicle velocity that provides the maneuvering system an added dimensionality, since low-drag control surfaces provide substantial forcing in this regime.

10.6 Future Directions

This chapter discussed the fundamentals of generating maneuvering forces from finite propulsive jets in biological and bioinspired engineering systems. The current state of the technology was summarized along with the vehicles developed to use this propulsion mechanism. The next stage of development is to subject the prototype thrusters to open ocean testing. The nearly instantaneous thrust output is predicted to substantially improve disturbance rejection and position tracking, especially for small-scale high-frequency disturbances like those found in shallow or coastal environments. However, improvements will need to be made of the internal cavity driving mechanism making it more suitable for corrosive environments.

There are also a number of more complicated vehicle control algorithms that become available through the use of this type of maneuvering system. Underwater vehicle dynamics are controlled by a combination of hydrodynamic and inertial forces acting on the vehicle. Traditionally, the hydrodynamic forces are decoupled into forces that act like inertia, added mass, and forces that act like linear drag terms. This allows the forces to be approximated from vehicle kinematic data but also requires extensive testing to get all the required drag and added mass coefficients. More recent control architectures propose using novel sensing systems to determine the exact hydrodynamic force distribution and then use the arrangement of ZNMF thrusters to instantaneously apply opposing forces and decouple the vehicle dynamics from the hydrodynamic forces (Xu and Mohseni 2013). In addition to negating the hydrodynamic forces, the thrusters can be used to change the nature of the hydrodynamic forces. The similarity between ZNMF thrusters and SJAs used in air suggests that this type of technology may be used for "hydroshaping" at high forward velocities. Instead of generating control forces strictly from the jetting momentum transfer, the ZNMF thruster can be used to inject energy into the vehicle boundary layer, altering the effective shape of the vehicle seen by the surrounding fluid. Aeroshaping is discussed in greater detail in Chapter 4.

Soft body robotics is also quickly gaining popularity in the robotics community, but it poses a difficult challenge since most actuation methods require rigid structures to provide necessary reaction forces. This challenge is certainly present for soft body marine robots, but the propulsive technology discussed in this chapter seems ideally suited for soft body robots, given that squid, jellyfish, and octopuses that inspire the technology are all invertebrates. Not surprisingly many of the first-generation soft body marine robots use this type of propulsion taking the body shapes of both octopus (Serchi et al. 2012) and jellyfish (Nawroth et al. 2012).

References

Anderson, E. J. and E. DeMont. 2000. The mechanics of locomotion in the squid *Loligo pealei*: Locomotory function and unsteady hydrodynamics of the jet and intramantle pressure. *Journal of Experimental Biology*, 203: 2851–2863.

Anderson, E. J. and M. A. Grosenbaugh. 2005. Jet flow in steadily swimming adult squid. *Journal of Experimental Biology*, 208: 1125–1146.

Barrett, D. S., M. S. Triantafyllou, D. K. P. Yue, M. A. Grosenbaugh, and M. Wolfgang. 1999. Drag reduction in fish-like locomotion. *Journal of Fluid Mechanics*, 392: 183–212.

Bartol, I. K., P. S. Krueger, W. J. Stewart, and J. T. Thompson. 2009. Hydrodynamics of pulsed jetting in juvenile and adult brief squid *Lolliguncula brevis*: Evidence of multiple jet "modes" and their implications for propulsive efficiency. *Journal of Experimental Biology*, 212: 1189–1903.

Bartol, I. K., P. S. Krueger, J. T. Thompson, and W. J. Stewart. 2008. Swimming dynamics and propulsive efficiency of squids throughout ontogeny. *Integrative and Comparative Biology*, 48(6): 1–14. doi:10.1093/icb/icn043.

Cantwell. B. J. 1986. Viscous starting jets. *Journal of Fluid Mechanics*, 173: 159–189.

Clark, T., P. Klein, G. Lake, S. Lawrence-Simon, J. Moore, B. Rhea-Carver, M. Sotola, S. Wilson, C. Wolfskill, and A. Wu. 2009. Kraken: Kinematically roving autonomously kontrolled electro-nautic. *The 47th AIAA Aerospace Sciences Meeting*, Orlando, FL, January 5–8.

Colin, S. P. and J. H. Costello. 2002. Morphology, swimming performance and propulsive mode of six co-occuring hyromedusae. *Journal of Experimental Biology*, 205: 427–437.

Dabiri, J. O., S. P. Colin, K. Katija, and J. H. Costello. 2010. A wake based correlate of swimming performance and foraging behavior in seven co-occurring jellyfish species. *Journal of Experimental Biology*, 213: 1217–1275.

Didden, D. 1979. On the formation of vortex rings: Rolling-up and production of circulation. *Journal of Applied Mathematics and Physics*, 30: 101–116.

Dunbabin, M., J. Roberts, K. Usher, G. Winstanley, and P. Corke. 2005. A hybrid AUV design for shallow water reef navigation. *Proceedings of the IEEE International Conference on Robotics and Automation*, Barcelona, Spain, April, pp. 2105–2110, DOI:10.1109/ROBOT.2005.1570424.

Ford, M. D. and J. H. Costello. 2000. Kinematic comparison of bell contraction by four species of hyromedusae. *Scientia Maria*, 64: 47–53.

Fossen, T. I. 1991. Nonlinear modelling and control of underwater vehicles. PhD thesis, Norwegian Institute of Technology, Trondheim, Norway.

Gharib, M., E. Rambod, and K. Shariff. 1998. A universal time scale for vortex ring formation. *Journal of Fluid Mechanics*, 360: 121–140.

Glezer, A. 1988. The formation of vortex rings. *Physics of Fluids*, 31(12): 3532–3542.

Gosline, J. and E. DeMont. 1985. Jet-propelled swimming in squids. *Scientific American*, 252: 96–103.

Hill, M. J. M. 1894. On a spherical vortex. *Philosophical Transactions of the Royal Society A*, 185: 213–245.

Krieg, M., C. Coley, C. Hart, and K. Mohseni. 2005. Synthetic jet thrust optimization for application in underwater vehicles. *Proceedings of the 14th International Symposium on Unmanned Untethered Submersible Technology (UUST)*, Durham, NH, August 21–24.

Krieg, M., P. Klein, R. Hodgkinson, and K. Mohseni. 2011. A hybrid class underwater vehicle: Bioinspired propulsion, embedded system, and acoustic communication and localization system. *Marine Technology Society Journal*, 45(4): 153–164.

Krieg, M. and K. Mohseni. 2008. Thrust characterization of pulsatile vortex ring generators for locomotion of underwater robots. *IEEE Journal of Oceanic Engineering*, 33(2): 123–132.

Krieg, M. and K. Mohseni. 2010. Dynamic modeling and control of biologically inspired vortex ring thrusters for underwater robot locomotion. *IEEE Transactions on Robotics*, 26(3): 542–554.

Krieg, M. and K. Mohseni. 2013. Modelling circulation, impulse, and kinetic energy of starting jets with non-zero radial velocity. *Journal of Fluid Mechanics*, 719: 488–526.

Krueger, P. S. 2008. Circulation and trajectories of vortex rings formed from tube and orifice openings. *Physica D*, 237: 2218–2222.

Krueger, P. S. and M. Gharib. 2005. Thrust augmentation and vortex ring evolution in a fully pulsed jet. *AIAA Journal*, 43(4): 792–801.

Lamb, H. 1945. *Hydrodynamics*. New York: Dover Publications.

Landau, L. D. and E. M. Lifshitz. 1959. Ideal Fluids (Energy Flux). In: *Fluid Mechanics*. eds. L. D. Landau and E. M. Lifshit, Oxford: Butterworth-Heinemann, pp. 9–10.

Licht, S. 2008. Biomimetic oscillating foil propulsion to enhance underwater vehicle agility and maneuverablity. PhD thesis, Massachusetts Institute of Technology, Cambridge, MA.

Licht, S., V. Polidoro, M. Flores, F. S. Hover, and M. S. Triantafyllou. 2004. Design and projected performance of a flapping foil AUV. *IEEE Journal of Ocean Engineering*, 29(3): 786–794.

Lipinski, D. and K. Mohseni. 2009a. Flow structures and fluid transport for the hydromedusa *Sarsia tubulosa. The 19th AIAA Fluid Dynamics Conference*, San Antonio, TX, June 22–25.

Lipinski, D. and K. Mohseni. 2009b. Flow structures and fluid transport for the hydromedusae *Sarsia tubulosa* and *Aequorea victoria*. *Journal of Experimental Biology*, 212: 2436–2447.

Maxworthy, T. 1972. The structure and stability of vortex rings. *Journal of Fluid Mechanics*, 51(1): 15–32.

Maxworthy, T. 1977. Some experimental studies of vortex rings. *Journal of Fluid Mechanics*, 80: 465–495.

Mclean, M. B. 1991. Dynamic performance of small diameter tunnel thrusters. PhD thesis, Naval Postgraduate School, Monterey, CA.

Mohseni, K. 2004. Pulsatile jets for unmanned underwater maneuvering. *The 3rd AIAA Unmanned Unlimited Technical Conference, Workshop and Exhibit*, Chicago, IL, September 20–23.

Mohseni, K. 2006. Pulsatile vortex generators for low-speed maneuvering of small underwater vehicles. *Ocean Engineering*, 33(16): 2209–2223.

Mohseni, K. and M. Gharib. 1998. A model for universal time scale of vortex ring formation. *Physics of Fluids*, 10(10): 2436–2438.

Mohseni, K., H. Ran, and T. Colonius. 2001. Numerical experiments on vortex ring formation. *Journal of Fluid Mechanics*, 430: 267–282.

Moslemi, A. A. and P. S. Krueger. 2009. Effect of duty cycle and stroke ratio on propulsive efficiency of a pulsed jet underwater vehicle. *The 39th AIAA Fluid Dynamics Conference*, San Antonio, TX, June.

Moslemi, A. A. and P. S. Krueger. 2010. Propulsive efficiency of a biomorphic pulsed-jet vehicle. *Bioinspiration & Biomimetics*, 5(3): 036003.

Moslemi, A. A. and P. S. Krueger. 2011. The effect of Reynolds number on the propulsive efficiency of a biomorphic pulsed jet underwater vehicle. *Bioinspiration & Biomimetics*, 6(2): 026001.

Nawroth, J. C., H. Lee, A. W. Feinberg, C. M. Ripplinger, M. L. McCain, A. Grosberg, J. O. Dabiri, and K. K. Parker. 2012. A tissue-engineered jellyfish with biomimetic propulsion. *Nature Biotechnology*, 30: 792–797.

O'Dor, R. K. 1988. The forces acting on swimming squid. *Journal of Experimental Biology*, 137: 421–442.

O'Dor, R. K. and D. M. Webber. 1991. Invertebrate athletes: Trade-offs between transport efficiency and power density in cephalopod evolution. *Journal of Experimental Biology*, 160: 93–112.

Pierson, W. J. and L. A. Moskowitz. 1964. Proposed spectral form for fully developed wind seas based on the similarity theory of S. A. Kitaigorodskii. *Journal of Geophysical Research*, 69(3): 5181–5190.

Pullin, D. I. 1979. Vortex ring formation in tube and orifice openings. *Physics of Fluids*, 22: 401–403.

Pullin, D. I. and W. R. C. Phillips. 1981. On a generalization of Kaden's problem. *Journal of Fluid Mechanics*, 104: 45–53.

Rosenfeld, M., K. Katija, and J. O. Dabiri. 2009. Circulation generation and vortex ring formation by conic nozzles. *Journal of Fluids Engineering*, 131(9): 091204.

Rosenfeld, M., E. Rambod, and M. Gharib. 1998. Circulation and formation number of laminar vortex rings. *Journal of Fluid Mechanics*, 376: 297–318.

Saffman, P. G. 1992. *Vortex Dynamics*. Cambridge: Cambridge University Press.

Sahin, M. and K. Mohseni. 2008.The numerical simulation of flow patterns generated by the hydromedusa *Aequorea victoria* using an arbitrary Lagrangian-Eulerian formulation. *The 38th AIAA Fluid Dynamics Conference and Exhibit*, Seattle, WA, June 23–26.

Sahin, M. and K. Mohseni. 2009. An arbitrary Lagrangian-Eulerian formulation for the numerical simulation of flow patterns generated by the hydromedusa *Aequorea victoria*. *Journal of Computational Physics*, 228: 4588–4605.

Sahin, M., K. Mohseni, and S. Colins. 2009. The numerical comparison of flow patterns and propulsive performances for the hydromedusae *Sarsia tubulosa* and *Aequorea victoria*. *Journal of Experimental Biology*, 212: 2656–2667.

Serchi, F. G., A. Arienti, and C. Laschi. 2012. A biomimetic, swimming soft robot inspired by *Octopus vulgaris*. *Proceedings of the 1st International Conference on Living Machines*, Barcelona, Spain, July 9–12, pp. 349–351.

Shariff, K. and A. Leonard. 1992. Vortex rings. *Annual Review of Fluid Mechanics*, 34: 235–279.

Shevstov, G. A. 1973. Results of tagging of the Pacific squid *Todarodes pacificus* Steenstrup in the Kuril Hokkaido region. *Izvestiya Tikhookean Nauchno-Issledovatel'skogo Instituta Rybnogo Khozyaistva i Okeanografii* [in Russian], 87: 120–126.

Torsiello, K. A. 1994. Acoustic positioning of the NPS autonomous underwater vehicle (AUV II) during hover conditions. Master's thesis, Naval Postgraduate School, Monterey, CA.

Triantafyllou, M. S., G. S. Triantafyllou, and D. K. P. Yue. 2000. Hydrodynamics of fishlike swimming. *Annual Review of Fluid Mechanics*, 32: 33–53.

Vogel, S. 2003. *Comparative Biomechanics: Life's Physical World*. Princeton, NJ: Princeton University Press.

Williamson, G. R. 1965. Underwater observations of the squid *Illex illecebrosus* Lesueur in Newfoundland waters. *Canadian Field Nature*, 79: 239–247.

Wilson, M., J. Peng, J. O. Dabiri, and J. D. Eldredge. 2009. Lagrangian coherent structures in low Reynolds number swimming. *Journal of Physics: Condensed Matter*, 21: 204105.

Xu, Y. and K. Mohseni. 2013. Bio-inspired real-time pressure feedforward for autonomous underwater vehicle control. *IEEE/ASME Transactions on Mechatronics*, 19(4): 1127–1137.

Yoerger, D., J. Cooke, and J.-J. Slotine. 1990. The influence of thruster dynamics on underwater vehicle behavior and their incorporation into control system design. *IEEE Journal of Oceanic Engineering*, 15(3): 167–178.

Index

Note: Locators "*f*" and "*t*" denote figures and tables in the text

A

B

C

D

E

F

G

H

Q

R

S

T